Progress in Mathematical Physics

Volume 64

For further volumes:
http://www.springer.com/series/4813

Bruno Cordani

Geography of Order and Chaos in Mechanics

Investigations of Quasi-Integrable
Systems with Analytical, Numerical,
and Graphical Tools

 Birkhäuser

Bruno Cordani
Italy
bruno.cordani@unimi.it

Please note that additional material for this book can be downloaded from http://extras.springer.com

ISBN 978-0-8176-8369-6   ISBN 978-0-8176-8370-2 (eBook)
DOI 10.1007/978-0-8176-8370-2
Springer New York Heidelberg Dordrecht London

Library of Congress Control Number: 2012948333

Mathematics Subject Classification (2010): 70F05, 70F10, 70F15, 70H08, 70H33, 70K65

Printed on acid-free paper

Springer is part of Springer Science+Business Media (www.birkhauser-science.com)

Alla mia famiglia,

che mi ha supportato e sopportato

Preface

The Newtonian program, well known by every student, is conceptually simple and attractive: given a mass distribution and the forces acting on it, write the differential equations arising from the fundamental law of dynamics and solve them in order to obtain the motion. Unfortunately, things are not so simple, and in the course of the program one encounters at least two essential and unavoidable obstacles.

First, we are not able in general to solve (technically, to integrate) a system of differential equations. Yes, every young student has learned how to tackle the harmonic oscillator, the two-body problem, or the free rigid body. But it is discouraging that these systems, along with a few others discovered mainly in the second half of the past century, exhaust the small list of integrable systems.

But even if one possessed magically an analytical formula giving exactly the time evolution, it would still be scarcely useful for various reasons. For example, the motion is in general very complicated, and following the solution in its wandering does not give valuable information about the nature of the phenomenon. What is more, a possible regularity in the motion is difficult to detect by simply inspecting the dynamical evolution of the physical coordinates. Another frequent difficulty is the extreme sensitivity to the initial conditions ("butterfly effect"), which in practice makes the concept of solution itself meaningless. But this should not come as a surprise: after all, everybody has felt a sense of frustration looking at the numerical solution of some three-dimensional systems, being unable to extract a meaning from the entangled trajectory appearing on the monitor.

The aim of this book is to show how to overcome these difficulties and grasp the essence of the dynamics in the particular but very important and significant case of quasi-integrable systems, i.e., integrable systems slightly perturbed by other forces. A paradigmatic case is the solar system, where the perturbations are the interactions among the planets. Besides their practical importance, these systems are also extremely interesting from a mathematical point of view, exhibiting an intricate and fascinating structure known as the "Arnold web."

In the book these systems will be studied both from the analytical and the numerical point of view. With regard to the first point, I think that it is impossible to overestimate the importance of the role played by the symplectic structure of the phase space or, in more traditional language, by the Hamiltonian form of the equations of motion. This structure is the natural one of the phase space, exactly as the Euclidean structure is the fundamental one of our physical space. It is the symplectic structure that forces the solutions of the integrable systems to evolve linearly on tori (products of circles) with some fundamental frequencies, providing the framework without which the two main theorems of perturbation theory, i.e., KAM and Nekhoroshev, could not even be enunciated. It is thus not surprising that, as already devised by the great founders of the analytical mechanics in the nineteenth century, one should constantly utilize the symplectic (canonical) coordinates adapted to the foliation in tori, the action-angle variables, which deeply reveal the hidden properties of the perturbed motions. Exploiting the advanced techniques of perturbation theory, many examples of reduction to normal form will be given, i.e., to an integrable, hence approximate form that however reproduces the true dynamics well.

In order to compare the approximate with the true dynamics one needs numerical methods. In the book I present some tools recently introduced: the Frequency Modified Fourier Transform (a refinement of the Discrete Fourier Transform), the Wavelets (which allow one to find the instantaneous frequency) and the Frequency Modulation Indicator (which detects the distribution of the resonances among the fundamental frequencies). The reader may also find many figures that well illustrate the effectiveness of the methods and, above all, the relative software. This is surely the main feature of the book: the reader himself can and is encouraged to reproduce the various figures of the book and experiment with other situations, exploring the details of various quasi-integrable systems. I am convinced that the union of theory and practice is the main route to try to master an argument that is considered difficult.

But a more profound motivation in resorting to numerical computations arises from some lack of reliability that to a certain extent every mathematician experiences when facing a theorem proof that is particularly lengthy and intricate, and that looks more like a rhetoric speech to persuade the reader than the granitic statement of an unquestionable truth. The numerical experiments become so an essential completion of the traditional proof, reversing Truesdell's thesis of "the computer: ruin of science and threat to mankind."

I'd like to make it clear that no knowledge on computer programming is needed in order to use the software: you only have to access directly to a MATLAB installation or, subordinately, to install a free reader. Indeed, the programs support a graphical user interface and require one only to click on buttons and menu having a hopefully clear meaning: see the final appendices to the book. The supplied programs in the accompanying CD can be downloaded as an iso image from the publisher's website by entering the book's ISBN (978-0-8176-8369-6) into http://extras.springer.com/ and are the following.

(i) POINCARE program analyzes symplectic maps with the aid of the Frequency Modulation Indicator.

(ii) HAMILTON program analyzes Hamiltonian systems with the Frequency Modulation Indicator.

(iii) LAGRANGE program regards the Lagrange points in the three-body problem.

(iv) KEPLER program studies the perturbations of the Kepler problem.

(v) LAPLACE program concerns the dynamics of a solar system.

I used part of the material presented here in some courses on Celestial Mechanics, Hamiltonian Systems, and Perturbation Theory, addressed to advanced undergraduate students. I think that the book may serve as an introduction to specialistic literature and to a serious study of perturbation theory, with particular emphasis on the KAM and Nekhoroshev theorems. The two theorems are proved in the book skipping some details, like the technical proof of bounding inequalities, which in a first approach (and also in a second) are more distracting than illuminating, and trying instead to stress the conceptual points. But I hope that professional researchers may also find this book useful, thanks to its enclosed software.

Briefly, the plan of the work is the following. In Chapter 1 a somewhat detailed account of the whole book is given, which should also help the reader to not lose the thread of the argument. Chapter 2 contains the basic concepts of differential geometry, Lie groups, and analytical mechanics, which Chapter 3 applies to perturbation theory. Chapter 4 deals with numerical integration of ordinary differential equations and Chapter 5 with some tools useful to numerically detect order and chaos. The final four chapters are devoted to the applications, i.e., to the perturbations of the Kepler problem, as the hydrogen atom in an electric and magnetic field, and to the planetary problem. These concrete applications are not only physically interesting but are also significant examples of how to investigate in general quasi-integrable Hamiltonian systems, combining the techniques of the reduction to normal form with the numerical analysis of how order, chaos, and resonances are distributed in phase space.

It is always useful to listen to several different voices on the same argument. Three books in particular are highly recommendable: Celletti (2010), Morbidelli (2002), and Ferraz-Mello (2007). More or less they cover the same topics of the present book, with a major emphasis on the applications but without including any software. A good introduction to this book is Tabor (1989).

I thank very much the two anonymous referees for all the useful comments and suggestions, which I've (almost) fully included.

Finally, I'd like to express my sincerest thanks to Tom Grasso and Ben Cronin at Birkhäuser for their highly professional and efficient handling of this project.

Milano, June 2012   B. CORDANI

Contents

List of Figures

CHAPTER 1

Introductory Survey

Excuse me if I alone introduce myself:
I am the Prologue.
— R. LEONCAVALLO, Pagliacci

Order and chaos, invariant tori, KAM theory, resonances, Arnold web, diffusion ..., these are "keywords" in the theory of dynamical systems. But for mathematicians who are not directly involved in this area they may sound a bit vague. To grasp what they are about, consider the following question: Given a conservative mechanical system, thus without dissipative forces, what is its ultimate fate? Without further information the answer probably cannot be given in general. However, if we restrict ourselves to the special, but very important, case of a *slightly perturbed integrable system*, we can claim that crucial progresses have been achieved. An example is our solar system, in which the perturbations are the interactions between planets. In this book we will suggest how to guess an answer, and on the way all these keywords will come into play.

The motion of an integrable system is totally ordered, in a sense that will be specified below. The central question is: Does a very small perturbation destroy this order completely? Before the fundamental work of Kolmogorov (1954), two completely opposite answers were given. For the astronomers, interested in the computation of perturbed orbits, the answer was (more or less tacitly) negative: for them the affirmative answer seemed to be a disaster, making meaningless their series expansions. On the contrary, for the statistical physicists, interested in the possibility of applying the ergodic theorem, the answer was decisively affirmative. That of Kolmogorov was, in some sense, a Solomon's verdict: the ultimate fate of a slightly perturbed orbit depends on the initial conditions, so that for the same perturbed system there coexist, in general, orbits that stay *forever* in the neighborhood of an unperturbed one, and other orbits that depart indefinitely.

What is the distinguishing feature that makes an orbit ordered or chaotic? The answer is given by the celebrated Kolmogorov–Arnold–Moser (KAM) theorem stated first in Kolmogorov (1954). Its proof is rather complicated and also the statement itself cannot be given without anticipating some concepts of symplectic geometry and analytical mechanics.

1.1 Configuration Space and Lagrangian Dynamics

Consider a mechanical system with n degrees of freedom, let $q^1, q^2, \ldots, q^n$ be the configuration coordinates, and denote with a dot their time derivative. The n Lagrange equations

$$\frac{d}{dt} \frac{\partial L(q, \dot{q})}{\partial \dot{q}} - \frac{\partial L(q, \dot{q})}{\partial q} = 0, \quad q = (q^1, q^2, \ldots, q^n)$$

are derived from the variational principle

$$\delta \int_{t_1}^{t_2} L(q, \dot{q}) \, dt = 0,$$

where the variations keep fixed the initial and final points. For natural systems, the Lagrangian L is usually given by the difference between the kinetic and potential energy.

The Lagrange equations admit a geometrical interpretation, which is the reason for their practical usefulness. In fact, in the particular case of a point constrained on a smooth surface, they are the projection of the fundamental law of dynamics onto the tangent plane, thus avoiding the introduction of reaction forces. For a mechanical system the surface is replaced by its natural generalization, an n-dimensional manifold Q, namely the *configuration space*. The kinetic energy is positive definite, homogeneous, and bilinear in the components $\dot{q}^k$ of the generalized velocity, thus defining a double symmetric tensor that equips Q with a metric (or Riemannian) structure and establishing an isomorphism between the tangent space and its dual, the cotangent space. This is referred to as the ability to raise or lower the indices. The manifold Q is *flat* or *Euclidean* if and only if a certain differential condition is satisfied (i.e., when the Riemann tensor, constructed with the first and second derivatives of the metric tensor, vanishes identically).

In this case it is possible to choose Cartesian coordinates, reducing the representation of the metric tensor to the identity matrix. However, in general Q is *not* flat.

In this context, the Lagrange equations represent a real cornerstone, like all the great intellectual achievements. They bring the Newtonian perspective to a natural conclusion and, having forces and accelerations as basic ingredients, convey the information in n equations of second order. At the same time, they are the starting point of a far-reaching path: the apparently unpretentious wish to express them as a set of $2n$ equations of the first order, also reveals that the tangent bundle TQ (i.e., the union of Q and its tangent spaces) has a sort of metric, namely the *symplectic structure*. The next section is a brief digression to state some definitions.

1.2 Symplectic Manifolds

A $2n$-dimensional manifold is said to be *symplectic* if it is equipped with a field of closed and regular 2-forms (i.e., double antisymmetric tensors). As in the Riemannian case, with the 2-form one can evaluate in any point the "scalar product" of two tangent vectors; moreover, the regularity property yields the isomorphism between the tangent and cotangent space and in turn the ability to raise or lower indices. The closure property is, in some sense, similar to the vanishing of the Riemann tensor: the Darboux theorem states that in an open neighborhood of a symplectic manifold the 2-form Ω takes a *canonical* expression in a suitable *canonical* coordinate system, its representation matrix being

$$\Omega = \begin{pmatrix} 0_n & -1_n \\ 1_n & 0_n \end{pmatrix}.$$

The canonical coordinates are therefore similar to the Cartesian coordinates of a Euclidean space. Usually, the first n canonical coordinates are named q^k, the remaining p_k, so that the canonical 2-form takes the expression $\Omega = \sum_{k=1}^{n} dp_k \wedge dq^k$; the wedge symbol $\wedge$ means *exterior product*, i.e., the antisymmetric part of the tensor product. Due to the closure property, a potential 1-form $\Theta = \sum_{k=1}^{n} p_k dq^k$ does exist at least locally and $\Omega = d\Theta$ is its exterior derivative.

The analogy between Cartesian and canonical coordinates can be taken further. The rigid transformations of a Euclidean space are rotations and translations and have the property to leave the metric invariant. Their counterparts, which leave the symplectic structure invariant, are the *symplectic transformations* that, in the canonical case, send the old q, p into the new $Q(q, p), P(q, p)$ through the relations

$$p_k = \frac{\partial W(q, Q)}{\partial q^k}, \quad P_k = -\frac{\partial W(q, Q)}{\partial Q^k},$$

where $W(q,Q)$ is a *generating function* satisfying the two relations

$$\sum_{k=1}^{n} p_k dq^k - P_k dQ^k = dW(q,Q), \quad \det\left(\frac{\partial^2 W}{\partial q^h \partial Q^k}\right) \neq 0.$$

Indeed, by exterior differentiating the first relation, one may check that the old canonical 2-form is still transformed into a canonical one, since $dd \equiv 0$; the latter relation simply ensures that the transformation is invertible.

A continuous one-parameter group of symplectic transformations can be generated as follows. Define the *Hamiltonian vector fields* as those obtained, first by differentiating a scalar function $f(q,p)$, called the *Hamiltonian*, and then by raising the indices with the symplectic structure; sometimes they are called the *symplectic gradient* of the Hamiltonian $f(q,p)$. In the class of generic vector fields defined on a symplectic manifold, the Hamiltonian ones occupy a privileged position, since a vector field generates a symplectic flow (i.e., leaving the symplectic structure invariant) if and only if it is Hamiltonian.

Lastly, let us define the *Poisson bracket* $\{f,g\}$ between two functions on a symplectic manifold as the scalar product of their gradients. Equivalently, it can be seen as the Lie (or directional) derivative of one of the two functions with respect to the vector field generated by the symplectic gradient of the other. In the canonical case we recover the well-known definition:

$$\{f,g\} = \sum_{k=1}^{n} \frac{\partial f}{\partial q^k}\frac{\partial g}{\partial p_k} - \frac{\partial g}{\partial q^k}\frac{\partial f}{\partial p_k}.$$

It is easy to see that the flows generated by the symplectic gradients of the two functions commute if and only if $\{f,g\} = 0$.

1.3   Phase Space and Hamiltonian Dynamics

In order to pass from the n equations of second order of the Lagrangian dynamics to $2n$ equivalent equations of first order, the most natural choice is to promote the components of the generalized velocity to independent variables by setting $\dot{q}^k = v^k$. Now we make a discovery. On TQ define the 1-form $\Theta_L = \sum_{k=1}^{n}\frac{\partial L}{\partial v^k}dq^k$ (this means that the coefficients of the dv^k terms are all null), so that the 2-form $\Omega_L = d\Theta_L$ is closed and, due to the regularity of the Riemannian structure, also regular: therefore the tangent bundle TQ becomes a symplectic manifold. Moreover, define the *Hamiltonian function* $H(q,v) = \sum_{k=1}^{n}\frac{\partial L}{\partial v^k}v^k - L(q,v)$. It is a simple matter of calculations to show that the symplectic flow generated by the symplectic gradient of the Hamiltonian $H(q,v)$ is equivalent to the Lagrange equations.

The definition of Θ_L and Ω_L shows that the symplectic structure of TQ is *not* in canonical form. In general, finding the canonical coordinates is not an easy task, but, fortunately, in the present case it is readily seen that the *Legendre transformation*

$$(q, v) \rightarrow (q, p) \quad \text{where } p_k(q, v) = \frac{\partial L(q, v)}{\partial v^k}$$

achieves our aim, sending Θ_L into the canonical $\Theta = \sum_{k=1}^{n} p_k dq^k$. In other words, by the Legendre transformation we pass from the tangent bundle TQ to the *cotangent bundle* T^*Q (the union of Q and all the dual spaces of the tangent spaces); this, in order to use the natural, or canonical, coordinates of the symplectic structure. The p_k terms are named *(canonical) momenta*, and the regularity of the Riemannian metric ensures that the relation $p_k(q, v)$ can be inverted, thus giving the velocities as functions of coordinates and momenta. T^*Q is usually referred as the *phase space* of the system.

With a little abuse of notation, we write $H(q, p) = H(q, v(q, p))$. Recalling that the Lagrange equations are expressed in terms of the symplectic flow generated by the symplectic gradient of the Hamiltonian, and that now the symplectic structure is the canonical one, we have that the celebrated *Hamilton equations*

$$\dot{q}^k = \frac{\partial H(q, p)}{\partial p_k}, \quad \dot{p}_k = -\frac{\partial H(q, p)}{\partial q_k}$$

are equivalent to the Lagrange equations. Hereafter we will work in canonical coordinates.

1.4 The Liouville and Arnold Theorems

Given a Hamiltonian, finding explicit solutions of the related, usually nonlinear, equations is in general a hopeless task. The very few cases in which this is possible share the property that the problem can be reduced to the quadrature, i.e., to invert functions and perform integrations. This is the case of *complete integrability*.

THEOREM 1.1 *(Liouville) A sufficient condition for the complete integrability of an n-dimensional Hamiltonian system is that there exist n first integrals $\Phi_k(q, p)$ that are independent and in involution, that is $\{\Phi_i, H\} = 0$ and $\{\Phi_i, \Phi_k\} = 0, \forall i, k$.*

For the proof, one basically seeks a canonical transformation sending the first integrals into the new momenta, so that the equations of the transformed Hamiltonian, which will depend only on the momenta, are trivially integrable. Clearly, the transformation exists if and only if the first integrals are in involution, since this holds true for any n-tuple of canonical momenta. For every n-tuple $\alpha_1, \ldots, \alpha_n$ of constants, the generating function appears as the potential of a vector field in Q that comes from inverting the relations $\Phi_k(q, p) = \alpha_k$ with respect to the momenta. This vector field is therefore known and finding its potential leads to performing n integrations.

Varying the constants, the n relations $\Phi_k(q, p) = \alpha_k$ determine a foliation of the $2n$-dimensional phase space T^*Q in n-dimensional level hypersurfaces. What is the topology of these hypersurfaces? At first glance one can say nothing on this topology, which depends on the analytical expression of the first integrals. But here the involutivity condition, which in turn is a direct consequence of the canonical structure, plays a key role.

THEOREM 1.2 *(Arnold) Given a completely integrable n-dimensional Hamiltonian system, the compact and connected components of the level surfaces of the first integrals are diffeomorphic to an n-dimensional torus. Moreover, there exist (locally) canonical coordinates called action-angle coordinates, such that the action variables parametrize the set of the tori whereas the angles parametrize the points on a torus. The Hamiltonian, expressed as a function of these coordinates, depends only on the actions, so that the dynamical evolution is a uniform rotation on an invariant torus.*

The key point in the proof consists of viewing the functions $\Phi_k(q, p)$ as Hamiltonians generating flows that, by involutivity, respect the foliation and commute with one another. It is natural to think (though this is the central point of the proof) that the sole n-dimensional compact hypersurface carrying n independent and commuting flows is the product of n circles, i.e., the torus $\mathbb{T}^n$. To find the action-angle variables I_j, φ^k, $j, k = 1, \ldots, n$, let γ_i be the cycles on the torus generated by I_i and φ^i the corresponding angles. In order that I_j, φ^k are canonical coordinates, we require that the two 1-forms $\sum_{k=1}^n p_k dq^k$ and $\sum_{k=1}^n I_k d\varphi^k$ differ by an exact 1-form, whose integral along a cycle is consequently zero. Hence

$$\sum_{k=1}^n \oint_{\gamma_i} I_k d\varphi^k = \sum_{k=1}^n \oint_{\gamma_i} p_k dq^k.$$

On the left-hand side $d\varphi^k = 0$, $\forall k \neq i$, and I_i is constant along the cycle since, by definition, it is just the Hamiltonian generating γ_i. Therefore we define

$$I_i = \frac{1}{2\pi} \sum_{k=1}^n \oint_{\gamma_i} p_k(q, \alpha) dq^k.$$

The action variables are therefore invertible functions of the first integrals Φ's only, thus arranged in a system of n independent first integrals in involution. We have therefore found a canonical transformation that sends the old Hamiltonian $H(p, q)$ into a new Hamiltonian $K(I)$, which now depends only on the actions. The Hamilton equations are

$$\dot{I}_k = -\frac{\partial K(I)}{\partial \varphi^k} = 0, \quad \dot{\varphi}^k = \frac{\partial K(I)}{\partial I^k},$$

i.e., the actions are first integrals (as already known) and the angles evolve linearly in time.

Notice that all the completely integrable systems with the same dimensions are *locally* isomorphic, being described by a foliation in tori, but differ for the singularity distribution. As an elementary example, let us consider a pendulum: the phase space is a cylinder and exhibits two equilibrium positions, the first is stable, the second unstable; this latter is a homoclinic point and is joined to itself by two separatrices, which are dynamically covered in infinite time (see Figure 3.3 on page 117). Cutting out these singularities, we are left with three disconnected components, each of them diffeomorphic to the product of a circle with an open interval of the real line: the inside of the two separatrices is the oscillatory or libration[1] zone, whereas the other two are the circulation zones. Comparing the harmonic oscillator with the pendulum, one sees that now the phase space is the plane minus the origin, and the system is isomorphic only to the libration zone of the pendulum.

1.5 Quasi-Integrable Hamiltonian Systems and KAM Theorem

Let us consider a completely integrable system to which we add a "small" nonintegrable perturbation or, in brief, a *quasi-integrable* system. The Hamiltonian will be of the type

$$H(I, \varphi) = H_0(I) + \varepsilon H_p(I, \varphi), \quad \varepsilon << 1. \tag{1.5.1}$$

As said previously, the central question is: Does this very small perturbation destroy the foliation in tori completely? Before proceeding, we consider a numerical experiment encompassing the essence of the problem, as will be clear later in the course of the book. Let us consider the *standard map*, introduced by Chirikov (1979) and regarding a symplectic transformation $S : x \rightarrow x'$ of the plane into itself:

$$x_1' = x_1 + \varepsilon \sin x_2, \ x_2' = x_2 + x_1'.$$

For $\varepsilon = 0$ the evolution of the two variables is very simple: x_1, which is of the action type, stays unchanged while x_2, an angular variable, grows

---

[1] From the Latin *libra*, i.e., balance.

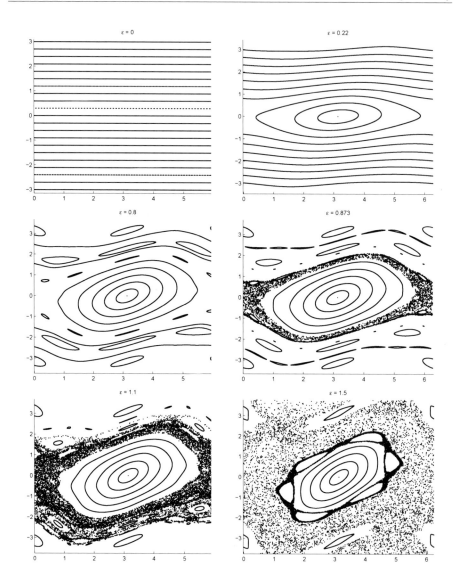

Figure 1.1: Rise of chaos in the standard map.

linearly: see Figure 1.1. Turning on the perturbation with $\varepsilon = 0.22$, the topology of the foliatiation changes abruptly, strongly resembling that of the pendulum, and what are called the *resonant* tori appear. Increasing the perturbation to $\varepsilon = 0.80$, some tori of the circulation zone are destroyed and replaced by a chain of adjacent resonant tori, even though the overall regularity of the motion seems preserved. However, increasing further to

$\varepsilon = 0.873$, 1.1, and 1.5 one sees that the foliation is progressively replaced by more and more wide zones of chaotic evolution, with the survival of some small islands of order. Notice, moreover, that also some tori of the libration zone break down and are replaced by a chain of resonant tori of second level, which in turn generate their own chaos.

Going back to the generic perturbed Hamiltonian (1.5.1), we see that the problem is no longer solvable through quadratures. Then, we proceed looking for a canonical transformation $I, \varphi \rightarrow I', \varphi'$, that differs from the identity by a quantity of order ε, such that the transformed Hamiltonian is integrable up to the second order terms. One may iterate this procedure, pushing the perturbation to the third order, and so on. If the process converges (but this is the key point), by increasing the order one obtains better and better approximations.

Let the canonical transformation $\varphi, I \mapsto \varphi', I'$ be generated by $W = \varphi \cdot I' + \varepsilon S(\varphi, I')$, that is

$$I = I' + \varepsilon \frac{\partial S}{\partial \varphi}, \quad \varphi' = \varphi + \varepsilon \frac{\partial S}{\partial I'},$$

where S is, for the moment, unknown. Define the *averaged* perturbation Hamiltonian

$$\overline{H_p}(I) = \frac{1}{(2\pi)^n} \int_0^{2\pi} \cdots \int_0^{2\pi} H_p(\varphi, I) \, d\varphi^1 \ldots d\varphi^n$$

and the *frequency vector*

$$\omega = \omega(I') = \left(\frac{\partial H_0}{\partial I} \right)_{I=I'}.$$

As one easily verifies, if we are able to find a function S that solves the *homological equation*

$$\omega \cdot \frac{\partial S}{\partial \varphi} + H_p(\varphi, I') - \overline{H_p}(I') = 0,$$

we succeed in pushing the perturbation to the second order. Iterating this procedure, we hope to end up with a canonical transformation $\text{Can}^\infty : I, \varphi \rightarrow I^\infty, \varphi^\infty$ and a completely integrable Hamiltonian $H_0^\infty(I^\infty)$. Therefore (but, we stress again, provided the procedure converges) the phase space of the perturbed Hamiltonian system would be foliated by n-dimensional hypersurfaces diffeomorphic to tori. The perturbation would simply cause a deformation of the original tori, i.e., those related to the unperturbed Hamiltonian, without destroying the well-ordered pattern.

The critical points in pursuing the outlined program are the following two: the solution of the homological equation and the convergence of the sequence of canonical transformations leading to $\text{Can}^\infty : I, \varphi \mapsto I^\infty, \varphi^\infty$.

Let us consider the first point. In order to solve the homological equation, we resort to Fourier series

$$H_p(\varphi, I') - \overline{H_p}(I') = \sum_{k \neq 0} H_k(I') e^{ik \cdot \varphi}, \quad S(\varphi, I') = \sum_{k \neq 0} S_k(I') e^{ik \cdot \varphi},$$

where $k = k_1, \ldots, k_n$ is a vector with integer components. We drop the term $k = 0, \ldots, 0$ since the mean value of S would be annihilated by the differentiation operator; this also imposes that the remaining part in the homological equation have null average. The *formal* solution of the homological equation is

$$S_k(I') = \frac{iH_k(I')}{\omega \cdot k},$$

which, however, shows that we are facing a serious convergence problem: clearly, there exist frequency vectors such that $\omega \cdot k = 0$ for some k, and this makes the formal solution meaningless. Such a frequency vector is called *resonant*, and it is characterized by the reciprocal rationality of its components. We must thus exclude such resonant terms and, moreover, those terms for which $\omega \cdot k$ is much smaller than the corresponding H_k in the numerator. This is the celebrated problem of the "small divisors" or "small denominators."

The situation may appear hopeless, but, fortunately, a classical result in Diophantine theory guarantees that $\omega \cdot k$ can be bounded from below, without yielding an empty set. More precisely, the inequality

$$|\omega \cdot k| \geq \frac{\gamma}{|k|^n} \quad \forall k \in \mathbb{Z}^n - \{0\}, \quad |k| \stackrel{\text{def}}{=} \sum_j |k_j|,$$

for some positive γ is satisfied by a set of real vectors ω of large relative measure, the complement of this set having Lebesgue measure $\mathcal{O}(\gamma)$. This inequality is referred to as the *Diophantine condition*. This is a key point. In fact, if H_p is analytic, it is easy to prove that its Fourier coefficients H_k decay exponentially with $|k|$, while $\frac{1}{|\omega \cdot k|}$ grows at most as a power, thanks to the Diophantine condition. This enables us to prove the convergence of the formal expansion.

The first point is thus overcome but at a price: the frequency vectors that do not satisfy the Diophantine condition, hence the corresponding tori, must be excluded, and the foliation of the phase space by invariant tori is lost: conserved and destroyed tori are mixed together, the first ones forming a complicated Cantor set.

The second point is technically more difficult. Basically, one fixes a torus to which there corresponds a frequency vector satisfying the Diophantine condition, then proves that, if the perturbation parameter is sufficiently small, the procedure converges to Can^∞. The proof, however, requires a further condition: in order to keep the frequency vector fixed when higher order terms of the perturbation come into play (a necessary condition, since in its, no matter how small, neighborhood there are other vectors *not* satisfying the Diophantine condition), one must slightly change the torus; hence, it is necessary that the *frequency map I → ω(I)* be at least locally invertible, thus $\det \left(\frac{\partial \omega_i}{\partial I_k} \right) = \det \left(\frac{\partial^2 H_0}{\partial I_i \partial I_k} \right) \neq 0$.

The above discussion is summarized by the famous KAM theorem.

THEOREM 1.3 *(KAM) Given the perturbed Hamiltonian*

$$H(\varphi, I) = H_0(I) + \varepsilon H_p(\varphi, I),$$

with $\det \left(\frac{\partial^2 H_0}{\partial I_h \partial I_j} \right) \neq 0$, *for every set I^* of the actions such that the unperturbed frequencies* $\omega(I^*) = \frac{\partial H_0}{\partial I}(I^*)$ *satisfy the Diophantine condition, the tori $I^* =$ constant survive, though slightly deformed, with respect to sufficiently small perturbations.*

Notice that the destroyed tori are not *completely* replaced by chaotic orbits. Indeed, inside a resonance we can find an adapted Hamiltonian (called *normal resonant*), which turns out to be the one corresponding to a slightly perturbed pendulum. Again, from the KAM theorem one expects the existence of regular resonant tori, which are obtained by deforming those of the unperturbed, thus integrable, pendulum; these, in turn, can develop secondary resonances with their related normal resonant Hamiltonians, then the pattern repeats itself endlessly. The chaos is restricted to the orbits starting in the thin stochastic layer surrounding the separatrices of the pendulum created by the resonances. These orbits "hesitate" among libration and clockwise or counterclockwise circulation, giving rise to chaotic dynamics through the mechanism of the *homoclinic tangle*.

1.6 Geography of the Phase Space

At this point, we are able to sketch the overall structure of the phase space of a quasi-integrable system, taking into account that the KAM theorem changes our point of view: from now on, our attention will be focused not on the single orbits but on the tori, since all the orbits on the same torus share the same destiny. Among other things, this drastically simplifies the work, reducing from $2n$ to n the number of the classifying parameters, for which we may use indifferently the actions or the relative frequency vector.

When the perturbative parameter grows, the nature of the phase space changes, covering, in ascending order, three different situations.

(i) *KAM*: essentially all points are regular, almost all unperturbed tori are conserved, and the dynamics is basically controlled by the KAM theorem. The system is in practice indistinguishable from a completely integrable one.

(ii) *Nekhoroshev*: the measure of the destroyed tori is small but not negligible. They form an *Arnold web*, which in the frequency space is given by frequencies satisfying the resonance relations $\omega \cdot k = 0$, along with a neighborhood decreasing exponentially with the order $\sum_{j=1}^{n} |k_j|$ of the resonance itself. The Arnold web is therefore the union of the neighborhoods of all the hyperplanes of codimension one through the origin and with rational slope. Assume for simplicity $n = 3$. In Figure 3.2 on page 116 a section with the plane $\omega_3 = 1$ in the 3-dimensional frequency space is shown, thus with equation $k_1 \omega_1 + k_2 \omega_2 + k_3 = 0$: the "skeleton" is formed by the lines whose slope and intersection with the axes take rational values, "fleshed out" by the resonance strips. The Arnold web is *connected, open, and dense* in the action space with, however, a relative small measure vanishing with the square root of the perturbative parameter. On a 2-dimensional energy surface of the action space an image of figure 3.2 appears, distorted under the diffeomorphism given by the local inverse of the frequency map $I \to \omega(I)$. The dynamics is still controlled almost everywhere by the KAM theorem except for the Arnold web, where it is controlled by the Nekhoroshev theorem. A point of a stochastic layer orbit (i.e., exactly on the border of a resonance) can in principle travel along the whole Arnold web, reaching the neighborhood of every point in action space but in a very long time, which grows exponentially with the inverse of the perturbative parameter. This phenomenon, whose existence is not in general proven, is known as *Arnold diffusion*.

(iii) *Chirikov*: the global measure of the resonances does not leave any place for invariant tori, and the dynamics is no longer controlled by the KAM and Nekhoroshev theorems but by the Chirikov overlapping criterion. When the resonances overlap, the motion can jump from one resonance to another, giving rise to large-scale diffusion with a time scale much shorter with respect to the Arnold diffusion. The system is fully chaotic.

1.7 Numerical Tools

To better understand the KAM theorem, it is useful to proceed with some numerical examples regarding case (ii), which is surely the most interesting. Several tools can be used.

The Poincaré section is a long-standing method, very effective for systems with two degrees of freedom, thus with a 4-dimensional phase space and a 3-dimensional hypersurface of constant energy. Sectioning with a plane and recording the points where it is crossed by an orbit, one can visualize the trace of the torus, if any, around which the orbit winds; a non-structured dust will denote instead a chaotic orbit. Two examples are given in figure 3.5 on page 121 and in Figure 5.1 on page 149.

As already pointed out, the usefulness of the method is clearly restricted to systems with just two degrees of freedom; moreover, if the perturbation is very small, the resonances are extremely thin and may escape from the visualization. In fact, the method is not in the spirit of the KAM theorem, since it focuses attention on the orbits, instead of on the tori. In contrast, the following methods are based on frequency analysis, and as such are tori-oriented.

The Fast Fourier Transform (FFT) is the implementation of the elementary Fourier transform and is applicable to the output of a numerical integration. If all the computed frequencies are a linear combination with integer coefficients of some n fundamental ones, the spectrum is regular and the motion winds around a KAM torus.

The Frequency Modified Fourier Transform (FMFT) allows one to find the spectrum of a "signal" $Z(t)$, but seeking numerically the maximum of the function

$$\phi(\omega) = \frac{1}{2\pi} \int_{-T}^{T} Z(t) e^{-i\omega t} dt.$$

The output is decisively more accurate, but nothing is perfect, and trying to resolve two very close frequencies yields a slightly imprecise result.

The Frequency Modulation Indicator (FMI) exploits just this imprecision to detect the resonances. It associates to each n-tuple of action values, hence to each torus, a number that measures how much the fundamental frequencies are frequency modulated. Indeed, for a KAM torus the n fundamental frequencies (i.e., those coming from the first component in the Fourier analysis) are time-constant; on the contrary, inside a resonance the superimposed pendulum causes a frequency modulation. Without going into detail on how this frequency modulation is numerically detected, we can reach the conclusion that a picture plotting the FMI values as a function of the actions will be able to represent the distribution of the resonances.

In Figure 8.11 on page 249 an example is given for a system with three degrees of freedom. The Arnold web in the action space shows up clearly, and distortions of the pattern in Figure 3.2 appear. The dark blue indicates negligible values of the FMI corresponding to KAM tori, whereas light blue, yellow, and red indicate intermediate and high values, i.e., resonances and chaos. Zooming into a resonance shows that the structure repeats over and over. The outcome of these numerical experiments is the concrete possibility of detecting more and more resonances, as long as we can afford to pay the price of computational complexity and time costs.

With regard to the Arnold diffusion, its existence has been proven to be possible, while its practical relevance is an open question; in particular it is unknown if the phenomenon is generic for every quasi-integrable Hamiltonian system. With the same software used for the computation of the FMI, one can numerically measure the possible drift of the values of the fundamental frequencies in a long-time orbit, thus recognizing a transition through different tori. Some preliminary results (*very* expensive in CPU time terms) seem to suggest that actually the phenomenon is generic: as expected, the points starting on the edge of a resonance, thus in the thin stochastic layer surrounding the separatrices, travel but very slowly along the resonance strips.

Equipped with such analytical and numerical tools, we can tackle some concrete examples regarding the perturbed Kepler (i.e., two-body) problem and the multi-body gravitational problem.

1.8 The Perturbed Kepler Problem

The starting point is the isomorphism between the regularized Kepler problem and the geodesic flow on the sphere. To get a geometrical insight, consider for simplicity the 2-dimensional Kepler problem and a geodesic circle on a 2-dimensional sphere which can be safely rotated into the position

$$X_1 = \sin s, \quad X_2 = -\cos\beta \cos s, \quad X_3 = \sin\beta \cos s.$$

The vector (X_1, X_2, X_3) is orthogonal to the vector $(0, \sin\beta, \cos\beta)$; β measures the angle between the equator $X_3 = 0$ and the circle in question, whereas s is the angle along the circle itself. The definition $Y_k = \frac{dX_k}{ds}$ yields

$$Y_1 = \cos s, \quad Y_2 = \cos\beta \sin s, \quad Y_3 = -\sin\beta \sin s.$$

Since the explicit form of the extended stereographic mapping (see figure 2.1 on page 19) is

$$x_k = \frac{X_k}{1 - X_{n+1}}, \quad y_k = Y_k (1 - X_{n+1}) + X_k Y_{n+1},$$

the image of the circle and its tangent vector under the extended stereographic projection is

$$x_1 = \frac{\sin s}{1 - \sin\beta \cos s}, \quad x_2 = -\frac{\cos\beta \cos s}{1 - \sin\beta \cos s},$$
$$y_1 = \cos s - \sin\beta, \quad y_2 = \cos\beta \sin s.$$

With $q_k = y_k$ and $E = \sin\beta$, this takes the form

$$q_1 = \cos s - E, \quad q_2 = \sqrt{1 - E^2} \sin s,$$

which is the representation of an ellipse of eccentricity $E = \sin\beta$ in terms of the eccentric anomaly s. Moreover, with $p_k = -x_k$ we get

$$p_1^2 + (p_2 - \tan\beta)^2 = 1 + \tan^2\beta,$$

which is the representation of a circle in the Cartesian plane $p_1 p_2$, which is the *hodograph curve*. Therefore, the trajectory of the moving point of the Kepler problem is the direct projection, followed by a translation, of a geodesic circle onto the equatorial plane, while the trajectory of the velocity is the stereographic projection.

Generalizing to the 3-dimensional case, some facts appear relevant. First, the group SO(4), which acts isometrically on the 3-dimansional sphere, is the symmetry group of the Kepler problem; then, the SO(2) group generates the motion on the geodesic circle; lastly, the dynamical evolution of position and velocity can be parametrized with two orthogonal vectors spanning the circle itself. Roughly speaking, the two groups and the couple of vectors fit together to form the dynamical group SO(2,4).

Taking the two orthogonal vectors as dynamical variables also turns out to be suited for studying the perturbed case, for example the hydrogen atom in electric and magnetic fields. The cotangent bundle to the 3-dimensional sphere, i.e., the phase space of the regularized Kepler problem, is twofold reduced. Thus we arrive to a 2-dimensional spheroid, on which the intersections of the level surfaces of the perturbation Hamiltonian describe globally the essence of the dynamics, obviously up to fast oscillations. See, e.g., Figures 8.1–8.4 on pages 240 and 241.

1.9   The Multi-Body Gravitational Problem

Deducing the motion of bodies interacting gravitationally is probably the most important mechanical problem but also the most difficult. Already the three-body problem is not integrable, even if the masses are very small but of comparable size, and this fact generally prevents the use of perturbative methods.

Some important exceptions are: the planar three-body problem, which admits a global treatment in its two limit, i.e., lunar and planetary cases; then the classical 3-dimensional planetary problem. By the *planetary problem* one means the mechanical system consisting of a body of large mass, the "Sun," and other bodies much smaller, the "planets," interacting through gravitational forces. By the *lunar problem* one means the system consisting of a small body, the "Moon," rotating around the "Earth," with a third body, the "Sun," much more distant.

Let us consider the planar case. The planar system is first reduced to four degrees of freedom thanks to its translational invariance; then, averaging along the unperturbed motion, it is further reduced to two degrees of freedom. The averaged Hamiltonian inherits, from the original one, the rotational plane invariance, and this symmetry results in a further reduction to a system with one degree of freedom, hence integrable.

In the 3-dimensional case, let us consider the three-body planetary problem, the extension to the generic case being straightforward. The system is easily reduced to two uncoupled Kepler problems plus a perturbative term proportional to the inverse of the distance $\triangle$ between the two planets. The secular Hamiltonian is obtained by averaging $\triangle^{-1}$ along the unperturbed motion, i.e., along the Keplerian ellipses. Unfortunately, this is a nontrivial task, which cannot be carried out in a closed form and requires two preliminary steps. With the first step the expression of the two position vectors is put in a suitable form, i.e., as a function of an evolutional parameter, closely related to the time, and of five constant parameters characterizing the ellipse. The second step consists in a series expansion of $\triangle^{-1}$ with respect to eccentricity and inclination.

To accomplish the first step, one could use the Keplerian elements of the orbit, but they suffer from the drawback of being singular for orbits that are circular and/or lying on the reference (ecliptic) plane. In contrast, the Poincaré variables are regular for orbits with small eccentricities and inclinations, and are thus well suited for studying the planetary problem.

The Keplerian elements of the orbit have a clear geometrical interpretation: semimajor axis and numerical eccentricity fix size and shape of the ellipse, while inclination, longitude of the ascending node, and argument of the pericenter are the three Euler angles fixing the spatial orientation of the ellipse. In contrast, the Poincaré variables are usually defined in a purely algebraic manner and lack a geometrical interpretation. This makes finding the expansion of the two position vectors somewhat involved and awkward, which surely does not simplify the subsequent series development and averaging process. It can be shown that exploiting the geometry of the SO(3) group allows one to write the expression of the Keplerian motion in a very suitable form. Then the development is performed in such a way that it is immediate, by direct inspection, to detect the terms that vanish under the averaging process. This produces a drastic simplification and allows us to smartly group the surviving terms in a reasonable and adequate manner, the final result being an even, real-valued polynomial in the Poincaré canonical variables. Taking into account only the first quadratic terms, one gets the classical Lagrange–Laplace planetary theory, whose dynamics is compared with the true one: see Figures 9.10 on page 298, 9.11 on page 299, 9.13 on page 301, and 9.14 on page 302.

Lastly, two numerical examples of the distribution of the resonances in our solar system are computed with the FMI method: see Figure 9.15 on page 303.

CHAPTER 2

Analytical Mechanics and Integrable Systems

*We are in the rarefied atmosphere of theories
of excessive beauty and we are nearing a high plateau
on which geometry, optics, mechanics, and wave
mechanics meet on common ground.*

— C. LANCZOS

Analytical mechanics is the basic tool we will utilize through the whole book. The aim of this chapter is to succinctly introduce and define some basic concepts, such as Hamilton equations, symplectic (i.e., canonical) transformations, symmetry and reduction, integrable systems, and action-angle variables. This will be done in the third section of the chapter. In the first two sections we recall the main ideas of differential geometry and Lie groups, which are the natural language of analytical mechanics.

2.1 Differential Geometry

The natural arena of Lagrangian dynamics is a Riemannian manifold, while that of Hamiltonian dynamics is a symplectic or Poisson manifold. In order to deal with these structures, we must state some definitions in advance: differentiable manifolds, the tangent and cotangent spaces, pullback and push-forward, tensors, and forms. For a more serious study, many books have been written, for example Kobayashi & Nomizu (1963–1968), Auslander & MacKenzie (1977), Helgason (1978), Choquet-Bruhat (1968), Von Westenholz (1978), Sternberg (1983), Dubrovin, Novikov & Fomenko (1982–1987), and the very readable Crampin & Pirani (1986) or Nash & Sen (1983). For topology, see Croom (1989).

2.1.1   Differentiable Manifolds

The concept of differentiable manifold generalizes the intuitive idea of surface, getting rid of the restriction to two dimensions and of the immersion in some ambient space. The central idea is that every surface can be constructed by assembling together some patches that are homeomorphic (i.e topologically equivalent) to open connected sets of $\mathbb{R}^n$.

DEFINITION 2.1 *A topological space M is a differentiable manifold of dimension n if*

   (i) *M is provided with a family of pairs $\{(U_\alpha, \phi_\alpha)\}$, where the U_α are a family of open connected sets that cover M, while the ϕ_α are homeomorphisms from U_α to an open set of $\mathbb{R}^n$;*

   (ii) *given U_α and U_β such that $U_\alpha \cap U_\beta \neq$ empty set, the map $\phi_\beta \circ \phi_\alpha^{-1}$ from the subset $\phi_\alpha(U_\alpha \cap U_\beta) \subset \mathbb{R}^n$ to the subset $\phi_\beta(U_\alpha \cap U_\beta) \subset \mathbb{R}^n$ is C^∞, thus infinitely differentiable (or smooth).*

Item (i) asserts that M is a space that is locally like $\mathbb{R}^n$; that is, M can be covered with patches U_α, which assign coordinates in $\mathbb{R}^n$ by the ϕ_α. Within one of these patches M looks like a subset of $\mathbb{R}^n$, but in general we do not expect M to be globally homeomorphic to $\mathbb{R}^n$, since this depends on how the patches fit together to form the whole M. Item (ii) asserts that, if two patches overlap, then in the overlap region $U_\alpha \cap U_\beta$ we have two sets of coordinates in $\mathbb{R}^n$, given by ϕ_α and ϕ_β, and that if we decide to change from one set of coordinates to the other, i.e., to use the function $\phi_\beta \circ \phi_\alpha^{-1}$, this can be done in a smooth manner.

EXAMPLE 2.1 Let us consider the sphere $S^n = \{X \in \mathbb{R}^{n+1} : X_1^2 + X_2^2 + \cdots + X_{n+1}^2 = 1\}$, which is in a natural way a space whose topology is induced from the usual topology of $\mathbb{R}^{n+1}$. The sphere can be covered with the two charts obtained by stereographic projection, one from the North pole $N \equiv (0, 0, \ldots, 1)$ and the other from the South pole $S \equiv (0, 0, \ldots, -1)$; see figure 2.1. The same point of the sphere is labeled by the two different coordinate sets $\{x_k^N\}$ and $\{x_k^S\}$, $k = 1, \ldots, n$, that is,

$$\phi^N : U_N = S^n - \{N\} \to \mathbb{R}^n, \quad \text{in coordinates } x_k^N = \frac{X_k}{1 - X_{n+1}},$$

$$\phi^S : U_S = S^n - \{S\} \to \mathbb{R}^n, \quad \text{in coordinates } x_k^S = \frac{X_k}{1 + X_{n+1}}.$$

In the overlap region $U_N \cap U_S = S^n - \{N\} - \{S\}$ the transition functions

$$\phi^S \circ (\phi^N)^{-1}, \quad \text{in coordinates } x_k^S = \frac{x_k^N}{(x_1^N)^2 + \cdots + (x_n^N)^2},$$

$$\phi^N \circ (\phi^S)^{-1}, \quad \text{in coordinates } x_k^N = \frac{x_k^S}{(x_1^S)^2 + \cdots + (x_n^S)^2},$$

are clearly C^∞.

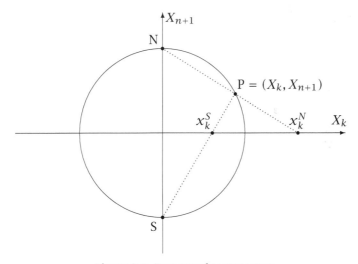

Figure 2.1: Stereographic projection.

EXAMPLE 2.2 Let us consider the 2-dimensional[1] real projective space

$$P_{\mathbb{R}}^2 = (\mathbb{R}^3 - \{0\})/\sim, \quad \vec{X} \sim \vec{X}' = \lambda \vec{X}, \ \vec{X} \in \mathbb{R}^3 - \{0\}, \ \lambda \in \mathbb{R} - \{0\}.$$

It is thus a space whose "points" are the straight lines through the origin of $\mathbb{R}^3$. A model is the sphere S^2 but with the diametrically opposed points

---

[1]We limit ourselves to two dimensions for simplicity, but the generalization to any dimension is straghtforward.

identified; it is obvious that it cannot be immersed homeomorphically into $\mathbb{R}^3$. We make $P_\mathbb{R}^2$ a topological space with the quotient topology. $P_\mathbb{R}^2$ may be covered with three charts:

$$\phi_x : U_x = P_\mathbb{R}^2 - \{X_1 = 0\} \to \mathbb{R}^2, \quad \text{in coordinates } x_1 = \frac{X_2}{X_1}, \ x_2 = \frac{X_3}{X_1},$$

$$\phi_y : U_y = P_\mathbb{R}^2 - \{X_2 = 0\} \to \mathbb{R}^2, \quad \text{in coordinates } y_1 = \frac{X_1}{X_2}, \ y_2 = \frac{X_3}{X_2},$$

$$\phi_z : U_z = P_\mathbb{R}^2 - \{X_3 = 0\} \to \mathbb{R}^2, \quad \text{in coordinates } z_1 = \frac{X_1}{X_3}, \ z_2 = \frac{X_2}{X_3},$$

where X_1, X_2, X_3 are Cartesian coordinates in $\mathbb{R}^3$. Consider for example the overlap region $U_x \cap U_y$; here the transition functions

$$\phi_x \circ \phi_y^{-1}, \quad \text{in coordinates } x_1 = \frac{1}{y_1}, \ x_2 = \frac{y_2}{y_1},$$

$$\phi_y \circ \phi_x^{-1}, \quad \text{in coordinates } y_1 = \frac{1}{x_1}, \ y_2 = \frac{x_2}{x_1},$$

are C^∞, and analogously for the other two overlap regions $U_x \cap U_z$ and $U_y \cap U_z$.

The pair (U_α, ϕ_α) for a fixed α is a *chart*. Two charts (U_α, ϕ_α) and (U_β, ϕ_β) are *compatible* if $\phi_\alpha \circ \phi_\beta^{-1} : \mathbb{R}^n \to \mathbb{R}^n$ and its inverse are smooth. A family of compatible charts $\{(U_\alpha, \phi_\alpha)\}$ such that $\bigcup_\alpha U_\alpha \supseteq M$ is an *atlas* for the differentiable manifold M.

Consider two differentiable manifolds M and N, respectively m and n-dimensional and let $f : M \to N$ be a map. Choosing local charts, ϕ for M and ψ for N say, the map f induces the map $\psi \circ f \circ \phi^{-1} : \mathbb{R}^m \to \mathbb{R}^n$, called the *coordinate presentation* of f. A coordinate presentation of a map f is therefore a map of open subsets of real number spaces, given explicitly by n functions of m variables. The map $f : M \to N$ is called *smooth* if its coordinate presentation is given by smooth functions for all charts of an atlas of both its domain and codomain. For the differentiability of M and N, this smoothness definition does not clearly depend on the choice of the particular local chart.

An important special case of a map of manifolds is a smooth bijective map with a smooth inverse. Such a map is called a *diffeomorphism*, and two manifolds connected by a diffeomorphism are said to be *diffeomorphic*. From the point of view of differential geometry, two diffeomorphic manifolds are the "same" manifold.

Tangent and Cotangent Spaces

A particular but very important case of a map between manifolds is that defining a curve. A *curve* in an n-dimensional manifold is a map $\sigma : \mathbb{R} \to M$, that is in local coordinates

$$\phi \circ \sigma : x^k = x^k(t), \quad k = 1.2. \dots, n \quad a < t < b, \ a, b \in \mathbb{R}.$$

Given a smooth function $f : M \to \mathbb{R}$, we define the derivative vf of the function along the curve σ :

$$vf = \frac{d}{dt}(f \circ \sigma) = \partial_k f \frac{dx^k}{dt}, \quad \partial_k = \frac{\partial}{\partial x^k}.$$

We can view v as an operator that maps functions into functions; it has the two properties (g is another function and $\lambda, \mu \in \mathbb{R}$)

(i) $v(\lambda f + vg) = \lambda vf + \mu vg,$

(ii) $v(fg) = (vf)g + f(vg),$

showing that v is a linear operator satisfying the Leibniz rule of the derivative of a product.

We define the *tangent vector* at a point p of a smooth manifold M as an operator on smooth functions that satisfies properties (i) and (ii). Denote the set of all tangent vectors to M at p by T_pM. We make this set a vector space defining the linear combination

$$(\lambda v + \mu w)f = \lambda vf + \mu wf, \quad v, w \in T_pM, \quad \lambda, \mu \in \mathbb{R}.$$

The linear space T_pM is called the *tangent space* to M at p. The definition of tangent space generalizes the intuitive idea of a plane tangent to a surface, but without resorting to the immersion of M into an ambient space.

A tangent vector v is written in coordinates as $v^k \partial_k$ and $\{\partial_k\}$ is a basis (more exactly, the *natural basis*) for the vector space T_pM. v^k are said to be the *contravariant components*. This shows, as is intuitively clear, that the dimension of the tangent space is equal to that of M. The adjective "contravariant" reminds us that if T_pM is submitted to a linear change of basis and if we pretend (as is sensible) that the tangent vector, which is intrinsically defined, does not change, the components v^k must change with the inverse transformation of the basis.[2]

As known from linear algebra, the set of the linear maps from a vector space V to $\mathbb{R}$ is again a vector space of the same dimension, which we denote V^* and call a *dual space*. In other words, if $v \in V$ and $w \in V^*$ then $w(v) \in \mathbb{R}$. One often writes this *pairing* as $\langle w, v \rangle$ instead of $w(v)$ to emphasize the linearity. If $\{e_k\}$ is a basis of V, the dual basis $\{e^h\}$ of V^* is by definition that satisfying $\langle e^h, e_k \rangle = \delta_k^h$. Moreover, as is easily shown, the dual of a dual space V^{**} is canonically, i.e., independently of the basis, isomorphic to V.

---

[2]As says one of the chief characters of the Italian novel *Il Gattopardo* by Tomasi di Lampedusa: "If we want things to stay as they are, things will have to change".

The dual space T_p^*M of the tangent space T_pM is called the *cotangent space* to M at p.

For fixed f and p, let us define a map $df : T_pM \to \mathbb{R}$ such that $v \mapsto (df)(v) = vf$. We call df the *differential* of f, where it is understood that every quantity is calculated at point p. It is immediate to verify that the linearity of the operator v entails that of df:

$$df(\lambda v + \mu w) = (\lambda v + \mu w)f = \lambda(df)(v) + \mu(df)(w).$$

The linearity ensures that $df \in T_p^*M$.

Take $f = x^k$, $k = 1, 2, \ldots, n$, where, as usual, $\{x^k\}$ are local coordinates on M, and let us examine the linear action of dx^k on the basis vectors ∂_h

$$dx^k : \partial_h \mapsto (dx^k)(\partial_h) = \frac{\partial x^k}{\partial x^h} = \delta_h^k,$$

which may be written as

$$\langle dx^k, \partial_h \rangle = \delta_h^k.$$

Thus, $\{dx^k\}$ is the dual basis of $\{\partial_h\}$.

An element $\omega \in T_p^*M$ is called a *covector* and is written in coordinates $\omega = \omega_k\, dx^k$. The ω_k are called *covariant* components, because they change like the natural basis.

Push-forward and Pull-back

Given two vector spaces V and W, not necessarily of the same dimension, and a linear map $\mathbf{T} : V \to W$, one defines the dual map $\mathbf{T}^* : W^* \to V^*$ as that satisfying

$$\langle \mathbf{T}^*\omega, v \rangle = \langle \omega, \mathbf{T}v \rangle, \quad \forall v \in V \text{ and } \forall \omega \in W^*.$$

Notice that the two pairings refer to two different couples of vector spaces, i.e., to V^*, V and W^*, W, respectively.

A map $\phi : M \to N$, where M and N are two differentiable manifolds, induces in a natural way a linear map $\mathbf{T}_\phi : T_pM \to T_{\phi(p)}N$, which is defined as that satisfying the intrinsic relation

$$(\mathbf{T}_\phi v)f = v(f \circ \phi), \quad \forall v \in T_pM \text{ and } \forall f : N \to \mathbb{R}.$$

Let us find the expression of the linear map $\mathbf{T}_\phi$ in local coordinates. Let $x = (x^1, \ldots, x^m)$ be the coordinates on M and $y = (y^1, \ldots, y^n)$ those on N; the map ϕ has coordinate presentation $y = \phi(x)$. From elementary calculus,

$$v(f \circ \phi) = v^k \frac{\partial}{\partial x^k} f(\phi(x)) = v^k \frac{\partial}{\partial y^h} f(y) \frac{\partial \phi^h}{\partial x^k} = (\mathbf{T}_\phi v)f, \qquad (2.1.1)$$

from which

$$(\mathbf{T}_\phi)^h_k = \frac{\partial \phi^h}{\partial x^k},$$

and thus the matrix representing $\mathbf{T}_\phi$ with respect to the standard basis is the Jacobian matrix of the map ϕ. We call $\mathbf{T}_\phi$ the *derivative map* of ϕ.

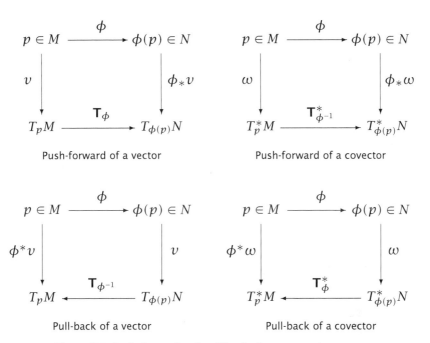

Figure 2.2: Push-forward and pull-back of vectors and covectors.

We can now define the image $\phi_* v$, or *push-forward*, of a vector v on M under a diffeomorphism ϕ as the vector on N that makes the relative diagram in Figure 2.2 commutative; then

$$\phi_* v = \mathbf{T}_\phi v \circ \phi^{-1},$$

which requires the inversion of the map ϕ. This explains our request that ϕ be a diffeomorphism.

Analogously, we define the inverse image, or *pull-back*, $\phi^* \omega$ of a covector on N. We start with a covector on N, instead of on M, because $\mathbf{T}^*$ reverses the direction of the map arrow with respect to $\mathbf{T}$. We have

$$\phi^* \omega = \mathbf{T}^*_\phi \omega \circ \phi.$$

Notice that the pull-back of a covector can also be defined if ϕ is not invertible. For example, if ω is a covector defined on N, it can be safely restricted to a covector on a submanifold $M \subset N$.

The definitions of the push-forward $\phi_* \omega$ of a covector and of the pull-back $\phi^* v$ of a vector are obvious;[3] see Figure 2.2, where $\mathbf{T}_{\phi^{-1}} = \mathbf{T}_\phi^{-1}$. Also in these two cases it is required that ϕ is invertible. If ϕ is not invertible, we may say, however, that two vectors v and w are ϕ-related if

$$w \circ \phi = \mathbf{T}_\phi v, \qquad (2.1.2)$$

which, in some sense, replaces the definition of pull-back and push-forward of a vector. The relation is natural with respect to the *Lie bracket*, i.e., $[v_1, v_2] \overset{\text{def}}{=} v_1 v_2 - v_2 v_1$, as elucidated in the following

PROPOSITION 2.2 *If v_1, v_2 are ϕ-related to w_1, w_2, respectively, then the Lie bracket $[v_1, v_2]$ is ϕ-related to $[w_1, w_2]$.*

A *vector field* on M is defined as the assignment to every point $p \in M$ of an element belonging to $T_p M$. Analogously, the same is true for a *covector field*, or *1-form*. The definition of pull-back and push-forward is straightforwardly extended to fields.

2.1.2 Tensors and Forms

An element of a dual space is a linear map from a vector space to the real numbers: the tensor definition is a natural generalization of this definition. First of all notice that the dual of the dual of a vector space V is canonically isomorphic to V itself; thus, we can define an element of V as a linear map from V^* to the real numbers. Despite the notation, there is therefore a perfect symmetry between V and V^*.

DEFINITION 2.3 *A tensor of (r, s)-type, or an r-fold contravariant and s-fold covariant tensor, is a multilinear (i.e., linear in every argument) map T of the type*

$$T : \overbrace{V^* \times \cdots \times V^*}^{r \text{ times}} \times \overbrace{V \times \cdots \times V}^{s \text{ times}} \to \mathbb{R}.$$

The set of all these maps is a vector space in a natural way, and is called $\mathcal{T}_s^r$. In particular, $\mathcal{T}_0^1 = V$ and $\mathcal{T}_1^0 = V^*$.

DEFINITION 2.4 *We define the* tensor product *of two tensors $T_1 \in \mathcal{T}_s^r$ and $T_2 \in \mathcal{T}_q^p$ the tensor $T_1 \otimes T_2 \in \mathcal{T}_{s+q}^{r+p}$ such that*

$$T_1 \otimes T_2(\omega_1, \ldots, \omega_{r+p}, v_1, \ldots, v_{s+q})$$
$$= T_1(\omega_1, \ldots, \omega_r, v_1, \ldots, v_s) T_2(\omega_{r+1}, \ldots, \omega_{r+p}, v_{s+1}, \ldots, v_{s+q}),$$
$$\forall \omega_k \in V^* \text{ and } \forall v_h \in V.$$

---

[3]Admittedly, the use of the asterisk in the various definitions is at this point somewhat inflated, but unfortunately this is the custom.

Tensor multiplication is associative and distributive with respect to addition, but not commutative.

It is easy to prove that, if $\{e_h\}$ is a basis of V and $\{e^k\}$ the dual basis of V^*, then

$$\left\{ e_{h_1} \otimes \cdots \otimes e_{h_r} \otimes e^{k_1} \otimes \cdots \otimes e^{k_s} \right\}$$

is a basis of $\mathcal{T}_s^r$, which has thus dimension n^{r+s}, where $\dim V = n$. For linearity, we can write for the generic tensor $T \in \mathcal{T}_s^r$

$$T = T_{k_1 \ldots k_s}^{h_1 \ldots h_r} \, e_{h_1} \otimes \cdots \otimes e_{h_r} \otimes e^{k_1} \otimes \cdots \otimes e^{k_s}$$

where

$$T_{k_1 \ldots k_s}^{h_1 \ldots h_r} = T(e^{h_1}, \ldots, e^{h_r}, e_{k_1}, \ldots, e_{k_s})$$

are the components of T in the chosen basis.

Let us consider a linear and invertible change of basis in V, that is, $e_h \mapsto e_k' = A_k^h e_h$. If we require that the transformed basis of V^* be still the dual basis, it induces the transformation $e^h \mapsto e'^k = (A^{-1})_h^k e^h$. Therefore the expression (which is at first sight terrifying)

$$e_{i_1}' \otimes \cdots \otimes e_{i_r}' \otimes e'^{j_1} \otimes \cdots \otimes e'^{j_s}$$

$$= A_{i_1}^{h_1} \ldots A_{i_r}^{h_r} (A^{-1})_{k_1}^{j_1} \ldots (A^{-1})_{k_s}^{j_s} e_{h_1} \otimes \cdots \otimes e_{h_r} \otimes e^{k_1} \otimes \cdots \otimes e^{k_s}$$

holds for the change of basis of $\mathcal{T}_s^r$, and

$$T'^{i_1 \ldots i_r}_{j_1 \ldots j_s} = (A^{-1})_{h_1}^{i_1} \ldots (A^{-1})_{h_r}^{i_r} A_{j_1}^{k_1} \ldots A_{j_s}^{k_s} T_{k_1 \ldots k_s}^{h_1 \ldots h_r}$$

for the corresponding change of the components. The basic point to keep in mind is that contravariant components (indices above) and covariant components (indices below) change with inverse law, so that *contracting* (or *saturating*, i.e., summing over) a pair of such indices leads to a quantity invariant under change of basis.

Sometimes, with a slight abuse of language, one calls for brevity $T_{j_1 \ldots j_s}^{i_1 \ldots i_r}$ "a tensor," instead of "the components of a tensor."

A *tensor field* is the assignment to every point $p \in M$ of a tensor whose underlying vector space is $T_p M$; thus $e_h = \partial_h$ and $e^k = dx^k$. The definition of push-forward and pull-back is naturally extended to the tensor fields, with $A_k^h = (\mathbf{T}\phi)_k^h$.

Forms and Exterior Derivative

Let us define the symmetry properties of a tensor. We say that a tensor is *symmetric* or *skew symmetric* in a pair of indices (of the same type) if

$$T_{\ldots i_r \ldots i_s \ldots} = T_{\ldots i_s \ldots i_r \ldots} \qquad \text{symmetric,}$$

$$T_{\ldots i_r \ldots i_s \ldots} = -T_{\ldots i_s \ldots i_r \ldots} \qquad \text{skew symmetric,}$$

and *completely* symmetric or skew symmetric if the property holds for every pair of indices. Analogously, we have the same for contravariant indices.

Given a generic tensor, we can extract from it the symmetric or skew symmetric part. Let P_r be the group of permutations of the r integers $1, \ldots, r$ and let $\pi \in P_r$. Define

$$(\pi T)_{i_1 \ldots i_r} = T_{\pi(i_1) \ldots \pi(i_r)}.$$

A symmetrization operator Sym and a skew symmetrization operator Sk can be defined as

$$\mathrm{Sym}(T) = \frac{1}{r!} \sum_{P_r} \pi T, \quad \mathrm{Sk}(T) = \frac{1}{r!} \sum_{P_r} \mathrm{sign}_\pi \, \pi T,$$

where $\mathrm{sign}_\pi = \pm 1$ according to whether the permutation π is even or odd. $\mathrm{Sym}(T)$ is a completely symmetric and $\mathrm{Sk}(T)$ a completely skew symmetric tensor, and analogously for contravariant indices.

DEFINITION 2.5 *The covariant completely skew symmetric tensors of order r are called r-covectors. The set of all r-covectors in an n-dimensional vector space V_n forms a vector space called $\bigwedge^r V_n$, $r \leq n$.*

DEFINITION 2.6 *The* exterior product $\wedge$ *(or* wedge product, *or* Grassmann product*) of an r-covector with an s-covector is a map*

$$\wedge : \left(\overset{r}{\bigwedge} V_n, \overset{s}{\bigwedge} V_n \right) \to \overset{r+s}{\bigwedge} V_n$$

defined by

$$\alpha \wedge \beta = \frac{(r+s)!}{r! s!} \mathrm{Sk}(\alpha \otimes \beta), \quad \alpha \in \overset{r}{\bigwedge} V_n, \ \beta \in \overset{s}{\bigwedge} V_n.$$

The exterior product is associative and distributive with respect to the sum, but in general not commutative

$$\alpha \wedge \beta = (-1)^{rs} \beta \wedge \alpha.$$

As one easily checks, a basis for $\bigwedge^r V_n$ is given by the set

$$\{ e^{i_1} \wedge \cdots \wedge e^{i_r}, \ i_1 < i_2 < \cdots < i_r \},$$

and $\dim \bigwedge^r V_n = \binom{n}{r}$.

DEFINITION 2.7 *The contravariant completely skew symmetric tensors of order r are called r-vectors; they form a vector space denoted $\bigwedge_r V_n$, and the exterior product is defined in an analogous way.*

The definition of *field* is naturally extended to r-vectors and r-covectors. In the latter case, we speak of r-*forms* or simply of *forms*.

DEFINITION 2.8 *The exterior derivative is a map from the space of the r-forms to the space of the $(r + 1)$-forms, given in local coordinates by*

$$d\omega = \frac{\partial \omega_{i_1 \dots i_r}}{\partial x^j} dx^j \wedge dx^{i_1} \wedge \cdots \wedge dx^{i_r}, \quad \forall j \text{ and } i_1 < i_2 < \cdots < i_r, \quad (2.1.3)$$

where

$$\omega = \omega_{i_1 \dots i_r} dx^{i_1} \wedge \cdots \wedge dx^{i_r}, \quad \text{with } i_1 < i_2 < \cdots < i_r,$$

is the expression of an r-form.

One checks that this definition is independent of the choice of coordinates. This is basically due to the fact that the skew symmetrization among j and the other indices kills the symmetric terms containing the second partial derivatives.

The exterior derivative satisfies the following properties:

(i) if f is a 0-form, the 1-form df is the differential already defined;

(ii) if $\omega = \omega_1 + \omega_2$, then $d\omega = d\omega_1 + d\omega_2$;

(iii) $dd\omega = 0, \ \forall \omega$;

(iv) $d(\omega_1 \wedge \omega_2) = d\omega_1 \wedge \omega_2 + (-1)^r \omega_1 \wedge d\omega_2$, ω_1 being an r-form and ω_2 any form.

The check of (i) and (ii) is immediate, that of (iii) is due to the symmetry of the second derivatives $\frac{\partial^2}{\partial x^i \partial x^k} = \frac{\partial^2}{\partial x^k \partial x^i}$, and that of (iv) to the skew symmetry of the exterior product of two 1-forms.

Vice versa, one can take the properties (i) – (iv) as defining the exterior derivative, and then show that the coordinate expression (2.1.3) follows uniquely.

Another useful property is given by the following

PROPOSITION 2.9 *The pull-back and the exterior derivative commute:*

$$\phi^*(d\omega) = d(\phi^*\omega).$$

Let σ be a r-dimensional domain and $\partial\sigma$ its $(r - 1)$-dimensional boundary; then we have the following

THEOREM 2.10 (GENERALIZED STOKES) *For every $(r - 1)$-form ω and r-dimensional domain σ the following beautiful formula holds:*

$$\int_\sigma d\omega = \int_{\partial\sigma} \omega.$$

Lie Derivative

Let Φ_τ be the flow generated by the vector field v: in some sense, v is the "velocity" field of the "motion" Φ_τ. We define the Lie derivative of a tensor field T with respect to v as

$$\mathcal{L}_v T = \lim_{\tau \to 0} \frac{1}{\tau} (\Phi_\tau^* T - T).$$

In other words, we compare the value taken by the tensor at the point $\Phi_\tau(p)$ (which is the image of the point p under the diffeomorphism Φ_τ) with that taken at the point p itself (i.e., at the position corresponding to $\tau = 0$). Obviously we are not allowed to directly compare the components of T at p and at $\Phi_\tau(p)$, because they belong to different tangent or cotangent spaces, but we must "pull-back" the tensor before.

From the definition one can immediately find the relation

$$\frac{d}{d\tau}(\Phi_\tau^* T) = \lim_{\triangle\tau \to 0} \frac{1}{\triangle\tau} (\Phi_{\tau+\triangle\tau}^* T - \Phi_\tau^* T)$$

$$= \lim_{\triangle\tau \to 0} \frac{1}{\triangle\tau} [\Phi_\tau^* (\Phi_{\triangle\tau}^* T - T)] = \Phi_\tau^* (\mathcal{L}_v T), \qquad (2.1.4)$$

which is sometimes taken as a definition of the Lie derivative.

In order to express the Lie derivative of a tensor field in local coordinates, we first prove the following two propositions.

PROPOSITION 2.11 *The Lie derivative is a derivation on the algebra of the differentiable tensor fields:*

$$\mathcal{L}_v(T_1 + T_2) = \mathcal{L}_v T_1 + \mathcal{L}_v T_2, \qquad \textit{(linearity)},$$
$$\mathcal{L}_v(T_1 \otimes T_2) = \mathcal{L}_v T_1 \otimes T_2 + T_1 \otimes \mathcal{L}_v T_2, \qquad \textit{(Leibniz rule)}.$$

Proof. The additive property can be readily verified. The Leibniz rule rests on the obvious fact that the transform of a tensor product is the tensor product of the transforms of its factors. Therefore,

$$\mathcal{L}_v(T_1 \otimes T_2) = \lim_{\tau \to 0} \frac{1}{\tau} (\Phi_\tau^* T_1 \otimes \Phi_\tau^* T_2 - T_1 \otimes T_2 + \Phi_\tau^* T_1 \otimes T_2 - \Phi_\tau^* T_1 \otimes T_2)$$

$$= \lim_{\tau \to 0} \frac{1}{\tau} [\Phi_\tau^* T_1 \otimes (\Phi_\tau^* T_2 - T_2) + (\Phi_\tau^* T_1 - T_1) \otimes T_2],$$

and the proposition follows.   **QED**

PROPOSITION 2.12 *The Lie and exterior derivatives commute:*

$$\mathcal{L}_v(d\omega) = d(\mathcal{L}_v \omega).$$

Proof. This is an immediate consequence of Proposition 2.9 and of the definition of $\mathcal{L}_v$. **QED**

An arbitrary tensor is a finite sum of tensor products of functions and of elements of the basis and the dual basis; hence we have to find the Lie derivative of dx^h and $\frac{\partial}{\partial x^k}$. Firstly,

$$\mathcal{L}_v(dx^h) = d(\mathcal{L}_v x^h) = d\left(v^k \frac{\partial x^h}{\partial x^k}\right) = \frac{\partial v^h}{\partial x^i} dx^i. \tag{2.1.5}$$

Then, to find $\mathcal{L}_v \frac{\partial}{\partial x^k}$ we require that the transforms of $\frac{\partial}{\partial x^k}$ and dx^h still be elements of the dual bases:

$$\left\langle dx^h + \tau \mathcal{L}_v dx^h + \mathcal{O}(\tau^2), \frac{\partial}{\partial x^k} + \tau \mathcal{L}_v \frac{\partial}{\partial x^k} + \mathcal{O}(\tau^2)\right\rangle = \delta_k^h + \mathcal{O}(\tau^2),$$

which entails

$$\left\langle \mathcal{L}_v dx^h, \frac{\partial}{\partial x^k}\right\rangle + \left\langle dx^h, \mathcal{L}_v \frac{\partial}{\partial x^k}\right\rangle = \mathcal{L}_v \left\langle dx^h, \frac{\partial}{\partial x^k}\right\rangle = 0.$$

Substituting the calculated expression of $\mathcal{L}_v dx^h$ and writing $\mathcal{L}_v \frac{\partial}{\partial x^k}$ as a linear combination $\lambda_k^i \frac{\partial}{\partial x^i}$ of the basis vectors, we find

$$\mathcal{L}_v \frac{\partial}{\partial x^k} = -\frac{\partial v^h}{\partial x^k} \frac{\partial}{\partial x^h}.$$

At this point we can calculate the Lie derivative of a generic tensor.

EXAMPLE 2.3 The Lie derivative of a vector u:

$$\mathcal{L}_v \left(u^h \frac{\partial}{\partial x^h}\right) = \left(v^k \frac{\partial}{\partial x^k} u^h - u^k \frac{\partial}{\partial x^k} v^h\right) \frac{\partial}{\partial x^h} = [v, u],$$

equals the Lie bracket of the two vectors.

Finally, let us view a formula connecting the Lie and the exterior derivative. Let us denote with the symbol $i_v \omega$ the contraction of a vector v with the first index of an r-form ω, which yields an $(r-1)$-form. Then

$$\mathcal{L}_v \omega = i_v \, d\omega + d(i_v \omega) \tag{2.1.6}$$

holds.

2.1.3 Riemannian, Symplectic, and Poisson Manifolds

Up to now, both the tangent and cotangent spaces have looked like separate worlds: a vector cannot be transformed into a covector, and vice versa. However, if a manifold is endowed with an additional, Riemannian or symplectic, structure, this becomes possible. These two structures differ for their symmetry properties. The Riemannian manifolds are typically the configuration spaces while the symplectic manifolds, with their Poisson generalization, are the phase spaces of mechanical systems.

Riemannian Manifolds

A *Riemannian manifold* is a smooth manifold M together with a covariant tensor field $g = g_{hk}dx^h \otimes dx^k$ of order two, called a *metric tensor*, such that

   (i) g is symmetric,

   (ii) for every point $p \in M$, the bilinear expression $g|_p$ is nondegenerate, that is, $g|_p(u,v) = 0$ for all $v \in T_pM$ if and only if $u = 0$.

A Riemannian manifold is called *proper* if

$$g|_p(u,u) > 0, \quad \forall u \in T_pM, \quad u \neq 0.$$

Otherwise, the manifold is called *pseudo-Riemannian* or is said to have an *indefinite metric*. The tensor g endows every vector space T_pM with an inner, or scalar product,

$$u \cdot v \overset{\text{def}}{=} g|_p(u,v) \quad \forall u,v \in T_pM.$$

For the linearity of the definition of the tensor map it follows that, in components,

$$u \cdot v = g_{hk}u^h v^k, \quad g_{hk} \overset{\text{def}}{=} g(e_h, e_k).$$

If we take a point $p' \in M$ "infinitesimally close" to p, the vector joining the two points belongs to T_pM, and the scalar product allows one to compute the distance between p and p'. One usually writes

$$ds^2 = g_{hk}dx^h dx^k,$$

without $\otimes$, for the symmetry of the tensor.

    Given a curve

$$y : \mathbb{R} \to M \quad \text{by } \tau \mapsto x^k = y^k(\tau), \quad a \leq \tau \leq b,$$

its length is given by

$$\int_\gamma ds \overset{\text{def}}{=} \int_a^b \sqrt{\gamma^*(ds^2)}.$$

The curve of minimal length joining two fixed points of M is called a *geodesic*.

An inner product on any vector space defines a canonical isomorphism between the space and its dual. For a fixed u the mapping

$$g|_p(u, \cdot) : T_pM \to \mathbb{R} \quad \text{by } v \mapsto u \cdot v$$

is by definition an element of T_p^*M. The canonical isomorphism is the mapping

$$T_pM \to T_p^*M \quad \text{by } u \mapsto g|_p(u, \cdot).$$

The same vector can be written as $u = u^h e_h$ or $u = u_k e^k$, the relation between contravariant u^h and covariant u_k components being given by

$$u_k = g_{kh}u^h, \quad u^h = g^{hk}u_k, \quad \text{where } g^{hk}g_{ki} = \delta_i^h.$$

The g^{ij} are the contravariant components of the metric tensor. One says that the indices of a generic tensor are raised or lowered by means of the metric tensor.

EXAMPLE 2.4 The most obvious example of Riemannian structure is the ordinary Euclidean space. Taking Cartesian coordinates, the metric tensor is represented by the unit matrix

$$(g_{hk}) = \begin{pmatrix} 1 & & \\ & \ddots & \\ & & 1 \end{pmatrix}.$$

EXAMPLE 2.5 A less trivial example is that of the induced metric. Let $f : M \to N$, $\dim M = m < \dim N = n$, be an immersion of rank m. When N (the ambient space) is endowed with a Riemannian metric g, the immersed manifold M acquires an induced metric f^*g, the pull-back of the ambient metric. If, for example, $N = \mathbb{R}^3$ with Euclidean metric and M is the sphere S^2 of radius R, then the induced metric is

$$(g_{hk}) = \begin{pmatrix} R^2 & 0 \\ 0 & R^2 \sin^2 \vartheta \end{pmatrix},$$

where the local coordinates $q^1 = \vartheta$, $q^2 = \varphi$ on the sphere are respectively colatitude and longitude.

Let M and M' be two smooth manifolds with Riemannian structures g and g', respectively. The mapping $f : M \to M'$ is called an *isometry* if f is a diffeomorphism and $f^*g' = g$. Two manifolds are said to be isometric if there exists an isometry of one onto the other. A manifold is said to be *flat* or *(pseudo)-Euclidean* if it is locally isometric to a manifold with a metric of the type $(g_{hk}) = \mathrm{diag}(-1, \ldots, -1, 1, \ldots, 1)$. A theorem, fundamental in Riemannian geometry and general relativity, states that a manifold is flat if and only if its curvature (or Riemann) tensor is identically zero; this tensor is constructed with the components of the metric tensor, along with the first and second derivatives.

A vector field v on M generates a one-parameter group of local isometries if and only if $\mathcal{L}_v g = 0$. For example, if M is the usual 3-dimensional Euclidean space, a family of continuous isometries is a composition of translations and rotations. If, on the contrary, the symmetric tensor $\mathcal{L}_v g$ of order two is different from zero, it is a measure of the deformation of the continuum, and is called the *strain tensor* in the theory of elasticity.

Symplectic Manifolds

A *symplectic manifold* is a $2n$-dimensional smooth manifold P endowed with a 2-form $\Omega = \Omega_{\mu\nu} dx^\mu \wedge dx^\nu$, called the *symplectic form*, such that

(i) Ω is closed: $d\Omega = 0$,

(ii) for every point $x \in P$ the symplectic form is nondegenerate, i.e.

$$\underbrace{\Omega \wedge \ldots \wedge \Omega}_{n \text{ times}} \neq 0.$$

Notice the differences and similarities with the Riemannian case: skew symmetry of the tensor $\Omega_{\mu\nu}$ versus symmetry of the tensor g_{hk}, but nondegeneracy in both cases. This last property ensures that a bivector $\Omega^{\alpha\beta}$, called a *Poisson bivector*, exists such that $\Omega_{\mu\nu}\Omega^{\nu\beta} = \delta_\mu^\beta$. As in the Riemannian case, the bilinear nondegenerate form Ω defines an inner product in $T_x P$, and thus a canonical isomorphism between $T_x P$ and its dual $T_x^* P$. This isomorphism is sometimes denoted with the symbols[4] $\sharp$ and $\flat$

$$\sharp : T_x^* P \to T_x P \quad \text{by } \omega \mapsto \omega^\sharp = v \quad \text{or } v^\mu = \Omega^{\mu\nu}\omega_\nu,$$
$$\flat : T_x P \to T_x^* P \quad \text{by } v \mapsto v^\flat = \omega \quad \text{or } \omega_\mu = \Omega_{\mu\nu}v^\nu.$$

The *symplectomorphisms* correspond to the isometries of the Riemannian case. Let P and P' be two $2n$-dimensional manifolds with symplectic

---

[4]The use of these symbols, exhibiting the great learning of the mathematicians, is inherited from music notation, where $\sharp$ (diesis or sharp) raises while $\flat$ (bemolle or flat) lowers a note by onehalf tone.

forms Ω and Ω', respectively. A map $f : P \rightarrow P'$ is called a symplectomorphism if f is a diffeomorphism and $f^*\Omega' = \Omega$. Two manifolds are said to be symplectomorphic if there exists a symplectomorphism of one onto the other.

Let v be a vector field on M. Clearly, the vector field v generates a one-parameter group of local symplectomorphisms if and only if $\mathcal{L}_v\Omega = 0$.

The closure property corresponds to the flatness of the Riemannian case, as will be clear in a moment by the Darboux theorem. In coordinates, the closure property reads as

$$\partial_\lambda\Omega_{\mu\nu} + \partial_\mu\Omega_{\nu\lambda} + \partial_\nu\Omega_{\lambda\mu} = 0,$$
$$\Omega^{\lambda\alpha}\partial_\alpha\Omega^{\mu\nu} + \Omega^{\mu\alpha}\partial_\alpha\Omega^{\nu\lambda} + \Omega^{\nu\alpha}\partial_\alpha\Omega^{\lambda\mu} = 0. \tag{2.1.7}$$

THEOREM 2.13 (DARBOUX) *For any $p \in U \subset P$ there is a chart such that, if $(x^1, \ldots, x^{2n})$ are local coordinates (called* canonical coordinates*) in U with $p = (0, \ldots, 0)$, the symplectic form acquires the canonical form*

$$\Omega|_U = \sum_{i=1}^n dx^{n+i} \wedge dx^i, \tag{2.1.8}$$

or, equivalently,

$$(\Omega_{\mu\nu}) = \begin{pmatrix} 0_n & -1_n \\ 1_n & 0_n \end{pmatrix}.$$

The Darboux theorem is a basic result, showing that all the symplectic manifolds are locally the "same" manifold, and look like $\mathbb{R}^{2n}$ endowed with the canonical form. They thus differ only for global, topological properties.

The canonical form (2.1.8) is natural for the phase space of mechanical systems, as the following example shows.

EXAMPLE 2.6 Let Q be an n-dimensional manifold, which will be the configuration space of a mechanical system in the applications. Define the *tangent* and *cotangent bundles,* respectively, as

$$TQ = \bigcup_{q \in Q} T_q Q, \quad T^*Q = \bigcup_{q \in Q} T_q^* Q.$$

They are $2n$-dimensional manifolds with local coordinates[5]

$$(q, v) = (q^1, \ldots, q^n, v^1, \ldots, v^n) \quad \text{and} \quad (q, p) = (q^1, \ldots, q^n, p_1, \ldots, p_n),$$

respectively; the first n's are local coordinates on Q, the other n's are the components, with respect to the natural bases $\left\{ \frac{\partial}{\partial q^k} \right\}$ and $\{dq^h\}$, of a tangent

---

[5]Hereafter p will denote a covector of T^*Q, no longer a generic point on a manifold.

and a cotangent vector, respectively. Tangent and cotangent bundles are instances of *fiber bundles*, i.e., manifolds that are *locally* the topological product of a *base* (here Q) and a *fiber* (here a tangent or cotangent space).

Any cotangent bundle carries a canonical 1-form Θ defined as follows. Let $\pi : T^*Q \rightarrow Q$, $x = (q,p) \mapsto q$ be the canonical projection and let $w \in T_x(T^*Q)$, $\Theta|_x \in T_x^*(T^*Q)$; then, taking advantage of the fact that a point $x \in T^*Q$ identifies a cotangent vector $p = p_k dq^k$ to Q, we define

$$\langle \Theta, w \rangle|_x = \langle p, \pi_* w \rangle|_q, \quad \forall w \in T_x(T^*Q).$$

Note that the pairing is between $T_x^*(T^*Q)$ and $T_x(T^*Q)$ on the left-hand side while that on the right-hand side is between T_q^*Q and T_qQ. In terms of local coordinates

$$\left\langle \Theta, \frac{\partial}{\partial q^k} \right\rangle = p_k, \quad \left\langle \Theta, \frac{\partial}{\partial p_k} \right\rangle = 0,$$

and therefore $\Theta = p_k dq^k$, from which $\Omega = d\Theta = dp_k \wedge dq^k$. We have reached the fundamental result that the cotangent bundle of a configuration space is endowed in a natural way with the canonical symplectic structure.

Obviously, not all the symplectic manifolds are cotangent bundles, as the following example shows.

EXAMPLE 2.7 Let us consider the sphere $S^2 = \{\vec{x} \in \mathbb{R}^3 : \|\vec{x}\| = R\}$. We can make the sphere a symplectic manifold, defining on it the area 2-form

$$\Omega|_{\vec{x}}(\vec{u}, \vec{v}) = \vec{x} \cdot \vec{u} \times \vec{v}, \quad \vec{u}, \vec{v} \in T_{\vec{x}}S^2.$$

This form is closed, because an $(n + 1)$-form on an n-dimensional manifold vanishes identically. This definition is clearly global, and no point on the sphere plays a privileged role. Now pick a polar axis and choose local coordinates ϑ and φ as in Example 2.5. Thus $\Omega = R^2 \sin \vartheta \, d\varphi \wedge d\vartheta$, and the symplectic form is undefined at the poles. Put

$$p = R \cos \vartheta, \quad q = R\varphi \quad \Rightarrow \quad \Omega = dp \wedge dq,$$

i.e., q, p are local canonical coordinates.

Poisson Manifolds

The Poisson manifolds are a generalization of the symplectic manifolds in the following sense. On a symplectic manifold P let us define the *Poisson bracket* $\{f, g\}$ between two functions $f, g : P \rightarrow \mathbb{R}$ by

$$\{f, g\} = \left\langle dg, df^\sharp \right\rangle = \frac{\partial g}{\partial x^\mu} \Omega^{\mu\nu} \frac{\partial f}{\partial x^\nu}. \tag{2.1.9}$$

In canonical coordinates,

$$\{f,g\} = \frac{\partial f}{\partial p_h}\frac{\partial g}{\partial q^h} - \frac{\partial g}{\partial p_h}\frac{\partial f}{\partial q^h}.$$

The closure property (2.1.7) of Ω entails (h is another function)

$$\{\{f,g\},h\} + \{\{h,f\},g\} + \{\{g,h\},f\} = 0,$$

which is known as the *Jacobi identity*.

From this basic identity a useful property follows, relating the Lie and Poisson bracket. Let us indicate the vector $(df)^\sharp$ with X_f; then, recalling that a vector is a derivation operator, we can write

$$\{f,g\} = X_f g = -X_g f.$$

Taking a third generic function h, one obtains

$$[X_f, X_g]h = X_f X_g h - X_g X_f h = \{f,\{g,h\}\} - \{g,\{f,h\}\}$$
$$\text{(for Jacobi identity)} = \{\{f,g\},h\} = X_{\{f,g\}}h,$$

therefore

$$[X_f, X_g] = X_{\{f,g\}}. \tag{2.1.10}$$

A *Poisson manifold* is a manifold on which a Poisson bracket is defined, or, equivalently, on which a bivector $\Omega^\sharp = \Omega^{\mu\nu}\partial_\mu\partial_\nu$ satisfying the closure property is defined. It is obvious that any symplectic manifold is also a Poisson manifold. However, if the matrix $(\Omega^{\mu\nu})$ is not invertible, there exists only the Poisson structure, without the corresponding symplectic structure. Notice that a Poisson manifold can be of odd dimension. Hereafter, "Poisson manifold" will denote in general a "true" one, which is not derived from a symplectic one.

At any point $x \in P$ only the map $\sharp$ still exists, and it is no longer surjective. The dimension r of its image is called the *rank*. The image of $\sharp$ is a subspace Δ_x of $T_x P$ of dimension r, and thus defines a distribution $\Delta = \bigcup \Delta_x$. Thanks to the closure property (2.1.7), the distribution Δ turns out to be integrable; in other words, there exists an r-dimensional (called *integral*) submanifold $N \subset P$ such that $T_x N = \Delta_x$, $\forall x \in N$.

We would like to restrict the bivector $\Omega^\sharp$ from the ambient space P to the submanifold N, but in general the pull-back of a k-vector is not defined. In the present case, however, we exploit the fact that the tangent space to N coincides with the image of $\Omega^\sharp$, and so we define the 2-form

$$\Omega(X_f, X_g) \overset{\text{def}}{=} \{f,g\}, \quad \forall f,g : P \to \mathbb{R}.$$

This 2-form is well defined on N (the vectors X_f and X_g are tangent to N) where it is obviously nondegenerate. Moreover, it is closed for the Jacobi identity; therefore, it endows N with a symplectic structure.

The integral manifolds of the distribution Δ are called *symplectic leaves* and are described by $(n - r)$ equations $C_a(x) =$ constant (n is the dimension of P). The functions $C_a(x)$ are called *Casimir functions* and have zero Poisson bracket with any functions, because

$$\{C_a(x), x^h\} = 0, \quad a = 1, \ldots, n - r, \quad h = 1, \ldots, n,$$

as is clear bearing in mind that dC_a belongs, by definition, to the kernel of $\sharp$.

EXAMPLE 2.8 Take $P = \mathbb{R}^3 - \{0\}$ with the Poisson structure

$$\{x^1, x^2\} = x^3, \quad \{x^2, x^3\} = x^1, \quad \{x^3, x^1\} = x^2,$$

or equivalently,

$$(\Omega^{\mu\nu}) = \begin{pmatrix} 0 & -x^3 & x^2 \\ x^3 & 0 & -x^1 \\ -x^2 & x^1 & 0 \end{pmatrix}.$$

The kernel of $\sharp$ is $(x^1 \ x^2 \ x^3)$, the rank is 2, the Casimir function is $C(x) = (x^1)^2 + (x^2)^2 + (x^3)^2$ and the symplectic leaves are the concentric spheres. Put $p = x^3$, $q = \arctan \frac{x^1}{x^2}$, from which $\{p, q\} = 1$, so that q, p are local canonical coordinates on a sphere. This symplectic structure on the sphere clearly coincides with that of example 2.7.

2.2 Lie Groups and Lie Algebras

A Lie group is a group whose elements are labeled by one or more parameters; taking these parameters as coordinates, a Lie group can be seen as a manifold, and the tangent space at the identity, once equipped with a natural composition law, as the corresponding Lie algebra. Lie groups are the fundamental tool when investigating the symmetries of mechanical systems.

For a deeper study the classical reference text is Chevalley (1946), while for a high readability, with applications to physics, see for example Cornwell (1989) or Gilmore (1974).

2.2.1 Definition and Properties

DEFINITION 2.14 *A Lie group $\mathfrak{G}$ is a manifold on which an analytic composition law $\mathfrak{G} \times \mathfrak{G} \to \mathfrak{G}$ is defined, satisfying the group multiplication law.*

This means that there exist in a neighborhood of the identity the *composition functions*

$$F : \mathfrak{G} \times \mathfrak{G} \to \mathfrak{G}, \quad (x, y) \mapsto F(x, y) = xy,$$

where xy is the group product, which are analytic and satisfy

(i) $F(x, 1) = F(1, x) = x$, where $1x = x$;

(ii) $F(x, x^{-1}) = F(x^{-1}, x) = 1$, where $x^{-1}x = 1$;

(iii) $F(x, F(y, z)) = F(F(x, y), z)$, i.e., the associative law.

These properties, in particular the third, endow the group manifold with a rich structure, which we want to study. We remark that, once given the composition functions in a neighborhood of the identity, the group is determined only locally, since the group manifold has in general a nontrivial topology; in general it cannot be covered with a single chart.

Let x^h, y^k, z^i, with $h, k, i = 1, \ldots, N$, be local coordinates of the points $x, y, z \in \mathfrak{G}$, and let the group multiplication have coordinate presentation $z^i = F^i(x, y)$. The most important actors on the scene are the $N + N$ vector fields defined by

$$R_a^k = \left. \frac{\partial F^k(x, y)}{\partial x^a} \right|_{x=1}, \quad L_a^k = \left. \frac{\partial F^k(y, x)}{\partial x^a} \right|_{x=1}. \tag{2.2.1}$$

As is clear from the definition, the vectors R_a, $a = 1, \ldots, N$ are the generators of the N flows of the *left translations*:

$$\lambda_x : \mathfrak{G} \to \mathfrak{G} \quad \text{by } \lambda_x y = xy,$$

while the vectors L_a are the generators of the N flows of the *right translations*:

$$\rho_x : \mathfrak{G} \to \mathfrak{G} \quad \text{by } \rho_x y = yx.$$

The maps λ_x and ρ_x, which are by assumption analytic, satisfy

$$\lambda_x \lambda_y = \lambda_{xy}, \quad \rho_x \rho_y = \rho_{yx},$$

and these relations may be taken as a definition of *left* and *right action*, respectively.

PROPOSITION 2.15 *The vectors R_a are invariant with respect to the right translations (or simply right invariant), while the vectors L_a are invariant with respect to the left translations (or simply left invariant).*

Proof. Differentiating the associative law with respect to x and z, we obtain

$$\frac{\partial F^h(x, yz)}{\partial x^k} = \frac{\partial F^h(xy, z)}{\partial (xy)^i} \frac{\partial F^i(x, y)}{\partial x^k},$$

$$\frac{\partial F^h(x, yz)}{\partial (yz)^i} \frac{\partial F^i(y, z)}{\partial z^k} = \frac{\partial F^h(xy, z)}{\partial z^k}.$$

Putting $x = 1$ in the first equation and $z = 1$ in the latter, we obtain

$$R_a^h(yz) = \frac{\partial F^h(y, z)}{\partial y^i} R_a^i(y),$$

$$\frac{\partial F^h(x, y)}{\partial y^i} L_a^i(y) = L_a^h(xy),$$

(2.2.2)

which may be rewritten as

$$R_a \circ \rho_z = \mathbf{T}_{\rho_z} R_a, \quad \text{that is,} \quad \rho_{z*} R_a = R_a,$$
$$\mathbf{T}_{\lambda_x} L_a = L_a \circ \lambda_x, \quad \text{that is,} \quad \lambda_{x*} L_a = L_a.$$

This is just the definition of invariance: the vector fields R_a and L_a are transformed into themselves by any right or left translation.   **QED**

The set of left (right) invariant vector fields constitutes a vector space, because of the linearity of $\lambda_{x*} (\rho_{x*})$. Moreover, a left (right) invariant vector field can be constructed by taking some vector at the point 1 and left (right) translating it at the generic point of the group manifold. The vector space of left (right) invariant vector fields may thus be identified with $T_1 \mathfrak{G}$, the tangent space to the group manifold at the identity, and denoted by $\mathfrak{g}$.

PROPOSITION 2.16 *The Lie bracket $[\cdot, \cdot]$ of two left (right) invariant vector fields is still a left (right) invariant vector field.*

The claim is a straightforward consequence of the following general property of the push-forward of vectors:

$$\phi_* [v, w] = [\phi_* v, \phi_* w]$$

(an analogous property holds for the pull-back). In turn, this simply derives from the naturalness of the ϕ-relation with respect to the Lie bracket.

As a basis for the vector space of the left (right) invariant vector fields we may take the N vectors $\{L_a\} (\{R_a\})$ previously defined. For the last proposition, the Lie bracket is still a left (right) invariant vector field and can thus be decomposed on the respective bases:

$$[L_a, L_b] = c_{ab}^c L_c, \quad [R_a, R_b] = d_{ab}^c R_c, \quad a, b, c = 1, \dots, N, \tag{2.2.3}$$

where c_{ab}^c and d_{ab}^c are constants. To find their value, substitute the definition (2.2.1) into the Lie brackets (2.2.3) and put $x = 1$. Taking into account that, for item (i) of the properties of the composition functions, the relations $R_a^k(1) = L_a^k(1) = \delta_a^k$ hold, we obtain

$$c_{ab}^c = \left(\frac{\partial^2 F^c}{\partial x^a \partial y^b} - \frac{\partial^2 F^c}{\partial x^b \partial y^a} \right)_{x=y=1}, \qquad (2.2.4)$$

$$d_{ab}^c = -c_{ab}^c.$$

Therefore, the two structures essentially coincide.

To complete the list of the Lie brackets among left and right invariant vector fields, we notice that left and right translations commute, so that

$$[L_a, R_b] = 0, \qquad \forall a, b.$$

Therefore, some constants c_{ab}^c, $a, b, c = 1, \dots, N$, called *structure constants*, are associated to any local Lie group. They cannot be completely arbitrary, but must satisfy

(i) skew symmetry: $[L_a, L_b] + [L_b, L_a] = 0 \quad \Rightarrow c_{ab}^c + c_{ba}^c = 0.$

(ii) Jacobi identity:

$$[[L_a, L_b], L_c] + [[L_b, L_c], L_a] + [[L_c, L_a], L_b] = 0$$
$$\Rightarrow c_{cd}^e c_{ab}^d + c_{ad}^e c_{bc}^d + c_{bd}^e c_{ca}^d = 0.$$

We have reached the fundamental result that the N-dimensional vector space of the left (right) invariant vector fields on the group manifold $\mathfrak{G}$ is naturally endowed with the composition laws (2.2.3). This vector space is identified with $\mathfrak{g} = T_1 \mathfrak{G}$, which is thus equipped with the same composition laws.

DEFINITION 2.17 *A vector space $\mathfrak{g}$ with a bilinear composition law*

$$\mathfrak{g} \times \mathfrak{g} \to \mathfrak{g} : (u, v) \mapsto [u, v]$$

satisfying the two properties

(i) $[u, v] = -[v, u],$

(ii) $[[u, v], w] + [[v, w], u] + [[w, u], v] = 0$

is said to be a Lie algebra.

We can rephrase the content of this section in terms of left (right) invariant dual forms. Consider the vector space $\mathfrak{g}^*$, the dual space of $\mathfrak{g}$, and let $\{\omega^a\}$ be the dual basis of $\{L_b\}$, that is, $\langle \omega^a, L_b \rangle = \delta_b^a$; then one can show that

$$d\omega^a + c_{bc}^a \omega^b \omega^c = 0, \quad a, b, c = 1, \ldots, N.$$

These equations are called the *Maurer-Cartan equations* for $\mathfrak{G}$. The ω^a are left invariant forms, as one easily deduces by applying the general relation $\langle \alpha, \phi_* v \rangle = \langle \phi^* \alpha, v \rangle$:

$$\langle \omega^a, L_b \rangle = \langle \omega^a, \lambda_{x*} L_b \rangle = \langle \lambda_x^* \omega^a, L_b \rangle \Rightarrow \lambda_x^* \omega^a = \omega^a.$$

Analogously, one could introduce the right invariant forms π^a, the dual forms of R_b, which satisfy

$$d\pi^a - c_{bc}^a \pi^b \pi^c = 0, \quad a, b, c = 1, \ldots, N.$$

2.2.2 Adjoint and Coadjoint Representation

Besides left and right translations, we may define a third action of a Lie group on itself: the *conjugation* or *inner automorphism*

$$\gamma_x : \mathfrak{G} \to \mathfrak{G} \quad \text{by } y \mapsto \gamma_x y = x y x^{-1}.$$

Since $\gamma_z \gamma_x = \gamma_{zx}$, the conjugation is a left action. The derivative map $\mathbf{T}_{\gamma_x}$ at the identity is a linear transformation of $T_1 \mathfrak{G} = \mathfrak{g}$ into itself, denoted $\mathbf{Ad}_x$, which satisfies $\mathbf{Ad}_z \mathbf{Ad}_x = \mathbf{Ad}_{zx}$. It is a linear representation of the group, called *adjoint representation*. Let us find its explicit expression in coordinates. By definition,

$$(\mathbf{Ad}_x)_b^a = \frac{\partial F^a(xy, x^{-1})}{\partial y^b} \bigg|_{y=1} = \frac{\partial F^a(xy, x^{-1})}{\partial (xy)^i} \frac{\partial F^i(x, y)}{\partial y^b} \bigg|_{y=1}.$$

Putting $z = y^{-1}$ in the first of (2.2.2), we find $\frac{\partial F^a(y, y^{-1})}{\partial y^i} = \pi_i^a(y)$, the right invariant forms. Hence

$$(\mathbf{Ad}_x)_b^a = \pi_i^a(x) L_b^i(x), \quad \text{and} \quad (\mathbf{Ad}_x^{-1})_b^a = \omega_i^a(x) R_b^i(x).$$

Besides the adjoint representation $\mathbf{Ad}_x : \mathfrak{g} \to \mathfrak{g}$ we can define by duality the *coadjoint representation* $\mathbf{Ad}_x^* : \mathfrak{g}^* \to \mathfrak{g}^*$ by

$$\langle \mathbf{Ad}_x^* \alpha, v \rangle = \langle \alpha, \mathbf{Ad}_x^{-1} v \rangle, \quad \forall \alpha \in \mathfrak{g}^*, \ \forall v \in \mathfrak{g}, \tag{2.2.5}$$

which satisfies $\mathbf{Ad}_x^* \mathbf{Ad}_y^* = \mathbf{Ad}_{xy}^*$.

REMARK 2.9 When we do not need to stress the particular group element x, we write $\mathbf{Ad}_\mathfrak{G}$ and $\mathbf{Ad}^*_\mathfrak{G}$, or even $\mathbf{Ad}$ and $\mathbf{Ad}^*$, if there is no risk of confusion.

The derivative map $\mathbf{ad}$ of the adjoint representation at the identity is strictly related to the structure constants. Indeed, differentiating at the identity the relation $\pi^a_k(x)R^k_b(x) = \delta^a_b$, one finds $\left.\frac{\partial \pi^a_b}{\partial x^c}\right|_{x=1} + \left.\frac{\partial R^a_b}{\partial x^c}\right|_{x=1} = 0$. Then, differentiate $\mathbf{Ad}_x$ at the identity and substitute. Remembering (2.2.4), one finds

$$\left.\frac{\partial}{\partial x^c}(\mathbf{Ad}_x)^a_b\right|_{x=1} = c^a_{cb}. \tag{2.2.6}$$

The derivative map of $\mathbf{Ad}_x$ at the identity is therefore

$$(\mathbf{ad}_{L_c})^a_b = c^a_{cb},$$

which may be intrinsically written as

$$\mathbf{ad}_u v = [u, v], \quad \forall u, v \in \mathfrak{g}.$$

$\mathbf{ad}_u$ is a linear operator $\mathfrak{g} \to \mathfrak{g}$, $\forall u \in \mathfrak{g}$, called the *adjoint map*.

In the linear space $\mathfrak{g}^*$ the submanifolds generated by the *coadjoint action* of $\mathfrak{G}$ over an element $\mu \in \mathfrak{g}^*$ are called *coadjoint orbits*,

$$\mathcal{O}_\mu \overset{\text{def}}{=} \{\mu' \in \mathfrak{g}^* : \mu' = \mathbf{Ad}^*_\mathfrak{G}\mu\},$$

and are symplectic manifolds in a natural way. The symplectic structure descends by the restriction to the symplectic leaves of a Poisson structure on $\mathfrak{g}^*$. Let us view it in detail.

We endow $\mathfrak{g}^*$ with a Poisson structure assigning a Poisson bivector, that is, for every $\mu \in \mathfrak{g}^*$, a linear map $\Omega^\sharp_{(\mu)} : T^*_\mu\mathfrak{g}^* \times T^*_\mu\mathfrak{g}^* \to \mathbb{R}$ satisfying the properties of skew symmetry and closure. For the linearity of the manifold $\mathfrak{g}^*$, we can identify $T^*_\mu\mathfrak{g}^*$ with $(\mathfrak{g}^*)^* = \mathfrak{g}$, and then define the Poisson structure

$$\Omega^\sharp_{(\mu)}(u, v) \overset{\text{def}}{=} -\langle \mu, [u, v]\rangle, \quad u, v \in \mathfrak{g}$$

(the minus sign is not essential and it is inserted only for convenience). Its skew symmetry is obvious, bearing in mind the skew symmetry of the Lie bracket. The closure property is checked by writing explicitly the Poisson bivector in components

$$(\Omega^\sharp_{(\mu)})_{ab} = -\mu_c c^c_{ab},$$

then the latter of (2.1.7) follows from the Jacobi identity on the structure constants. Assigning the Poisson structure $\Omega^\sharp_{(\mu)}$ makes $\mathfrak{g}^*$ a Poisson manifold, endowed with the Poisson bracket

$$\{v_a, v_b\} = c^c_{ab}v_c, \quad v \in \mathfrak{g}^*, \ v_a = \langle v, L_a\rangle.$$

The rank of the Poisson structure is not constant on $\mathfrak{g}^*$, but only on each coadjoint orbit. We show indeed that the image of $\Omega^{\sharp}_{(\mu)}$, considered for each fixed μ as a linear operator $\mathfrak{g} \to \mathfrak{g}^*$, coincides with $T_\mu \mathcal{O}_\mu$, the tangent space in μ to the coadjoint orbit through μ. To this end, let us consider, in the definition of coadjoint orbit, elements of $\mathfrak{G}$ very close to the identity. Then, remembering that the derivative map at the identity of **Ad** is **ad**, we may write

$$\mathbf{Ad}^*_{\mathfrak{G}}\mu = \mu - \langle \mu, \mathbf{ad}_{\mathfrak{g}} \cdot \rangle + \cdots$$

while, by the definition of a Poisson bivector,

$$\Omega^{\sharp}_{(\mu)}(u, \cdot) = - \langle \mu, \mathbf{ad}_u \cdot \rangle, \quad u \in \mathfrak{g}.$$

This proves the above statement. We know that the image of a Poisson bivector is an integrable distribution; therefore, we can conclude that the symplectic integral manifold through μ coincides with $\mathcal{O}_\mu$. The coadjoint orbits are therefore symplectic manifolds: in some sense, the symplectic form is the inverse of the restriction, to the orbit, of the Poisson bivector and is called the *Kirillov form*, or the *Kirillov-Kostant-Souriau form*.

EXAMPLE 2.10 Take $\mathfrak{G} = SO(3)$ and identify its Lie algebra $so^*(3)$, i.e., the linear space of the skew symmetric 3×3 matrices, with $\mathbb{R}^3$. Then the coadjoint action coincides with the similarity action on the matrices and induces a rotation on $\mathbb{R}^3$. The coadjoint orbits are therefore the concentric spheres S^2 equipped with the symplectic structure of Example 2.7.

2.2.3   Action of a Lie Group on a Manifold

For what concerns us, Lie groups arise as transformation groups of manifolds and describe the symmetries of mechanical systems, whose configuration or phase space is just the manifold on which the group acts. On this manifold some vector fields are naturally defined, which are, in some sense, the velocity fields of the incipient motion caused by the group action. Not surprisingly, the Lie algebra of the group turns out to be homomorphic to the Lie algebra of these fields.

Indeed, let us consider an n-dimensional manifold M with local coordinates $q = \{q^\mu\}$ and let

$$\Phi_x : M \to M \quad \text{by } q \mapsto \Phi_x q, \quad \text{in coordinates } q^\mu \mapsto \Phi^\mu(x, q),$$

be a left action of the N-dimensional Lie group $\mathfrak{G}$:

(i) $\Phi_1 q = q$,   i.e., $\Phi^\mu(1, q) = q^\mu$,

(ii) $\Phi_x \Phi_y q = \Phi_{xy} q$,   i.e., $\Phi^\mu(x, \Phi(y, q)) = \Phi^\mu(F(x, y), q)$.

We define N vector fields on M by $X_a^\mu(q) = \left. \dfrac{\partial \Phi^\mu}{\partial x^a} \right|_{x=1}$. They are called *infinitesimal generators* of the action.

PROPOSITION 2.18 *The map $L_a \mapsto -X_a$ is a Lie algebra homomorphism:*

$$[X_b, X_c] = -c_{bc}^a X_a.$$

If none of the vectors X_a is identically zero (thus if the action of $\mathfrak{G}$ is effective; see below), the map is an isomorphism.

Starting instead with a right action: $\Phi_x \Phi_y q = \Phi_{yx} q$, one finds that the homomorphism is given by $L_a \mapsto X_a$.

Let us view some general features of a group action. The set of the elements of $\mathfrak{G}$ leaving fixed a chosen point $q \in M$ is a subgroup of $\mathfrak{G}$, called the *isotropy (sub)group* of q and is denoted by $\mathfrak{G}_q$. Thus $\mathfrak{G}_q = \{x \in \mathfrak{G} : \Phi_x q = q\}$. If the isotropy group of every point of M is the identity, then $\mathfrak{G}$ is said to act *freely* on M; in this case no element except the identity leaves *some* point fixed, i.e., *all* points are moved by the nontrivial group action. Less strongly, $\mathfrak{G}$ is said to act *effectively* on M if no element, except the identity, leaves *every* point fixed, i.e., not all but surely *some* points are moved by the group action. The orbit $\mathcal{O}_q$ is defined as the set of points of M that can be reached, starting from q, by the action of the whole group $\mathfrak{G}$; thus $\mathcal{O}_q = \{\Phi_x q, \, x \in \mathfrak{G}\}$. If $\mathcal{O}_q = M$, the group $\mathfrak{G}$ is said to act *transitively*, and M is called a *homogeneous manifold* of $\mathfrak{G}$.

EXAMPLE 2.11 Let us consider the action of the rotation and translation group on the 3-dimensional Euclidean space. The group of translations acts freely and transitively. The group of rotations, around a given point O, acts effectively but neither freely nor transitively. The orbit for a point P, different from O, is the sphere with center O and radius $\overline{OP}$; the restriction of the action of the rotation group to this sphere is a homogeneous non-free action, the isotropy subgroup for a point P being the subgroup of the rotations around the axis OP.

An action Φ of $\mathfrak{G}$ on a manifold M defines an equivalence relation among the points of M belonging to the same orbit. Explicitly, for $q, q' \in M$, we write $q \sim q'$ if there exists an $x \in \mathfrak{G}$ such that $\Phi_x q = q'$, that is, if $q' \in \mathcal{O}_q$. We let $M/\mathfrak{G}$ be the set of these equivalence classes, that is, the set of orbits, sometimes called the *orbit space*. We give to $M/\mathfrak{G}$ the quotient topology. However, to guarantee that the orbit space also has a smooth manifold structure, further conditions on the action are required. An action Φ is called *proper* if the mapping $\widetilde{\Phi} : \mathfrak{G} \times M \to M \times M$ defined by $\widetilde{\Phi}(x, q) = (q, \Phi_x q)$ is proper; that is, for $K \subset M \times M$ compact, then $\widetilde{\Phi}^{-1}(K)$ also is compact. We notice that if $\mathfrak{G}$ is compact, the property is automatically satisfied. The next proposition gives a useful sufficient condition for $M/\mathfrak{G}$ to be a smooth manifold.

PROPOSITION 2.19 *If* $\Phi : \mathfrak{G} \times M \to M$ *is a free and proper action, then* $M/\mathfrak{G}$ *is a smooth manifold and* $\pi : M \to M/\mathfrak{G}$ *a smooth submersion.*

For the proof see Abraham & Marsden (1978, proposition 4.1.23).

EXAMPLE 2.12 Let M be the punctured plane and $\mathfrak{G}$ the rotation group about the puncture point. The action of $\mathfrak{G}$ is proper (the rotation group is compact) and free (the point fixed under rotations has been removed). The orbits are concentric circles, and the value of the radius $R > 0$ parametrizes the set of the orbits. Then $M/\mathfrak{G} = \mathbb{R}^+$.

An important particular case is the following. Consider the action on $\mathfrak{G}$ of a closed Lie subgroup $\mathfrak{H}$ of $\mathfrak{G}$ itself, given by restriction of the left action. The orbits of $\mathfrak{G}$ under this action are called *right cosets* of $\mathfrak{H}$ in $\mathfrak{G}$; they are the sets $\mathfrak{H}x = \{hx, h \in \mathfrak{H}, x \in \mathfrak{G}\}$ (unfortunately, some authors call them left cosets). One verifies that this action is proper and free, see Abraham & Marsden (1978, proposition 4.1.23). Therefore, $\mathfrak{G}/\mathfrak{H}$ is a smooth manifold, M say.

One may reverse the argument. Let us consider a transitive non-free action of a Lie group $\mathfrak{G}$ on a manifold M, and let $\mathfrak{G}_q$ be the isotropy subgroup relative to any point $q \in M$ (notice that all $\mathfrak{G}_q$, $\forall q \in M$, are isomorphic for the homogeneity of M); thus $M = \mathfrak{G}/\mathfrak{G}_q$. The closure of any isotropy subgroup is always assured, since it is the preimage of a point q (which is a closed set) of the smooth map $\Phi_q : \mathfrak{G} \to M$, given by $\Phi_q(x) = \Phi(x, q)$.

EXAMPLE 2.13 The 3-dimensional rotation group $SO(3)$ acts transitively on the sphere S^2. Choose a point of S^2: its isotropy subgroup is the group of the rotations about the axis through this point. Thus $S^2 = SO(3)/SO(2)$.

As seen, some structure constants satisfying the two properties of skew symmetry and Jacobi identity are associated to any local Lie group. Vice versa, it is possible to prove that, given some constants satisfying the two properties, or, in other words, given an abstract Lie algebra $\mathfrak{g}$, there is a local Lie group $\mathfrak{G}$ whose Lie algebra is isomorphic to $\mathfrak{g}$. The group is determined by $\mathfrak{g}$ up to a local isomorphism. This is the content of the converse of the third Lie theorem.

In general there are homomorphic Lie groups with isomorphic Lie algebras. We remember that a subgroup $\mathfrak{D}$ of $\mathfrak{G}$ is *discrete* if every $d \in \mathfrak{D}$ has a neighborhood in $\mathfrak{G}$ that contains no element of $\mathfrak{D}$, apart from d. $\mathfrak{D}$ is *normal* if $xdx^{-1} \in \mathfrak{D}$, $\forall d \in \mathfrak{D}$, $\forall x \in \mathfrak{G}$. It is thus possible to prove that the general situation is the following: if $\mathfrak{G}$ is a simply connected Lie group (that is, every closed curve on the group manifold can be smoothly shrunk to a point), then any other Lie group $\mathfrak{G}'$, such that $\mathfrak{g}' = \mathfrak{g}$, is covered with $\mathfrak{G}$, that is, $\mathfrak{G}' = \mathfrak{G}/\mathfrak{D}$. A typical example (see the following section for the definitions) is the simply connected Lie group $SU(2)$, which is the twofold covering of $SO(3)$, thus $\mathfrak{D} = \mathbb{Z}_2$.

2.2.4 Classification of Lie Groups and Lie Algebras

We recall some definitions.

Let $\mathfrak{G}_1$ and $\mathfrak{G}_2$ be any two Lie groups. Consider the set of pairs (x_1, x_2), where $x_1 \in \mathfrak{G}_1$ and $x_2 \in \mathfrak{G}_2$, and define the product of two such pairs (x_1, x_2) and (y_1, y_2) by $(x_1, x_2)(y_1, y_2) = (x_1 y_1, x_2 y_2)$. Then the set of pairs (x_1, x_2) forms a group, as one easily checks, called the *direct product* and denoted by $\mathfrak{G}_1 \times \mathfrak{G}_2$. The corresponding Lie algebras satisfy

$$[\mathfrak{g}_1, \mathfrak{g}_1] \subseteq \mathfrak{g}_1, \quad [\mathfrak{g}_2, \mathfrak{g}_2] \subseteq \mathfrak{g}_2, \quad [\mathfrak{g}_1, \mathfrak{g}_2] = 0.$$

A group $\mathfrak{G}$ is said to be a *semidirect product* group if it possesses two subgroups $\mathfrak{G}_1$ and $\mathfrak{G}_2$ such that

(i) $\mathfrak{G}_1$ is an invariant subgroup of $\mathfrak{G}$, i.e., $g^{-1} g_1 g \subseteq \mathfrak{G}_1$;

(ii) $\mathfrak{G}_1$ and $\mathfrak{G}_2$ have only the identity element in common;

(iii) any element of $\mathfrak{G}$ can be written as the product of an element of $\mathfrak{G}_1$ and an element of $\mathfrak{G}_2$.

In this case, we will write $\mathfrak{G} = \mathfrak{G}_2 \times_S \mathfrak{G}_1$. The corresponding Lie algebras satisfy

$$[\mathfrak{g}_1, \mathfrak{g}_1] \subseteq \mathfrak{g}_1, \quad [\mathfrak{g}_2, \mathfrak{g}_2] \subseteq \mathfrak{g}_2, \quad [\mathfrak{g}_1, \mathfrak{g}_2] \subseteq \mathfrak{g}_1.$$

One says that $\mathfrak{g}_1$ is an *invariant subalgebra* of $\mathfrak{g} = \mathfrak{g}_1 \oplus \mathfrak{g}_2$, or that it is an *ideal*.

EXAMPLE 2.14 The Euclidean group of the continuous isometries of the 3-dimensional Euclidean space is the semidirect product of $\mathfrak{G}_1$ = translations and $\mathfrak{G}_2$ = rotations.

A nonabelian (that is noncommutative) Lie algebra is said to be *simple* if it does not posses an invariant subalgebra. Hence $[\mathfrak{g}, \mathfrak{g}] = \mathfrak{g}$; that is, the Lie brackets among all the elements generate the whole algebra. A Lie group is said to be *simple* if its Lie algebra is simple. A Lie group is said to be *semisimple* if it is the direct product of simple groups.

A Lie algebra is said to be *solvable* if the k-derived algebra $\mathfrak{g}^{(k)}$ is zero for some $k \geq 0$, where the *k-derived algebra* is defined recursively as

$$\mathfrak{g}^{(1)} = [\mathfrak{g}, \mathfrak{g}], \quad \mathfrak{g}^{(i+1)} = [\mathfrak{g}^{(i)}, \mathfrak{g}^{(i)}].$$

A Lie group is *solvable* if its algebra is solvable.

Simple and solvable Lie algebras are the two basic constituents of any Lie algebra. Indeed, any Lie algebra $\mathfrak{g}$ can be decomposed into the sum of a semisimple $\mathfrak{s}$ and a solvable $\mathfrak{h}$ Lie algebra: $\mathfrak{g} = \mathfrak{s} \oplus \mathfrak{h}$ (this is known as the *Levi-Malcev decomposition*), satisfying the relations

$$[\mathfrak{s}, \mathfrak{s}] = \mathfrak{s}, \quad [\mathfrak{s}, \mathfrak{h}] \subseteq \mathfrak{h}, \quad [\mathfrak{h}, \mathfrak{h}] = \mathfrak{h}^{(1)} \subset \mathfrak{h}.$$

Clearly, the classification of all Lie groups is a fundamental problem. Unfortunately, an exhaustive classification of the solvable groups does not exist, while, fortunately, that of the simple groups does: it has been performed by Killing and Cartan. Without entering the beautiful but lengthy analysis (which is however easily understandable, requiring only linear algebra), we proceed to the conclusions.

The result of this classification is that a compact simple Lie group is necessarily one of the "classical groups."

(i) The orthogonal or rotation groups $SO(n)$ with $n = 3, 5, 6, 7 \ldots$, which are the groups of the $n \times n$ matrices $\mathbf{R}$ such that

$$\mathbf{R}^t \mathbf{R} = 1_n, \quad \det \mathbf{R} = 1 \quad (\mathbf{R}^t = \text{transposed matrix}).$$

To find the corresponding Lie algebras $so(n)$, we write

$$\mathbf{R} = 1_n + \mathbf{r} + \mathcal{O}(\mathbf{r}^2),$$

then

$$(1_n + \mathbf{r}^t)(1_n + \mathbf{r}) = 1_n + \mathbf{r} + \mathbf{r}^t + \mathcal{O}(\mathbf{r}^2) = 1_n,$$

showing that $\mathbf{r} \in so(n) \Rightarrow \mathbf{r}$ is skew symmetric. The case $n = 2$ is excluded since $SO(2)$ is abelian, as well as the case $n = 4$ since $SO(4) = SO(3) \times SO(3)$.

(ii) The special unitary groups $SU(n)$ with $n = 2, 3, 4, 5 \ldots$, which are the groups of the $n \times n$ matrices $\mathbf{U}$ such that

$$\mathbf{U}^\dagger \mathbf{U} = 1_n, \quad \det \mathbf{U} = 1 \quad (\mathbf{U}^\dagger = \text{Hermitian conjugate}).$$

To find the corresponding Lie algebras $su(n)$, we write

$$\mathbf{U} = 1_n + \mathbf{u} + \mathcal{O}(\mathbf{u}^2),$$

then $\det \mathbf{U} = 1 \Rightarrow \text{Tr}\,\mathbf{u} = 0$, and

$$(1_n + \mathbf{u}^\dagger)(1_n + \mathbf{u}) = 1_n + \mathbf{u} + \mathbf{u}^\dagger + \mathcal{O}(\mathbf{u}^2) = 1_n,$$

showing that $\mathbf{u} \in su(n) \Rightarrow \mathbf{u}$ is skew Hermitian.

(iii) The symplectic groups $Sp(n)$, which are the groups of the $2n \times 2n$ matrices $\mathbf{S}$ such that

$$\mathbf{S}^t\mathbf{E}_n\mathbf{S} = \mathbf{E}_n, \quad \mathbf{E}_n = \begin{pmatrix} \mathbf{0}_n & \mathbf{1}_n \\ -\mathbf{1}_n & \mathbf{0}_n \end{pmatrix}.$$

A matrix $\mathbf{s}$ belongs to the Lie algebra $sp(n)$ if $\mathbf{s}^t\mathbf{E} + \mathbf{E}\mathbf{s} = \mathbf{0}_{2n}$.

To the groups of these three series one must add the five "exceptional groups," which are isolated.

Besides the above compact simple groups, there exist some other non-compact forms. They are obtained by complexifying the compact forms (that is, allowing the assumption of complex values to the group coordinates), then finding all the possible real structures. For example, one finds the noncompact simple Lie groups $SO(p, q)$, which are the isometry groups of pseudo-Euclidean metric, or the special (that is with det $= 1$) linear groups $SL(n)$: their corresponding compact forms are $SO(p + q)$ and $SU(n)$, respectively.

The simple groups obtained in this way are all distinct, but the same statement does not hold for the corresponding algebras. For example, we will show now that $SO(3) = \frac{SU(2)}{\mathbb{Z}_2}$, $\mathbb{Z}_2 = \{1, -1\}$, from which $so(3) = su(2)$. Let

$$\psi = \begin{pmatrix} \psi_1 \\ \psi_2 \end{pmatrix} \in \mathbb{C}^2, \quad \psi^\dagger\psi = \overline{\psi}_1\psi_1 + \overline{\psi}_2\psi_2 = 1$$

be, by definition, a *spinor*. $\psi\psi^\dagger$ is a 2×2 Hermitian matrix with $\operatorname{Tr}\psi\psi^\dagger = \psi^\dagger\psi = 1$ and $\det \psi\psi^\dagger = 0$; hence we may write

$$\psi\psi^\dagger = \frac{1}{2}\begin{pmatrix} 1 - X_3 & X_1 + iX_2 \\ X_1 - iX_2 & 1 + X_3 \end{pmatrix}, \quad \det \psi\psi^\dagger = 0 \Rightarrow X_1^2 + X_2^2 + X_3^2 = 1.$$

We see that if $\psi' = g\psi$, $g \in SU(2)$, then $\psi'\psi'^\dagger$ is still a Hermitian matrix with trace 1 and null determinant; thus $SU(2)$ induces an action on $\mathbb{R}^3$ that leaves the Euclidean metric invariant. A homomorphism between $SU(2)$ and $SO(3)$ is so established, whose kernel is discrete because the dimension of the two groups is equal. Let us investigate the topology of the two group manifolds. Any matrix $g \in SU(2)$ can be written as

$$g = \begin{pmatrix} \alpha & \beta \\ -\overline{\beta} & \overline{\alpha} \end{pmatrix}, \quad \det g = \alpha\overline{\alpha} + \beta\overline{\beta} = 1, \quad \alpha, \beta \in \mathbb{C},$$

thus $SU(2) \simeq S^3$, the 3-dimensional sphere. On the other hand, all the elements of $SO(3)$ are rotations. There is a correspondence between rotations and vectors $\vec{x} \in \mathbb{R}^3$, $\|\vec{x}\| \leq \pi$, which is 1-1 when $\|\vec{x}\| < \pi$, while one must make the identification $\vec{x} = -\vec{x}$ when $\left\|\vec{x}\right\| = \pi$. Therefore, $SO(3) \simeq P_\mathbb{R}^3 \simeq \frac{S^3}{\mathbb{Z}_2}$.

A very important property of the semisimple Lie groups is that the corresponding group manifold is endowed with a natural (pseudo-)Riemannian metric, invariant under right and left translation. Indeed it is possible to prove that the symmetric matrix $F_{ab} = -c_{ad}^c c_{bc}^d$ is nonsingular if and only if the group is semisimple, and positive definite if it is compact. Then one defines

$$F_{hk}(x) = F_{ab}\omega_h^a \omega_k^b = F_{ab}\pi_h^a \pi_k^b, \tag{2.2.7}$$

bringing the metric defined at the identity around on the group manifold. One obtains the same metric using either left or right invariant forms, because the structure constants are invariant under the action of the adjoint representation, and one passes from right to left invariant forms (and vice versa) just with the action of the adjoint representation. The metric defined in (2.2.7) is called the *Cartan-Killing metric*.

2.3   Lagrangian and Hamiltonian Mechanics

By Lagrange equations we mean the equations of dynamics derived from a function L, the Lagrangian, which is a function of generalized coordinates q^μ and generalized velocities $\dot{q}^\mu$ (and in general an explicit function of time). We shall interpret q^μ as local coordinates on a manifold Q, called configuration space of the mechanical system, and $\dot{q}^\mu$ as the corresponding fiber coordinates on TQ. A Lagrangian is then a function on TQ, the tangent bundle of the configuration space.

Analogously, Hamilton equations are the equations of dynamics derived from a function H, the Hamiltonian, which is a function of generalized coordinates q^μ and generalized momenta p_μ (and in general an explicit function of time). We interpret q^μ as local coordinates on a manifold Q and p_μ as the corresponding fiber coordinates on T^*Q. A Hamiltonian is then a function on T^*Q, the cotangent bundle of the configuration space.

Many books have been written on analytical mechanics. We quote Whittaker (1917), Goldstein (1980), Sommerfeld (1964), and Landau & Lifchitz (1960) or, for those who like the conceptual aspects, the beautiful Lanczos (1970), all written in traditional language. But in the last decades the matter has changed its language, employing more and more the tools of differential geometry; see Arnold (1989), Abraham & Marsden (1978), Souriau (1997), Libermann & Marle (1987), Cushman & Bates (1997), and Marsden & Ratiu (1994).

2.3.1 Lagrange Equations

The Lagrange equations are a powerful tool that allows one to write, in an automatic way, as many pure (i.e., without forces of reaction) equations of the second order as degrees of freedom.

Let $\vec{r}_1, \ldots, \vec{r}_N$ be the position vectors of the N particles of our system[6] of mass $m_1, \ldots, m_N$. Let $\vec{v}_k = \frac{d\vec{r}_k}{dt}$ and $\vec{a}_k = \frac{d^2\vec{r}_k}{dt^2}$ be velocity and acceleration vectors of the generic kth particle. The total force acting on this particle is divided into the sum of two forces: $\vec{F}_k$, the active force, and $\vec{\Phi}_k$, the reaction force due to the constraints. The fundamental law of dynamics gives

$$\vec{F}_k + \vec{\Phi}_k = m_k \vec{a}_k.$$

We state the following basic postulate concerning $\vec{\Phi}_k$: the virtual work of the forces of reaction is always zero for any virtual displacement that is in harmony with the given kinematic constraints. We recall that a virtual displacement is infinitesimal and that we suppose the constraints frozen at the time when the virtual displacement is imposed. Plainly, the postulate generalizes the idea that the forces of reaction are orthogonal to the constraints. Let $\delta\vec{r}_k$ be the virtual displacement of the generic point; for the above postulate we can write

$$\sum_{k=1}^{N} \vec{F}_k \cdot \delta\vec{r}_k = \sum_{k=1}^{N} m_k \vec{a}_k \cdot \delta\vec{r}_k,$$

which is an equation from which the forces of reaction have disappeared. If the constraints are holonomic, we can obtain as many equations as degrees of freedom by projecting the above equation along all the independent virtual displacements.

Let $q^1, \ldots, q^n$ be local coordinates on the configuration manifold Q of the mechanical system; hence the position vector of any point can be expressed as a function $\vec{r}_k(q, t)$. If the time t does not appear explicitly, the constraints are said to be *fixed*. Any virtual displacement impresses to the generic point the displacement

$$\delta\vec{r}_k = \frac{\partial\vec{r}_k}{\partial q^\mu}\delta q^\mu, \quad \mu, \nu \ldots = 1, \ldots, n.$$

Notice that instead for a *real* displacement: $d\vec{r}_k = \delta\vec{r}_k + \frac{\partial\vec{r}_k}{\partial t}dt$. While the vectors $\delta\vec{r}_k$ are not in general independent (they take the constraints of

---

[6]We suppose the material system to be discrete. Otherwise, it is understood that the sums over the particles are replaced by integrals.

the system into account), the δq^μ are independent, if the constraints are holonomic as we suppose. We can thus write the n independent equations

$$\sum_{k=1}^{N} \vec{F}_k \cdot \frac{\partial \vec{r}_k}{\partial q^\mu} = \sum_{k=1}^{N} m_k \vec{a}_k \cdot \frac{\partial \vec{r}_k}{\partial q^\mu}.$$

The left-hand side is the sum of the projections of all the active forces onto the vectors of the natural basis and is denoted by Q_μ, or (the covariant components of the) *generalized force*. The right-hand side is rearranged as follows. Write

$$m_k \vec{a}_k \cdot \frac{\partial \vec{r}_k}{\partial q^\mu} = \frac{d}{dt}\left(m_k \vec{v}_k \cdot \frac{\partial \vec{r}_k}{\partial q^\mu}\right) - m_k \vec{v}_k \cdot \frac{d}{dt}\frac{\partial \vec{r}_k}{\partial q^\mu}.$$

Then, from the definition of real velocity

$$\vec{v}_k = \frac{\partial \vec{r}_k}{\partial q^\mu}\dot{q}^\mu + \frac{\partial \vec{r}_k}{\partial t},$$

we derive the two identities

$$\frac{\partial \vec{r}_k}{\partial q^\mu} = \frac{\partial \vec{v}_k}{\partial \dot{q}^\mu} \quad \text{and} \quad \frac{d}{dt}\frac{\partial \vec{r}_k}{\partial q^\mu} = \frac{\partial^2 \vec{r}_k}{\partial q^\nu \partial q^\mu}\dot{q}^\nu + \frac{\partial^2 \vec{r}_k}{\partial t \partial q^\mu} = \frac{\partial \vec{v}_k}{\partial q^\mu}.$$

Defining the total kinetic energy $T = \sum_k \frac{1}{2}m_k \vec{v}_k \cdot \vec{v}_k$, we can write

$$\sum_{k=1}^{N} m_k \vec{a}_k \cdot \frac{\partial \vec{r}_k}{\partial q^\mu} = \frac{d}{dt}\frac{\partial T}{\partial \dot{q}^\mu} - \frac{\partial T}{\partial q^\mu},$$

from which the Lagrange equations follow:

$$\frac{d}{dt}\frac{\partial T}{\partial \dot{q}^\mu} - \frac{\partial T}{\partial q^\mu} = Q_\mu.$$

If the active forces admit a potential energy V, that is, if $Q_\mu = -\frac{\partial V}{\partial q^\mu}$, then the Lagrange equations may be written as

$$\frac{d}{dt}\frac{\partial L}{\partial \dot{q}^\mu} - \frac{\partial L}{\partial q^\mu} = 0,$$

where $L = T - V$ is the *Lagrangian* of the mechanical system.

The general expression of the kinetic energy is

$$T = \frac{1}{2}g_{\mu\nu}\dot{q}^{\mu}\dot{q}^{\nu} + A_{\mu}\dot{q}^{\mu} + V,$$

$$g_{\mu\nu}(q,t) = \sum_{k=1}^{N} m_k \frac{\partial \vec{r}_k}{\partial q^{\mu}} \cdot \frac{\partial \vec{r}_k}{\partial q^{\nu}},$$

$$A_{\mu}(q,t) = \sum_{k=1}^{N} m_k \frac{\partial \vec{r}_k}{\partial q^{\mu}} \cdot \frac{\partial \vec{r}_k}{\partial t},$$

$$V(q,t) = \frac{1}{2}\sum_{k=1}^{N} m_k \frac{\partial \vec{r}_k}{\partial t} \cdot \frac{\partial \vec{r}_k}{\partial t}.$$

If the constraints are fixed, the kinetic energy is a quadratic homogeneous function of the velocities

$$T = \frac{1}{2}g_{\mu\nu}(q)\dot{q}^{\mu}\dot{q}^{\nu},$$

and endows the configuration manifold Q with a Riemannian metric $g_{\mu\nu}$.

2.3.2 Hamilton Principle

Before discussing the very important fact that the Lagrange equations can be derived from a variational principle,[7] we state a basic analytic identity.

Let us consider a generic smooth function $F : TQ \times \mathbb{R} \rightarrow \mathbb{R}$ by $(q, v, t) \mapsto F(q, v, t)$. Define the *action integral*

$$S = \int_{t_1}^{t_2} F(q(t), \dot{q}(t), t)\,dt,$$

which is a functional; i.e., it yields a number in correspondence to every curve $q(t)$, $t_1 \le t \le t_2$ of the configuration space. We want to calculate the "infinitesimal" variation of this action integral when passing from a curve y_0 to other curves y_{α} "infinitesimally close." These curves are labeled with a real parameter α:

$$y_{\alpha} : t \mapsto q(t, \alpha).$$

Substituting into the action integral, this becomes a function of the parameter α. The usual methods of elementary calculus, in particular the derivation under the integral sign, give

$$\delta S[\alpha] \stackrel{\text{def}}{=} \frac{\partial S[\alpha]}{\partial \alpha}\delta\alpha = \int_{t_1}^{t_2} \left(\frac{\partial F}{\partial q^{\mu}}\frac{\partial q^{\mu}}{\partial \alpha} + \frac{\partial F}{\partial \dot{q}^{\mu}}\frac{\partial \dot{q}^{\mu}}{\partial \alpha} \right)\delta\alpha\,dt. \qquad (2.3.1)$$

---

[7] In this context, they are sometimes called *Euler–Lagrange equations*.

We invert the order of the differentiations with respect to the two parameters t and α, then perform an integration by parts on the second term:

$$\int_{t_1}^{t_2} \frac{\partial F}{\partial \dot{q}^\mu} \frac{\partial \dot{q}^\mu}{\partial \alpha} \delta \alpha \, dt = \frac{\partial F}{\partial \dot{q}^\mu} \frac{\partial q^\mu}{\partial \alpha} \delta \alpha \Big|_{t_1}^{t_2} - \int_{t_1}^{t_2} \frac{d}{dt} \frac{\partial F}{\partial \dot{q}^\mu} \frac{\partial q^\mu}{\partial \alpha} \delta \alpha \, dt.$$

Now put $\alpha = 0$, so that δS measures the variation of the integral action when passing from y_0 to y_α, as required. Put, for brevity, $\delta q^\mu = \dfrac{\partial q^\mu}{\partial \alpha} \delta \alpha \Big|_{\alpha=0}$. The final formula we seek is

$$\delta S = \int_{t_1}^{t_2} \left(\frac{\partial F}{\partial q^\mu} - \frac{d}{dt} \frac{\partial F}{\partial \dot{q}^\mu} \right) \delta q^\mu \, dt + \frac{\partial F}{\partial \dot{q}^\mu} \delta q^\mu \Big|_{t_1}^{t_2}, \qquad (2.3.2)$$

which, as we stress again, is a pure analytical identity, i.e., not involving any physical principle.

    We can now state the Hamilton principle. Let us restrict ourselves to the class of curves y_α that start from a point P_1 at time t_1 and reach another point P_2 at time t_2. Because $\delta q^\mu = 0$ at time t_1 and t_2, the last term in (2.3.2) vanishes. Therefore, the fundamental identity (2.3.2) shows that

(i) if we choose y_0 to be the real trajectory of the mechanical system satisfying the physical laws, then the Lagrange equations hold, and we can infer that $\delta S = 0$. In other words, the natural motion of a mechanical system makes stationary the action integral;

(ii) if we assume as a primitive physical law that the action integral is stationary and suppose the constraints holonomic, then the Lagrange equations follow.

The Hamilton principle may be summarized as

$$\textit{Stationary action integral} \Leftrightarrow \textit{Lagrange equations.}$$

2.3.3 Noether Theorem

Several conservation laws of mechanics (for example of the linear or angular momentum and of the energy) are particular cases of the same general theorem, due to Noether: to each one-parameter group of diffeomorphisms of the configuration space, leaving invariant the Lagrangian, corresponds a first integral of the equations of motion. We recall that a first integral is a function $I : TQ \to \mathbb{R}$, whose total time derivative vanishes when calculated along a solution of the Lagrange equations. In this section we consider a time-independent Lagrangian.

Before proving the theorem, we must define exactly what the invariance condition of the Lagrangian means. Let us consider a one-parameter group of diffeomorphisms acting on the configuration space: $\Phi_\alpha : q \mapsto \Phi_\alpha q$, and let $\Phi_{\alpha*}v$ be the induced push-forward of a tangent vector field. We say that a Lagrangian $L(q, \dot{q})$ is invariant under the action of the group if

$$L(\Phi_\alpha q, \Phi_{\alpha*}v) = L(q, v), \quad \forall \alpha.$$

We also say that the Lagrangian has a symmetry.

To prove the theorem, we start from the fundamental identity (2.3.2), where we put $F = L$. Let γ_0 be the real curve satisfying the Lagrange equations and γ_α be the curve transformed under the action of the symmetry group. Therefore, $\delta q^\mu = \left. \dfrac{\partial \Phi_\alpha^\mu}{\partial \alpha} \delta \alpha \right|_{\alpha=0} = X^\mu(q)\,\delta\alpha$, where X^μ is the infinitesimal generator of the action of the group. In (2.3.2), $\delta S = 0$ holds, for the assumption on the invariance of the Lagrangian, while the integral is zero, because only curves satisfying Lagrange equations are considered. Therefore, the difference $\left. \dfrac{\partial L}{\partial \dot{q}^\mu} X^\mu \right|_{t_2} - \left. \dfrac{\partial L}{\partial \dot{q}^\mu} X^\mu \right|_{t_1}$ also vanishes. Since t_2 and t_1 are generic, we infer that

$$I(q.\dot{q}) \stackrel{\text{def}}{=} \frac{\partial L}{\partial \dot{q}^\mu} X^\mu$$

is a first integral.

EXAMPLE 2.15 If the Lagrangian does not depend on some coordinate, say q^n, it is invariant under the group of translations along this coordinate, which is called *ignorable*. Then $X = (0, \ldots, 0, 1)$ and $I = \frac{\partial L}{\partial \dot{q}^n}$.

EXAMPLE 2.16 Let Q be the 2-dimensional Euclidean plane, with Cartesian coordinates x and y, and let the Lagrangian be invariant under rotations around the origin. Thus

$$\Phi_\alpha \begin{pmatrix} x \\ y \end{pmatrix} = \begin{pmatrix} \cos\alpha & \sin\alpha \\ -\sin\alpha & \cos\alpha \end{pmatrix} \begin{pmatrix} x \\ y \end{pmatrix} \Rightarrow X = \begin{pmatrix} y \\ -x \end{pmatrix}.$$

If $L = T - V$ with $T = \frac{1}{2}m(\dot{x}^2 + \dot{y}^2)$, then $I = m(\dot{x}y - \dot{y}x)$, which is the angular momentum with respect to the origin.

It is possible to completely eliminate the ignorable coordinate q^n from the Lagrangian, reducing the problem to $n-1$ degrees of freedom. This is a first elementary example of *reduction by symmetry*. Later, the idea of reducing the number of the degrees of freedom by exploiting the symmetries of the system will be generalized. Let us indeed consider the differential (2.3.1) of the action integral and, *before* performing the integration by parts, take into account that now $\frac{\partial L}{\partial q^n} = 0$ and $\frac{\partial L}{\partial \dot{q}^n} = c_n$ hold. Obviously, $c_n \delta \dot{q}^n = \delta(c_n \dot{q}^n)$, and we may write, after an integration by parts on the first $n-1$ coordinates only,

$$\delta \int_{t_1}^{t_2} (L - c_n \dot{q}^n) dt = \int_{t_1}^{t_2} \left(\frac{\partial L}{\partial q^\mu} - \frac{d}{dt} \frac{\partial L}{\partial \dot{q}^\mu} \right) \delta q^\mu \, dt + \frac{\partial L}{\partial \dot{q}^\mu} \delta q^\mu \Big|_{t_1}^{t_2},$$

$$\mu = 1, \ldots, n-1.$$

We can now argue as in the previous section, and deduce that the equations of motion for the first $n-1$ coordinates are derived from the reduced Lagrangian $L_R = L - c_n \dot{q}^n$.

EXAMPLE 2.17 Let $L = \frac{1}{2} m (\dot{r}^2 + r^2 \dot{\theta}^2) - V(r)$ be the Lagrangian of a central motion on the plane with polar coordinates r, θ. The angle θ is ignorable, $\frac{\partial L}{\partial \theta} = mr^2 \dot{\theta}$ is a constant c and $\dot{\theta} = \frac{c}{mr^2}$, then $L_R = \frac{1}{2} m \dot{r}^2 - V(r) - \frac{c^2}{2mr^2}$. The motion is now 1-dimensional, but a new force with potential energy $\frac{c^2}{2mr^2}$ appears.

2.3.4 From Lagrange to Hamilton

Hereafter, Lagrangian functions are assumed to nonexplicitly depend on the time. It is tempting to write the n Lagrange equations of the second order as $2n$ equations of the first order on the tangent bundle TQ. The natural local coordinates on TQ are $\{q^\mu, v^\nu\}$, so that we may write

$$\frac{d}{dt} q^\mu = v^\mu, \quad \frac{d}{dt} \frac{\partial L(q, v)}{\partial v^\nu} = \frac{\partial L(q, v)}{\partial q^\nu}, \quad \mu, \nu = 1, \ldots, n. \qquad (2.3.3)$$

It is a very remarkable fact that TQ is naturally endowed with a symplectic structure and that a Hamiltonian form can be given to the system (2.3.3).

DEFINITION 2.20 *A vector field on a symplectic manifold P is said to be Hamiltonian if it is the symplectic gradient of a Hamiltonian function $H : P \to \mathbb{R}$; that is, has the form $\Omega^\sharp(dH)$, where Ω is the symplectic 2-form of P. A dynamical system is said to be Hamiltonian if the relative vector field is Hamiltonian.*

Let u^α, $\alpha, \beta, \gamma = 1, \ldots, 2n$ be local coordinates on P and

$$\Omega = \Omega_{\alpha\beta} \, du^\alpha \wedge du^\beta$$

the symplectic form; a Hamiltonian system can therefore be written as

$$\dot{u}^\alpha = \Omega^{\alpha\beta} \frac{\partial H}{\partial u^\beta}, \quad \Omega^{\alpha\beta} \Omega_{\beta\gamma} = \delta^\alpha_\gamma.$$

To show that (2.3.3) is a Hamiltonian system, first we endow TQ with the symplectic form $\Omega_L = d\Theta_L$, where $\Theta_L = \dfrac{\partial L}{\partial v^\mu} dq^\mu$. This ensures that $d\Omega_L = 0$. Then, we suppose that $\det\left(\dfrac{\partial^2 L}{\partial v^\mu \partial v^\nu}\right) \neq 0$, which ensures the invertibility of the matrix $(\Omega_L^{\alpha\beta})$: a Lagrangian satisfying this property is said to be *regular*. Explicitly

$$\Omega_L = \frac{\partial^2 L}{\partial q^\mu \partial v^\nu} dq^\mu \wedge dq^\nu + \frac{\partial^2 L}{\partial v^\mu \partial v^\nu} dv^\mu \wedge dq^\nu.$$

Finally, putting $(u^1, \ldots, u^{2n}) = (q^1, \ldots, q^n, v^1, \ldots, v^n)$, define the Hamiltonian

$$H(u) \stackrel{\text{def}}{=} \frac{\partial L}{\partial v^\mu} v^\mu - L(q, v).$$

Then, it is a simple matter of calculation to check that the equations (2.3.3) are equivalent to

$$\Omega_L(\dot{u}) = dH,$$

which, for the invertibility of the matrix of Ω_L, proves the statement.

From the Darboux theorem we know that there exists a chart in which a symplectic structure is put in canonical form. In the Lagrangian case it is immediate to find it by simply inspecting Θ_L; it is indeed natural to define the map $\text{Leg} : TQ \to T^*Q$ by $(q, v) \mapsto (q, p)$ where $p_\mu = \dfrac{\partial L}{\partial v^\mu}$, so that T^*Q becomes equipped with the canonical 1-form $\Theta = p_\mu dq^\mu$.

The Leg map, called the *Legendre transformation*, is therefore a symplectomorphism between TQ, equipped with the symplectic structure Θ_L, and T^*Q, equipped with the canonical structure $\Omega = dp_\mu \wedge dq^\mu$:

$$\text{Leg}^*\Omega = \Omega_L.$$

Let us sum up, in the traditional language, how to pass from Lagrangian to Hamiltonian formulation.

(i) Invert the definition of the momenta $p_\mu = \dfrac{\partial L}{\partial \dot{q}^\mu}$, finding the generalized velocities as functions of the momenta; this is possible if the Lagrangian is regular.

(ii) Define the Hamiltonian $H(q, p) = p_\mu \dot{q}^\mu - L(q, \dot{q})$; then the equations of motion are the Hamilton equations in canonical form

$$\dot{q}^\mu = \frac{\partial H(q, p)}{\partial p_\mu}, \quad \dot{p}_\nu = -\frac{\partial H(q, p)}{\partial q^\nu}.$$

A particular but important case is that of the natural systems: the momenta $p_\mu = g_{\mu\nu} \dot{q}^\nu$ are the covariant components of the velocity, and the Hamiltonian $H = \frac{1}{2} g^{\mu\nu} p_\mu p_\nu + V$ is the total energy.

2.3.5 Canonical Transformations

A chart in T^*Q, with coordinates $u = (u^1, \ldots, u^{2n})$, is said to be *canonical* if the symplectic 2-form has the expression $\Omega = \sum_\mu du^{n+\mu} \wedge du^\mu$. A change of chart is said to be *canonical* if it preserves the canonical form. A canonical transformation is thus a (in general local) symplectomorphism of T^*Q into itself. Making the identifications[8] $u^\mu = q^\mu$, $u^{n+\mu} = p_\mu$, the change of chart is given by assigning the $2n$ functions

$$Q^\mu = Q^\mu(q, p), \quad P_\nu = P_\nu(q, p).$$

These functions are not completely arbitrary. Take the definition (2.1.9) of the Poisson bracket into account, then $\Omega^{\alpha\beta} = -\{u^\alpha, u^\beta\}$, $\alpha, \beta = 1, \ldots, 2n$ holds, so that the change of chart defines a canonical transformation $\Gamma :$ $(q, p) \mapsto (Q, P)$ if and only if

$$\{Q^\mu, Q^\nu\} = 0, \quad \{P_\mu, P_\nu\} = 0, \quad \{P_\mu, Q^\nu\} = \delta_\mu^\nu. \tag{2.3.4}$$

Given a Hamiltonian $H(q, p)$, put $K = \Gamma_* H$; then, for the canonicity of Γ, the Hamilton equations in the new variables are

$$\dot{Q}^\mu = \frac{\partial K(Q, P)}{\partial P_\mu}, \quad \dot{P}_\nu = -\frac{\partial K(Q, P)}{\partial Q^\nu}.$$

Sometimes this property, i.e., to leave invariant in form *any* Hamiltonian system, is taken as the definition of a canonical transformation.

The conditions (2.3.4) allow us to check if a given transformation is canonical, but they do not tell us how to construct one. To this end, we notice that a transformation $(q, p) \mapsto (Q, P)$ such that

$$p_\mu dq^\mu - P_\mu dQ^\mu = dF(q, Q) \tag{2.3.5}$$

is surely canonical, since $ddF = 0$ implies that $dp_\mu \wedge dq^\mu = dP_\mu \wedge dQ^\mu$. In particular, F may be identically zero, then $p_\mu = \frac{\partial Q^\nu}{\partial q^\mu} P_\nu$, which is the transformation rule of a covector under a coordinate transformation $q \mapsto Q$. Such transformations are called *punctual*, and exhaust the class of transformations allowed in the Lagrangian formalism.

A function $F(q, Q)$, satisfying $\det\left(\frac{\partial^2 F}{\partial q^\mu \partial Q^\nu}\right) \neq 0$, defines, at least locally, an invertible canonical transformation, and is called a *generating function*. Indeed, expanding its differential, one finds

$$p_\mu = \frac{\partial F(q, Q)}{\partial q^\mu}, \quad P_\mu = -\frac{\partial F(q, Q)}{\partial Q^\mu}.$$

---

[8]Unfortunately, there is a certain problem with the position of the indices. We put them above, like for the u's coordinates, looking at the symplectic structure, while we put them below, like for the p's, looking at the momenta as the components of the 1-form Θ but *defined on the configuration space*.

Due to the condition on the Hessian, these two relations can be inverted, expressing the new as functions of the old canonical coordinates and vice versa.

The most simple canonical transformation of this type is generated by $F = \sum_\mu q^\mu Q^\mu$:

$$p_\mu = Q^\mu, \quad P_\mu = -q^\mu,$$

that is, the exchange between coordinates and momenta. However, the identity transformation cannot be generated by $F(q, Q)$.

We may consider other classes of generating functions; for example

$$p_\mu dq^\mu - P_\mu dQ^\mu = dF(q, P) - d(P_\mu Q^\mu)$$

gives surely a canonical transformation

$$p_\mu = \frac{\partial F(q, P)}{\partial q^\mu}, \quad Q^\mu = \frac{\partial F(q, P)}{\partial P_\mu}, \tag{2.3.6}$$

if the condition $\det\left(\frac{\partial^2 F}{\partial q^\mu \partial P_\nu}\right) \neq 0$ is satisfied. The identity is generated by $F = P_\mu q^\mu$. The punctual, or *extended point transformation*, can be generated by $F(q, P) = f^\mu(q)P_\mu$, with f^μ arbitrary functions:

$$Q^\mu = f^\mu(q), \quad p_\mu = \frac{\partial f^\nu}{\partial q^\mu} P_\nu.$$

Moreover, one may also consider generating functions of the type $F(p, P)$ or $F(p, Q)$, although they are never used.

2.3.6 Hamilton–Jacobi Equation

The basic idea of the integration method of Hamilton–Jacobi is to find a canonical transformation Γ such that, in the new coordinates, the transformed Hamiltonian $K = \Gamma_* H$ is easily integrable.

As a typical example, let us seek a Γ such that $K(Q, P) \equiv P_1$, so that

$$\dot{P}_\mu = -\frac{\partial K}{\partial Q^\mu} = 0 \quad \Rightarrow P_\mu = \alpha_\mu, \quad \mu = 1, \ldots, n,$$

$$\dot{Q}^1 = \frac{\partial K}{\partial P_1} = 1 \quad \Rightarrow Q^1 = t + \beta^1,$$

$$\dot{Q}^\nu = \frac{\partial K}{\partial P_\nu} = 0 \quad \Rightarrow Q^\nu = \beta^\nu, \quad \nu = 2, \ldots, n,$$

where α_μ, β^μ are $2n$ integration constants; in particular $\alpha_1 = E$, the total energy. After inversion of Γ, the general integral can be written as

$$q^\mu = q^\mu(t, \alpha, \beta), \quad p_\mu = p_\mu(t, \alpha, \beta),$$

and the dynamical problem is solved.

Of course, there is nothing magical in all that, the problem now being deferred to find such a canonical transformation. Let $W(q, P)$ (but one could take any of the other three types) be the unknown generating function of Γ. Taking the first of (2.3.6) into account, consider the partial differential equation

$$H\left(q^1, \ldots, q^n, \frac{\partial W}{\partial q^1}, \ldots, \frac{\partial W}{\partial q^n}\right) = \alpha_1. \tag{2.3.7}$$

The meaning of (2.3.7) should be clear: it imposes that the canonical transformation Γ we seek be such that the transformed Hamiltonian $K = \Gamma_* H$ coincides with the first of the new momenta. If one is able to find a *complete integral* of this equation, that is, a solution $W(q, \alpha)$ such that $\det\left(\frac{\partial^2 W}{\partial q^\mu \partial \alpha_\nu}\right) \neq 0$, $\mu, \nu = 1, \ldots, n$, then the dynamical problem is solved. The partial differential equation of first order (2.3.7) is called a *(reduced) Hamilton–Jacobi equation*.

The full name (without "reduction") is reserved for the time-dependent Hamiltonian, whose Hamilton–Jacobi equation

$$H\left(q^1, \ldots, q^n, \frac{\partial W}{\partial q^1}, \ldots, \frac{\partial W}{\partial q^n}, t\right) + \frac{\partial W}{\partial t} = 0,$$

imposes that the transformed Hamiltonian will vanish identically.

2.3.7 Symmetries and Reduction

In this section we will generalize the Noether theorem to the case of an N-dimensional Lie group $\mathfrak{G}$ of symmetries.

Let $\Phi : M \to M$ be a symplectomorphism of a $2n$-dimensional symplectic manifold (M, Ω). Obviously, Φ leaves invariant the Poisson brackets:

$$\Phi^* \Omega = \Omega \iff \{\Phi^* f, \Phi^* g\} = \Phi^* \{f, g\}. \tag{2.3.8}$$

Let us now consider the left action $u \mapsto \Phi_x u$, $u \in M$, $x \in \mathfrak{G}$ of $\mathfrak{G}$ on M. Let

$$X_a = \left.\frac{\partial \Phi_x}{\partial x^a}\right|_{x=1}, \quad a = 1, \ldots, N$$

be a basis for the vector space of the infinitesimal generators of the action. We know from Proposition 2.18 that they form a Lie algebra that is homomorphic to the Lie algebra $\mathfrak{g}$. In other words, if $\xi = \xi^a L_a$, $\eta = \eta^a L_a$ are two elements of $\mathfrak{g}$, and $X_\xi = \xi^a X_a$, $X_\eta = \eta^a X_a$ the corresponding vector fields on M, we have

$$[X_\xi, X_\eta] = -X_{[\xi, \eta]}, \quad \xi, \eta \in \mathfrak{g}. \tag{2.3.9}$$

Now suppose that the $\mathfrak{G}$-action on M is a symplectomorphism for every $x \in \mathfrak{G}$. Recalling the relation (2.1.6), we can write

$$\mathcal{L}_{X_\xi}\Omega = i_{X_\xi}\, d\Omega + d(i_{X_\xi}\Omega).$$

But $\mathcal{L}_{X_\xi}\Omega = 0$ holds, which is the "infinitesimal version" of (2.3.8); moreover, $d\Omega = 0$, by the definition of symplectic manifold. Therefore, the 1-form $i_{X_\xi}\Omega$ is closed, hence locally exact: for a fixed ξ, there exists locally a function $J_\xi : M \to \mathbb{R}$ such that

$$i_{X_\xi}\Omega = -dJ_\xi, \quad \text{or } X_\xi = dJ_\xi^\sharp.$$

Thus, the infinitesimal generators of the action are local Hamiltonian vectors, with Hamiltonian J_ξ. If, as we will suppose hereafter, the vector fields X_ξ are globally Hamiltonian, the action Φ is said to be *Hamiltonian*.

For a fixed point $u \in M$, the dependence of $J_\xi(u)$ from the element of the Lie algebra is linear. Indeed, since the basis vector fields X_a on M are Hamiltonian, $X_a = dJ_a^\sharp$ holds for some functions J_a, $a = 1,\ldots,N$. If $\xi = \xi^a L_a$, then X_ξ is Hamiltonian, with Hamiltonian $J_\xi = \xi^a J_a$. Since the dual of a vector space is the space of the linear functions, we can define a function $J : M \to \mathfrak{g}^*$ such that

$$J_\xi(u) = \langle J(u), \xi \rangle .$$

J is called a *moment map*.

Let us find the basic property of the moment map. In expression (2.1.10) take $f = J_\xi$ and $g = J_\eta$; since $X_\xi = X_{J_\xi}$ and $X_\eta = X_{J_\eta}$ we obtain, by comparison with (2.3.9), that

$$X_{\{J_\xi, J_\eta\}} = -X_{J_{[\xi,\eta]}}.$$

The operator $\sharp$ is an isomorphism, and the above relation implies that the two functions $-J_{[\xi,\eta]}$ and $\{J_\xi, J_\eta\}$ have the same gradient, and thus differ in a quantity constant on M

$$\{J_\xi, J_\eta\} = -J_{[\xi,\eta]} + C(\xi, \eta).$$

Projecting this relation on the basis $\{L_a\}$, we can write

$$\{J_a, J_b\} = -c_{ab}^c J_c + C_{ab}, \quad a, b, c = 1,\ldots,N.$$

The skew symmetric bilinear function $C(\xi_1, \xi_2)$ is called a *2-cocycle of the Lie algebra*, and must satisfy the relation

$$C([\xi_1, \xi_2], \xi_3) + C([\xi_2, \xi_3], \xi_1) + C([\xi_3, \xi_1], \xi_2) = 0,$$

due to the Jacobi identity. When the cocycle is zero (which happens, for example, when the Lie group is semisimple), there is an isomorphism between the Lie algebra of the infinitesimal generators X_a and the Lie algebra of the Hamiltonians J_a with respect to the Poisson brackets: the moment map is said to be *Ad*-equivariant*. This name is due to the fact that

$\{J_a, J_b\} = -c_{ab}^c J_c$ is the "infinitesimal" version of the statement that the following diagram is commutative:

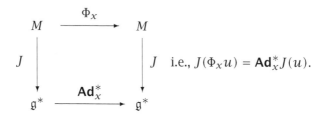

$$J(\Phi_X u) = \mathbf{Ad}_x^* J(u).$$

Indeed, projecting this commutativity relation on the basis $\{L_a\}$, we obtain

$$J_a(\Phi_X u) = (\mathbf{Ad}_x^{-1})_a^c J_c(u) \tag{2.3.10}$$

by definition (2.2.5) of coadjoint representation. Differentiating with respect to x^b and putting $x = 1$, then taking into account (2.2.6), along with the skew symmetry of the Poisson bracket and of the structure constants, we find

$$\{J_a, J_b\} = -c_{ab}^c J_c. \tag{2.3.11}$$

This shows that (2.3.11) holds if the diagram is commutative.

For the connected component of the identity of $\mathfrak{G}$, also the reverse holds. Equation (2.3.11) asserts that both members of (2.3.10) have the same derivative at the identity. Moreover, the two members of (2.3.10) agree at the identity: for the homogeneity of the group, the statement follows. Hereafter, we will consider only Ad*-equivariant maps.

EXAMPLE 2.18 An important example is that arising when a group acts on the configuration space Q and the symplectomorphism Φ is an extended point transformation. The infinitesimal generator X_ξ is a vector field on Q and the relative Ad*-equivariant moment map is $J_\xi = \langle \Theta, X_\xi \rangle$; in coordinates $J_\xi(q, p) = p_\mu X_\xi^\mu(q)$. Indeed, a straightforward calculation shows that

$$\{J_\xi, J_\eta\} = [X_\xi, X_\eta]^\mu p_\mu,$$

which ensures that (2.3.11) holds. See the following two examples.

EXAMPLE 2.19 Take $\mathfrak{G} = SO(3)$ acting on $Q = \mathbb{R}^3$; thus the moment map $T^*\mathbb{R}^3 \to so^*(3)$ is given by the angular momentum $\vec{G} = \vec{q} \times \vec{p}$ of a point in $\mathbb{R}^3$. It is easy to check that

$$\{G_1, G_2\} = -G_3, \ \{G_2, G_3\} = -G_1, \ \{G_3, G_1\} = -G_2, \tag{2.3.12}$$

according to (2.3.11).

EXAMPLE 2.20 Take $\mathfrak{G} = SO(3)$ acting on itself through left and right translations; thus the two relative moment maps concern a rigid body with a fixed point, and they are given by the projection of the angular momentum on the axes fixed in the space and with the moving body, respectively. The three components of each one of the projections satisfy (2.3.12), while the mutual Poisson brackets are zero. See Marsden & Ratiu (1994, Chapter XV).

If the Φ action on M is transitive, then for the Ad*-equivariance the coadjoint action of $\mathfrak{G}$ on $J(M)$ also is transitive, and M can be identified with (a cover of) a coadjoint orbit of $\mathfrak{G}$ on $\mathfrak{g}^*$. This is the content of the Kirillov–Kostant–Souriau theorem (Guillemin & Sternberg 1990).

EXAMPLE 2.21 Consider the action of $SO(3)$ on the sphere S^2 equipped with the symplectic structure of Example 2.7. The action is clearly symplectic and transitive, and in effect S^2 is a coadjoint orbit on $so^*(3)$; see Example 2.10.

Let us now suppose that on the symplectic manifold M a dynamics is given, which is invariant with respect to the group $\mathfrak{G}$. In other words, a Hamiltonian $H : M \to \mathbb{R}$ is assigned such that

$$H \circ \Phi_x = H, \quad \forall x \in \mathfrak{G}.$$

Differentiating with respect to the elements of $\mathfrak{G}$ at the identity, and noticing that the right-hand side does not depend on $\mathfrak{G}$, we find

$$\left.\frac{\partial}{\partial x^a} H(\Phi_x u)\right|_{x=1} = \left\langle dH, dJ_a^\sharp \right\rangle = \{J_a, H\} = 0 \quad a = 1, \ldots, N.$$

The components of the moment map relative to the symmetry group $\mathfrak{G}$ are therefore first integrals. The expression

$$\{J_a, H\} = \mathcal{L}_{dJ_a^\sharp} H = -\mathcal{L}_{dH^\sharp} J_a = 0$$

can be interpreted in two "complementary" ways. In the first one, the Hamiltonian is invariant along the flows generated by the $\mathfrak{G}$-action; thus the Hamiltonian is invariant with respect to the group action. In the latter, the components of the moment map are invariant along the time flow generated by the Hamiltonian; thus the J_a's are first integrals. The relation between symmetries and conservation laws constitutes the *generalized Noether theorem* and frequently allows us to guess the first integrals of a mechanical system without calculations.

EXAMPLE 2.22 The Hamiltonian $p^2/2 + V(r)$ in $T^*\mathbb{R}^3$ is manifestly invariant under the extended action of the rotation group; hence, from Example 2.19 it follows that the angular momentum is a first integral.

But we can go even further. In the Lagrangian environment a one-para-meter group of symmetry allows us to reduce the number of degrees of freedom of a mechanical system by one. Also, in the Hamiltonian environment an N-dimensional Lie group of symmetries allows us to proceed to a reduction. This is the content of the Marsden–Weinstein theorem (Marsden & Weinstein 1974), concerning the reduction in the symplectic case. The theorem can be generalized to the Poisson case (Marsden & Ratiu 1986).

In fact, let us define

$$J^{-1}(\mu) = \{u \in M : J(u) = \mu, \ \mu \in \mathfrak{g}^*\}.$$

We say that μ is a *regular value* if $dJ|_u$ is surjective for every $u \in J^{-1}(\mu)$; in such a case $J^{-1}(\mu)$ is a differentiable submanifold of M. The action of $\mathfrak{G}$ on M in general moves this submanifold, while the action of the isotropy subgroup

$$\mathfrak{G}_\mu \overset{\text{def}}{=} \{x \in \mathfrak{G} : \mathbf{Ad}_x^*\mu = \mu\},$$

leaves it invariant. Indeed, if $J(u) = \mu$ and $x \in \mathfrak{G}_\mu$, for the Ad*-equivari-ance we have $J(\Phi_x u) = \mathbf{Ad}_x^* J(u) = \mathbf{Ad}_x^*\mu = \mu$. If the action of $\mathfrak{G}_\mu$ on $J^{-1}(\mu)$ is free and proper, then the space of the orbits $M_\mu = J^{-1}(\mu)/\mathfrak{G}_\mu$ is a dif-ferentiable manifold. The restriction of the symplectic 2-form Ω to $J^{-1}(\mu)$ is still a closed 2-form, which is, however, in general degenerate: the in-duced linear map $\flat : TM \to T^*M$ has a nontrivial kernel. The content of the Marsden–Weinstein theorem is essentially that this kernel is spanned by the tangent vectors to the orbit of the $\mathfrak{G}_\mu$-action on $J^{-1}(\mu)$, so that, passing to the quotient, we obtain that M_μ is symplectic. M_μ is said to be the *symplec-tic reduced manifold* of M. Moreover, if the Hamiltonian on M is invariant with respect to $\mathfrak{G}$-action, the dynamics projects down to M_μ.

THEOREM 2.21 (MARSDEN–WEINSTEIN) *Let (M, Ω) be a symplectic mani-fold on which the Lie group $\mathfrak{G}$ acts symplectically and let $J : M \to \mathfrak{g}^*$ be an Ad*-equivariant moment map for this action. Assume $\mu \in \mathfrak{g}^*$ is a regular value of J and that the action of the isotropy subgroup $\mathfrak{G}_\mu$ on $J^{-1}(\mu)$ is free and proper. Then $M_\mu = J^{-1}(\mu)/\mathfrak{G}_\mu$ has a unique symplectic form Ω_μ with the property*

$$\pi_\mu^* \Omega_\mu = i_\mu^* \Omega, \tag{2.3.13}$$

where $\pi_\mu : J^{-1}(\mu) \to M_\mu$ is the canonical projection and $i_\mu : J^{-1}(\mu) \to M$ the inclusion.

The restriction of a $\mathfrak{G}$-invariant Hamiltonian H to $J^{-1}(\mu)$ induces a Hamil-tonian H_μ on the reduced manifold M_μ, called the reduced Hamiltonian, *de-fined by*

$$\pi_\mu^* H_\mu = i_\mu^* H.$$

Then, the vector field X_H on $J^{-1}(\mu)$ is π_μ-related to the reduced vector field X_{H_μ} on M_μ, that is,

$$\mathbf{T}_{\pi_\mu} X_H = X_{H_\mu} \circ \pi_\mu.$$

As for the dimension of the reduced manifold, $\dim M_\mu = \dim M - \dim \mathfrak{G} - \dim \mathfrak{G}_\mu$.

In practice the theorem does not provide an explicit recipe, ready-made for concrete applications. Rather, it unifies and clarifies several different procedures that emerged historically, orienting the wayfarer in the calculation desert.

EXAMPLE 2.23 Consider again the Hamiltonian $p^2/2 + V(r)$ in $T^*(\mathbb{R}^3 - \{0\})$. In this case, the reduced manifold concides with the cotangent bundle of $(\mathbb{R}^3 - \{0\})/SO(3) = \mathbb{R}_+$, the positive half-line. The reduced Hamiltonian is

$$ H = \frac{1}{2} \left(p_r^2 + \frac{G^2}{r^2} \right) + V(r), \ G = \left\| \vec{G} \right\| = \text{const.} $$

The system has been reduced to one degree of freedom, but the new potential G^2/r^2 appears.

EXAMPLE 2.24 The configuration manifold of an asymmetric free rigid body with a fixed point is $SO(3)$, which is also the symmetry group. A general construction shows that reducing $T^*\mathfrak{G}$ under the action of the group $\mathfrak{G}$ itself gives a coadjoint orbit in $\mathfrak{g}^*$; see Arnold (1989, appendix 5). Therefore, in the present case we expect $M_\mu = S^2$. The angular momentum of the free rigid body is a first integral; taking its projections on the orthogonal principal axes of inertia, fixed with the body, the Hamiltonian is $H = \sum_{k=1}^{3} G_k^2/2I_k$, where I_k are the principal momenta of inertia. Taking into account (2.3.12), the Euler equations result:

$$ \frac{d\vec{G}}{dt} = \{H, \vec{G}\} = \vec{\omega} \times \vec{G}, \text{ where } \omega_k = \frac{G_k}{I_k}. $$

The vector $\vec{G}$ moves with respect to the axes fixed with the body and spans a sphere S^2, which is just the reduced symplectic manifold.

EXAMPLE 2.25 The completely integrable systems of the next section are a particular case, with $\mathfrak{G} = \mathfrak{G}_\mu = \mathbb{R}^n$, while the moment map is given by the first integrals in involution and $\dim M_\mu = 0$.

2.3.8 Liouville Theorem

Given a Hamiltonian, finding explicit solutions of the related, usually nonlinear, equations is in general a hopeless task. The very few cases in which this is possible share the property that the problem can be reduced to the quadrature, i.e., to invert functions and to perform integrations: this is the case of *complete integrability*.

THEOREM 2.22 *(Liouville) A sufficient condition for the complete integrability of an n-dimensional Hamiltonian system is that*

   (i) *it admits n first integrals $\Phi_k(q, p) = \alpha_k$, $k = 1, \ldots, n$,*

   (ii) *that are in involution, i.e., $\{\Phi_h, \Phi_k\} = 0$, $\forall h, k$*

   (iii) *and independent, i.e., $\det\left(\frac{\partial \Phi_h}{\partial p_k}\right) \neq 0$.*

Proof. One basically seeks for a canonical transformation sending the first integrals into the new momenta, so that the transformed Hamiltonian depends only on the momenta and is trivially integrable. Clearly, the transformation does exist if and only if the first integrals are in involution, since this holds true for any n-tuple of canonical momenta. In order to find such a canonical transformation, notice that the n relations $\Phi_k(q, p) = \alpha_k$ can be inverted, thanks to (iii), giving $p_k = P_k(q, \alpha)$ from which

$$\Phi_k(q, P(q, \alpha)) \equiv \alpha_k \quad \Rightarrow \quad \frac{\partial \Phi_k}{\partial q^i} + \frac{\partial \Phi_k}{\partial p_j}\frac{\partial P_j}{\partial q^i} \equiv 0.$$

Replacing inside the Poisson brackets, a simple calculation gives

$$\{\Phi_h, \Phi_k\} = \frac{\partial \Phi_h}{\partial p_j}\frac{\partial \Phi_k}{\partial p_i}\left(\frac{\partial P_i}{\partial q^j} - \frac{\partial P_j}{\partial q^i}\right).$$

Involutivity condition (ii) and independence condition (iii) implies that

$$\frac{\partial P_i}{\partial q^j} - \frac{\partial P_j}{\partial q^i} = 0 \quad \Rightarrow \quad \exists\, W(q, \alpha) : P_k = \frac{\partial W}{\partial q^k}.$$

Therefore, $W(q, \alpha)$ is the generating function of the sought canonical transformation $p, q \to \alpha, \beta$ and can be found through n integrations. The transformed Hamiltonian K will depend on the variables α only, since

$$\frac{\partial K}{\partial \beta} = -\frac{d\alpha}{dt} = 0,$$

and the Hamilton equations are trivially integrable.   **QED**

The given proof is purely computational but can be rephrased in more geometric terms. The key idea is that of *Lagrangian submanifold*.

DEFINITION 2.23 *A* Lagrangian submanifold *of a 2n-dimensional symplectic manifold, with Ω the symplectic 2-form, is an n-dimensional submanifold Λ such that $\Omega|_\Lambda = 0$.*

With $\Omega|_\Lambda$ we mean the evaluation of Ω on all the pairs of tangent vectors to Λ or, in other words, the pull-back of Ω to Λ. Notice that n is the maximal dimension of a submanifold for which the defining property can hold.

Consider the cotangent bundle T^*Q (the generalization to a generic symplectic manifold is immediate owing to the Darboux theorem) and let $\pi : T^*Q \to Q$, $u \mapsto q$ be the canonical projection. Let Λ be a section of T^*Q; it is an n-dimensional submanifold of T^*Q with a one-to-one projection on the base manifold Q. But, by definition, a section of T^*Q is also a 1-form (i.e., a covector field) λ on Q, so that Λ and λ are two different symbols for the same object, according to whether we consider it as a submanifold of T^*Q or as a 1-form on Q. Moreover, λ may also be regarded as a map $Q \to \Lambda$.

We need the following basic property of the canonical 1-form Θ of T^*Q :

$$\lambda^*\Theta = \lambda, \quad \forall \lambda.$$

On the left-hand side, $\lambda : Q \to \Lambda \subset T^*Q$ is viewed as a map, while on the right-hand side λ is viewed as a 1-form on Q. In canonical coordinates, if $\lambda = \lambda_\mu(q)dq^\mu$ and $\Theta = p_\mu dq^\mu$, then $\lambda^*\Theta$ is obtained by simply replacing $p_\mu = \lambda_\mu(q)$ into Θ, which gives back λ, so $\lambda^*\Theta = \lambda$. To prove the property intrinsically, we recall (Example 2.6) that the canonical 1-form Θ is defined by

$$\langle \Theta, w \rangle|_u = \langle \lambda, \pi_*w \rangle|_q, \quad \Theta|_u \in T_u^*(T^*Q), \quad \forall w \in T_u(T^*Q).$$

Therefore, $\forall v \in T_qQ$, and taking into account that $\pi \circ \lambda = $ identity in Q,

$$\langle \lambda^*\Theta, v \rangle\big|_q = \langle \Theta, \lambda_*v \rangle|_u = \langle \lambda, \pi_*\lambda_*v \rangle|_q$$
$$= \langle \lambda, (\pi \circ \lambda)_*v \rangle|_q = \langle \lambda, v \rangle|_q,$$

holds, from which $\lambda^*\Theta = \lambda$.

We can now prove the following

THEOREM 2.24 *A section Λ of T^*Q is Lagrangian if and only if the corresponding 1-form λ is closed.*

Proof. From the above property, and bearing in mind that the pull-back and the exterior derivative commute, one finds

$$d\lambda = d(\lambda^*\Theta) = \lambda^*d\Theta = \lambda^*\Omega = \Omega|_\Lambda,$$

from which the theorem follows. $\hspace{4cm}$ QED

An immediate consequence is that Λ is Lagrangian if and only if there exists a scalar function $W : Q \to \mathbb{R}$, called the *generating function* of the

Lagrangian section Λ, such that, at least locally, $\lambda = dW$. But we may argue in another way: since $\Omega|_\Lambda = d\Theta|_\Lambda = 0$, thus, at least locally, $\Theta|_\Lambda = dW_\Lambda$, where $W_\Lambda : \Lambda \to \mathbb{R}$ is a scalar function defined on the Lagrangian submanifold Λ. The relation between the two functions is, up to an additive constant,

$$W = \lambda^* W_\Lambda = W_\Lambda \circ \lambda,$$

indeed

$$d(\lambda^* W_\Lambda) = \lambda^* dW_\Lambda = \lambda^* \Theta = \lambda = dW.$$

Therefore, W is the projection of W_Λ on the base Q: even if W_Λ is a well-defined function on Λ, W fails to be so on the critical points of the projection.

Let us consider a *Lagrangian foliation* of a symplectic manifold,[9] that is, by definition, a family λ_α of 1-forms continuously depending on n parameters $\alpha = \alpha_1, \ldots, \alpha_n.$. There is another way to describe a Lagrangian foliation. Consider n functions $\Phi_k : T^*Q \to \mathbb{R}$, $k = 1, \ldots, n$; if their differentials $d\Phi_k$ are independent, i.e., if the matrix $\frac{\partial(\Phi_1, \ldots, \Phi_n)}{\partial(q^1, \ldots, p_n)}$ has rank n (the condition (iii) of the Liouville theorem is a particular case), then the n equations

$$\Phi_1(q, p) = \alpha_1, \ldots, \Phi_n(q, p) = \alpha_n \tag{2.3.14}$$

describe an n-dimensional foliation Λ_α of T^*Q when varying the parameters α.

THEOREM 2.25 *The foliation Λ_α is Lagrangian if and only if the functions Φ_k are in* involution, *that is, $\{\Phi_h, \Phi_k\} = 0$, $h, k = 1, \ldots, n$.*

Proof. From definition (2.1.9) of the Poisson bracket, one immediately deduces

$$\{f, g\} = \Omega(df^\sharp, dg^\sharp).$$

If the two functions are in involution, then the vector $df^\sharp$ is tangent to the manifold $g = $ constant and $dg^\sharp$ to $f = $ constant, since

$$\{f, g\} = \mathcal{L}_{df^\sharp} g = -\mathcal{L}_{dg^\sharp} f = 0.$$

Therefore, the theorem follows from

$$\{\Phi_h, \Phi_k\} = \Omega(d\Phi_h^\sharp, d\Phi_k^\sharp)$$

and from the independence of $d\Phi_1^\sharp, \ldots, d\Phi_n^\sharp$, which span the tangent space to the level manifold Λ_α of (2.3.14).   **QED**

---

[9]A *foliation* of a manifold M is a decompositiion of M into disjoint connected immersed submanifolds of constant dimension less than that of M.

The relation between the functions Φ_k, the 1-forms λ_α, and the generating function $W(q, \alpha)$ should now be clear. If $\det\left(\frac{\partial \Phi_k}{\partial p_h}\right) \neq 0$, one can invert, at least locally, the n Equations (2.3.14) to find $p_k = \mathcal{P}_k(q, \alpha)$, with which one constructs the 1-forms $\lambda_\alpha = \mathcal{P}_k(q, \alpha)\, dq^k$. Varying the parameters α_k, these 1-forms describe a foliation Λ_α of sections of T^*Q that are Lagrangian, since by hypothesis the functions Φ_k are in involution. Thus λ_α is closed and a generating function $W_\alpha \equiv W(q, \alpha)$ exists such that $\lambda_\alpha = dW_\alpha$.

Let us look at some examples of integrable Hamiltonians, bearing in mind, however, that a general analytical method able to detect such systems is not known.

EXAMPLE 2.26 Every one-dimensional time-independent Hamiltonian is integrable. The solutions are given by the level lines of $H(q, p) = $ constant in the two-dimensional phase space.

EXAMPLE 2.27 The two systems of Examples 2.23 and 2.24 admit the angular momentum as first integral, but its three Cartesian components are not in involution; however, $\{G, G_3\} = 0$ holds, since the norm of a vector is invariant under the rotations generated by a component of the angular momentum. Thus G, G_3 and the Hamiltonian itself are three independent first integrals in involution. In Example 2.24, the dynamical solutions are given by the intersection lines between the reduced phase space S^2 and the ellipsoid $H = $ const.

EXAMPLE 2.28 The (Lagrange) spinning top is a heavy rigid body with rotational symmetry, suspended at a point of the symmetry axis different from the barycenter. The system is invariant for rotations about the vertical axis and about the axis of material symmetry; therefore, G_3^{space} and G_3^{body} are first integrals, which moreover are in involution: see Example 2.20.

EXAMPLE 2.29 The variable separation in the Hamilton–Jacobi equation is historically the richest source of integrable systems. A Hamiltonian is said to be *separable* if, when one attemps a solution of the type

$$S(q_1, \ldots, q_n, \alpha_1, \ldots, \alpha_n) = S_1(q_1, \alpha_1, \ldots, \alpha_n) + \cdots + S_n(q_n, \alpha_1, \ldots, \alpha_n),$$

the Hamilton–Jacobi equation splits into n ordinary differential equations of first order, thus integrable through quadrature: see Cordani (2003, Chapter VIII) for a summary and bibliography. In the Appendix to this chapter we briefly study the Euler problem of the two fixed centers. Another celebrated system is the Stark problem, a constant electric field acting on the hydrogen atom. The Stark problem can be seen as the limit of the Euler problem when one of the two fixed points goes to infinity.

2.3.9 Arnold Theorem

A natural question is: What is the topology of the hypersurfaces generated by the foliation Λ_α? At first glance one can say nothing about this topology, which depends on the analytical expression of the first integrals Φ. But here the involutivity condition, which in turn is a direct consequence of the canonical structure, plays a key role.

THEOREM 2.26 *(Arnold) Given a completely integrable n-dimensional Hamiltonian system, the compact and connected components of the level surfaces of the first integrals are diffeomorphic to an n-dimensional torus, i.e., the product of n circles. Moreover, there exist canonical coordinates called action-angle coordinates, such that the action variables parametrize the set of the tori whereas the angles parametrize the points on a torus. The Hamiltonian, expressed as a function of these coordinates, depends only on the actions, so that the dynamical evolution is a uniform rotation on an invariant torus.*

The key point in the proof consists of viewing the functions $\Phi_k(q, p)$ as Hamiltonians generating flows that, by the involutivity, respect the foliation and commute with one another. It is natural to think (though this is the central point of the proof) that the sole n-dimensional compact hypersurface carrying n independent and commuting flows is the product of n circles, i.e., the torus $\mathbb{T}^n$.

The Arnold theorem will be proved through a series of lemmas, following Arnold (1989) and Fasano & Marmi (1994). Let Λ_α be one of the n-dimensional differentiable compact manifolds of the foliation (2.3.14) of the $2n$-dimensional symplectic manifold M. On Λ_α act n commuting vector fields $X_{\Phi_k} = d\Phi_k^\sharp$, inducing n commuting flows $\phi_1^{t_1}, \ldots, \phi_n^{t_n}$. The commutativity of the vector fields X_{Φ_k} is with respect to the Lie brackets and is inherited from the Poisson involutivity of the first integrals. We can therefore define an action on Λ_α of the n-dimensional abelian group $\mathbb{R}^n$ by

$$\mathbb{R}^n \times \Lambda_\alpha \to \Lambda_\alpha : (t_1, \ldots, t_n, u) \mapsto \phi_1^{t_1} \circ \ldots \circ \phi_n^{t_n} u,$$

which will be simply denoted by $u \mapsto \phi^t u, \ t \in \mathbb{R}^n$.

LEMMA 2.27 *For any fixed $u_0 \in \Lambda_\alpha$ the map $\phi : \mathbb{R}^n \to \Lambda_\alpha$ given by $\phi^U u_0 = V$ is a local diffeomorphism; here U is a neighborhood of $0 \in \mathbb{R}^n$ and V a neighborhood of u_0.*

Proof. For the independence of the differentials $d\phi_k$ and the invertibility of $\sharp$, the n commuting vectors $X_{\Phi_k} = d\Phi_k^\sharp$ are independent. Integrating along their directions, it is possible to parametrize all points of V through an element of U (note that $\phi^0 u_0 = u_0$). Thus the transformation $t \mapsto u(t)$ is defined, and the rows of the Jacobian matrix of this transformation, evaluated at $t = 0$, are just the n vectors generating the flows. Therefore, the Jacobian matrix is nonsingular, and the transformation is invertible for U and V sufficiently small. **QED**

REMARK 2.30 The map $\phi : \mathbb{R}^n \to \Lambda_\alpha$ cannot be bijective since Λ_α is compact, contrary to $\mathbb{R}^n$.

LEMMA 2.28 *The action ϕ of $\mathbb{R}^n$ over Λ_α is transitive, that is*

$$\forall u_0, u \in \Lambda_\alpha, \ \exists t \in \mathbb{R}^n : \phi^t u_0 = u.$$

Proof. Connect u to u_0 with a curve on Λ_α. For the compactness of Λ_α the curve can be covered with a finite number of neighborhoods of the previous lemma, then define t as the sum of the various translations corresponding to the neighborhoods of the covering. **QED**

The preceding lemma says that Λ_α is a homogeneous manifold under the action of the Lie group $\mathbb{R}^n$. But, in general, if a Lie group $\mathfrak{G}$ acts transitively on a manifold M, the isotropy subgroups $\mathfrak{G}_u$, relative to any point $u \in M$, are all isomorphic (owing to the homogeneity of the $\mathfrak{G}$-action) and M can be identified with $\mathfrak{G}/\mathfrak{G}_u$. We are therefore led to study the isotropy subgroup, say Γ, of the $\mathbb{R}^n$-action on Λ_α, that is, the set of inverse images of $u \in \Lambda_\alpha$ under the map $\phi^t : \mathbb{R}^n \to \Lambda_\alpha$.

LEMMA 2.29 *The isotropy subgroup Γ of $\mathbb{R}^n$ is a discrete group.*

Proof. Since ϕ^t is a local diffeomorphism, the origin $0 \in \mathbb{R}^n$ is an isolated point of Γ, so that there is a neighborhood $U \subset \mathbb{R}^n$ of the origin such that $\Gamma \cap U = \{0\}$. This imposes that any other point of Γ is isolated. Indeed, let us suppose on the contrary that $t \neq 0$ is an accumulation point of Γ. Then

$$t + U = \{t + u : u \in U\}$$

is a neighborhood of t; therefore, an element $s \in (t + U) \cap \Gamma$, with $s \neq t$, should exist. But $s - t \neq 0$ and $s - t \in \Gamma \cap U$, since $s, t \in \Gamma$, and this is a contradiction. **QED**

All the discrete subgroups of $\mathbb{R}^n$ are generated by k linearly independent vectors $e_1, \ldots, e_k \in \mathbb{R}^n$, $0 \leq k \leq n$; that is, every subgroup Γ is the set of all their linear combinations with integer coefficients:

$$\Gamma : \{m_1 e_1 + \cdots + m_k e_k, \ (m_1, \ldots, m_k) \in \mathbb{Z}^k\}.$$

A basis $\{e_1, \ldots, e_k\}$ of Γ is not unique; indeed we have the following

LEMMA 2.30 *Let* **M** *be a unimodular matrix, i.e., a square matrix with integer entries and* $\det \mathbf{M} = \pm 1$. *Then any other set* $\{e'_1, \ldots, e'_k\} \subset \Gamma$ *of* k *independent vectors is a basis of* Γ *if and only if there exists a* $k \times k$ *unimodular matrix* (M_{ij}) *such that*

$$e'_i = \sum_{j=1}^{k} M_{ij} e_j, \quad i = 1, \ldots, k. \tag{2.3.15}$$

Proof. Let us suppose that the condition (2.3.15) is satisfied. Then the set $\{e'_i\}$ generates a subgroup $\Gamma' \subseteq \Gamma$, since its vectors are independent and generated with integer coefficients by a basis of Γ. Inverting (2.3.15) and noticing that the inverse of a unimodular matrix is still a unimodular matrix, we can also infer, with the same argument, that $\Gamma \subseteq \Gamma'$. Therefore $\Gamma' = \Gamma$. Let us now suppose that $\{e'_1, \ldots, e'_k\}$ is a basis of Γ. Then there exists an invertible matrix **M** transforming the basis $\{e_1, \ldots, e_k\}$ to the basis $\{e'_1, \ldots, e'_k\}$. Since $e'_j \in \Gamma$, the matrix **M** must have integer entries. On the other hand, since $\{e'_1, \ldots, e'_k\}$ is a basis of Γ, the inverse matrix also must have integer entries. That implies that both **M** and $\mathbf{M}^{-1}$ have an integer determinant, and by $\det \mathbf{M} \cdot \det \mathbf{M}^{-1} = 1$ we get $\det \mathbf{M} = \pm 1$. **QED**

REMARK 2.31 If $\det \mathbf{M} = -1$, one can recover the case $\det \mathbf{M} = 1$ by simply exchanging two basis vectors. Hereafter, we will always take $\mathbf{M} \in \mathrm{SL}(n, \mathbb{Z})$.

We notice that k independent vectors of Γ do not form in general a basis, since they do not generate the whole Γ but only a subgroup. In order to find such a basis, let us associate the k-dimensional parallelogram $D(e_1, \ldots, e_k)$, defined as

$$D(e_1, \ldots, e_k) = \{\mu_1 e_1 + \cdots + \mu_k e_k, \ 0 \le \mu_j < 1, \ j = 1, \ldots, k\},$$

to a set $\{e_1, \ldots, e_k\} \subset \Gamma$ of independent vectors. We claim that this set is a basis of Γ if and only if

$$D(e_1, \ldots, e_k) \cap \Gamma = \{0\}. \tag{2.3.16}$$

Indeed, if the condition is not satisfied, the vectors belonging to $D \cap \Gamma$ cannot be generated (with integer coefficients) by the set, which therefore is not a basis of Γ. Vice versa, if the condition is satisfied, all the elements of Γ are generated, since any element of Γ is external to D and by assumption the set $\{e_1, \ldots, e_k\}$ is a basis of the vector space $\mathbb{R}^k$ containing Γ. Giving to $\mathbb{R}^k$ a Euclidean metric, we can speak of the area (or volume) of D. If the condition (2.3.16) is not satisfied, one can substitute an element of $\{e_1, \ldots, e_k\}$ with an element belonging to D, in such a way that the new set is still a basis of $\mathbb{R}^k$ and, moreover, the area of the new parallelogram is strictly smaller. With a finite number of steps we can construct a set satisfying the condition (2.3.16), which therefore forms a basis of Γ.

From the previous lemmas we deduce that $\Lambda_\alpha \simeq \mathbb{R}^n/\Gamma \simeq \mathbb{R}^{n-k} \times \mathbb{T}^k$. If Λ_α is compact, $k = n$ and we reach the basic conclusion that Λ_α is diffeomorphic to the n-dimensional torus $\mathbb{T}^n$. Moreover, we can now make a semiglobal statement about the topology of the symplectic manifold M.

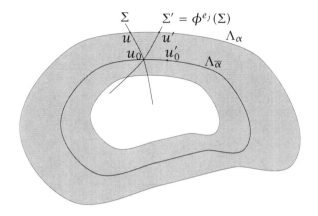

Figure 2.3: Construction of the coordinates α, ψ. The shaded area is the neighborhood $V \subset M$ of $\Lambda_{\overline{\alpha}}$.

LEMMA 2.31 *There exists an open neighborhood $V \subset M$ of any Λ_α that is diffeomorphic to the product of an open neighborhood $U \subset \mathbb{R}^n$ and a torus, i.e., $V \simeq U \times \mathbb{T}^n$, and can be covered with coordinates $\alpha_1, \ldots, \alpha_n$ on U and angular coordinates $\psi^1, \ldots, \psi^n$ on $\mathbb{T}^n$.*

Proof. See Figure 2.3. For a fixed $\overline{\alpha}$ take any point $u_0 \in \Lambda_{\overline{\alpha}}$ and construct a differentiable manifold Σ transversal to $\Lambda_{\overline{\alpha}}$ in u_0: by definition this implies that the tangent space to Σ is complementary in $T_{u_0}M$ to that of $\Lambda_{\overline{\alpha}}$, thus $\dim \Sigma = n$. Consider the basis vectors e_j of the isotropy subgroup of u_0, i.e., $\phi^{e_j} u_0 = u_0$, $j = 1, \ldots, n$. We can write, for any $u_0' \in \Lambda_{\overline{\alpha}}$, that $u_0' = \phi^t u_0$, $t \in \mathbb{R}^n$ and give to this point angular coordinates ψ, by putting $t = \sum_{i=1}^n \frac{\psi^i}{2\pi} e_i$. Let the surface Σ evolve under the action of ϕ^{e_j}. Another surface $\Sigma' = \phi^{e_j}(\Sigma)$ is obtained, different in general from Σ but surely containing u_0. If U is sufficiently small, Σ' is transversal to Λ_α, $\forall \alpha \subset U$; therefore, for every $u' \in \Sigma'$, uniquely determined by a value α, there exists a differentiable function $\tau_j(\alpha)$ such that $\tau_j(\overline{\alpha}) = 0$ and, if $u \in \Sigma \cap \Lambda_\alpha$, such that $\phi^{\tau_j(\alpha)} u' = u$. Hence, $\phi^{e_j + \tau_j(\alpha)} u = u$. For every $\alpha = (\alpha_1, \ldots, \alpha_n)$ in U, the periods $e_j(\alpha) = e_j + \tau_j(\alpha)$ that determine the isotropy subgroup of u on any Λ_α, and allow us to give angular coordinates ψ to any point of Λ_α (as above to u_0'), can thus be defined. **QED**

The previous lemma is global with regard to the parametrization of points on a torus, but only local with regard to the parametrization of the tori themselves. The lemma says that M is a fiber bundle whose base is covered with a family of open sets like U and the fiber is an n-dimensional torus. In this sense we say that it is a semiglobal statement, since nothing is said about the global topology of the bundle, which in general may be nontrivial.

The coordinates $\alpha_1, \ldots, \alpha_n, \psi^1, \ldots, \psi^n$ are *not* in general canonical coordinates. However, it is possible to construct a set of canonical coordinates as follows. Let γ_i be a cycle on the torus, parametrized by the angle φ^i. In order to impose that $I_j, \varphi^k, \ j, k = 1, \ldots, n$ are canonical coordinates, we require, as in (2.3.5), that the two 1-forms $p_k dq^k$ and $I_k d\varphi^k$ differ for an exact 1-form, whose integral along a cycle is therefore zero. Thus

$$\oint_{\gamma_i} I_k d\varphi^k = \oint_{\gamma_i} p_k dq^k.$$

On the left-hand side $d\varphi^k = 0$, $\forall k \neq i$, and I_i is constant along the cycle since, by definition, it is just the Hamiltonian generating γ_i. Therefore,

$$I_i = \frac{1}{2\pi} \oint_{\gamma_i} \mathcal{P}_k(q, \alpha) dq^k. \qquad (2.3.17)$$

The action variables are therefore functions of the first integrals Φ's only. For any foliation these functions are invertible, so that the action variables are a system of n independent first integrals in involution.

We remark that this definition does not depend, to some extent, on the particular cycle γ. If we take, indeed, any other cycle $\gamma' = \gamma + \partial\sigma$, $\sigma \subset \Lambda_\alpha$, homologous to γ, the value of the integral does not change, owing to the Stokes theorem and to the fact that $\Omega|_{\Lambda_\alpha} = d\Theta|_{\Lambda_\alpha} = 0$:

$$\oint_{\gamma'} \Theta = \oint_{\gamma} \Theta + \oint_{\partial\sigma} \Theta = \oint_{\gamma} \Theta + \oint_{\sigma} d\Theta = \oint_{\gamma} \Theta.$$

This definition, however, must be taken with some care when (in practice: always) the first integrals are generated by separation of variables, so that the cycle γ_k is assumed to be the closed curve described by $\Phi_k(q_k, p_k) = $ constant on the phase plane $q_k p_k$. Varying the value of the first integrals, γ_k also varies: the subtle point is that, coming back to the starting value of the first integrals, the cycle γ_k does *not* in general agree with the initial one. Rather, if the torus bundle is not trivial,

$$\gamma_k \mapsto \gamma'_k = \sum_h M_{kh}\gamma_h, \quad \text{Identity} \neq (M_{kh}) \in \text{SL}(n, \mathbb{Z}).$$

This implies that the definition (2.3.17) is only local; i.e., it is meaningful as long as the value of the first integrals varies in a simply connected open set of the base space. Globally, the action variables can be in general multivalued functions of the first integrals: in this case, the system is said to have a nontrivial *monodromy*. See Cushman & Sadovskií (2000) or Cordani (2003, pages 312 and 417–422) for a concrete example.

The canonical transformation $(p, q) \mapsto (I, \varphi)$ transforms a completely integrable Hamiltonian $H(p, q)$ into a new Hamiltonian $K(I)$, which depends only on the actions. The Hamilton equations are

$$\dot{I}_k = -\frac{\partial K(I)}{\partial \varphi^k} = 0, \quad \dot{\varphi}^k = \frac{\partial K(I)}{\partial I^k};$$

i.e., the actions are first integrals (as already known) and the angles evolve linearly with the time.

The action-angle variables are not defined univocally. New coordinates *on the same torus* can be generated by the following canonical transformations.

(i) Translations of the action variables:

$$I'_k = I_k + c_k, \quad \varphi'^h = \varphi^h, \quad (c_1, \ldots, c_n) \in \mathbb{R}^n;$$

the actions appear indeed in the formulae only through their gradients.

(ii) Translations of the origin of the angles by quantities depending on the torus:

$$I'_k = I_k, \quad \varphi'^h = \varphi^h + \frac{\partial f(I)}{\partial I_h},$$

where f is an arbitrary scalar function; this reflects the arbitrariness in the choice of the point u_0 and of the surface Σ, see figure 2.3.

(iii) A linear transformation by a unimodular matrix:

$$\varphi' = \mathbf{M}^{-1}\varphi, \quad I' = \mathbf{M}^t I,$$

which reflects the arbitrariness in the choice of the basis of the isotropy subgroup Γ: see Lemma 2.30.

2.3.10 Action-Angle Variables: Examples

Let us study some examples of action-angle variables that will be relevant in the sequel.

EXAMPLE 2.32 Consider the 1-dimensional harmonic oscillator, with Hamiltonian (in rescaled form)

$$H = \frac{1}{2}\omega_0(p^2 + q^2).$$

The action is given by dividing by 2π the area enclosed by the circles $H =$ const., thus with radius $\sqrt{p^2 + q^2}$:

$$I = \frac{1}{2\pi} \cdot \pi(p^2 + q^2), \quad \varphi = -\arctan\frac{p}{q}, \quad H = \omega_0 I.$$

Notice that $\{I, \varphi\} = 1$.

The extreme simplicity of this example is deceptive. Finding explicitly the action-angle variables is equivalent to solving the equations of motion, and in general requires the evaluation of nontrivial integrals, even involving nonelementary transcendental functions, as the following examples show.

EXAMPLE 2.33 The vibrational motion of diatomic molecules can be described by the Morse oscillator

$$H = \frac{1}{2}p^2 + \frac{1}{2}(e^{-q} - 1)^2.$$

The potential has a minimum for $q = 0$ and tends to infinity for $q \to -\infty$ and to $1/2$ for $q \to +\infty$. To restrict to non equilibrium bounded orbits, $0 < H < 1/2$ must hold. To calculate the action, one must evaluate the integral

$$I(H) = \frac{1}{\pi} \int_{q_1}^{q_2} \sqrt{2H - (e^{-q} - 1)^2} dq, \quad q_{1,2} = -\log(1 \pm \sqrt{2H}).$$

The final result is

$$I = 1 - \sqrt{1 - 2H},$$
$$\cos\varphi = \frac{1 - e^q(1 - 2H)}{\sqrt{2H}}.$$

EXAMPLE 2.34 The pendulum is described by the Hamiltonian

$$H = \frac{1}{2}p^2 - \omega_0^2 \cos q,$$

where ω_0 is the frequency for the linearized oscillation about the stable equilibrium point. Define the elliptic integral of first and second kind

$$F(\varphi, k) = \int_0^\varphi \frac{d\varphi'}{\sqrt{1 - k^2 \sin^2\varphi'}}, \quad E(\varphi, k) = \int_0^\varphi \sqrt{1 - k^2 \sin^2\varphi'}\, d\varphi'$$

respectively; define also

$$k = \sqrt{\frac{1}{2}\left(1 + \frac{H}{\omega_0^2}\right)}, \quad \sin\eta = \frac{1}{k}\sin\frac{q}{2}.$$

Clearly, for $k < 1$ the motion is libratory and for $k > 1$ it is oscillatory. Then one finds

$$I = \frac{8\omega_0}{\pi} \begin{cases} E(\pi/2, k) - (1 - k^2)F(\pi/2, k) & k < 1 \\ kE(\pi/2, 1/k) & k > 1 \end{cases}$$

$$\varphi = \frac{\pi}{2} \begin{cases} F(\eta, k)/F(\pi/2, k) & k < 1 \\ 2F(q/2, 1/k)/F(\pi/2, 1/k) & k > 1 \end{cases}.$$

The following last example is a nontrivial generalization of the harmonic oscillator. It regards a Hamiltonian quadratic in $2n$ canonical coordinates, which gives rise to *linear* equations of motion.

Before we consider the example, we recall that trying a solution of the type $\underline{x}(t) = \underline{c}\, e^{\lambda t}$ for the linear equation $\underline{\dot{x}} = \mathbf{A}\underline{x}$, one finds that λ and $\underline{c}$ must be an eigenvalue and an n-dimensional eigenvector, respectively, of the $n \times n$ matrix $\mathbf{A}$. If, as we always suppose, the eigenvalues are distinct, hence the eigenvectors independent, the general integral is

$$\begin{pmatrix} x_1(t) \\ \vdots \\ x_n(t) \end{pmatrix} = \begin{pmatrix} T_{11} & \cdots & T_{1n} \\ \cdots & \cdots & \cdots \\ T_{n1} & \cdots & T_{nn} \end{pmatrix} \begin{pmatrix} e^{\lambda_1 t} \\ \vdots \\ e^{\lambda_n t} \end{pmatrix},$$

where the columns of the $n \times n$ invertible matrix $\mathbf{T}$ are the eigenvectors, each one multiplied by an arbitrary integration constant. Define the diagonal matrix $\Lambda = \text{diag}(\lambda_1, \ldots, \lambda_n)$; thus $\mathbf{A}\mathbf{T} = \mathbf{T}\Lambda$ and the matrix $\mathbf{T}$ diagonalizes $\mathbf{A}$. From $\underline{\dot{x}} = \mathbf{T}\Lambda\mathbf{T}^{-1}\underline{x}$, it follows that $\underline{x}' \stackrel{\text{def}}{=} \mathbf{T}^{-1}\underline{x}$ satisfies the equation $\underline{\dot{x}}' = \Lambda\underline{x}'$. The origin $\underline{x} = 0$ is an equilibrium position that is (not asymptotically) stable if the eigenvalues are imaginary numbers: in such a case, the equilibrium point is called *elliptic*.

EXAMPLE 2.35 Let $\underline{x}^t = (q_1, p_1, \ldots, q_n, p_n)$ be the transpose of a $2n$ dimensional column vector $\underline{x}$ and $\mathbf{H}$ a $2n \times 2n$ symmetric matrix. Define the Hamiltonian $H = \frac{1}{2}\underline{x}^t\mathbf{H}\underline{x}$. Let[10]

$$\Omega = \text{diag}\left[\begin{pmatrix} 0 & 1 \\ -1 & 0 \end{pmatrix}, \ldots, \begin{pmatrix} 0 & 1 \\ -1 & 0 \end{pmatrix} \right]$$

be the inverse of the $2n \times 2n$ canonical matrix, then $\mathbf{A} = \Omega\mathbf{H}$.

PROPOSITION 2.32 *If λ is an eigenvalue of $\Omega\mathbf{H}$, then also $-\lambda, \lambda^*, -\lambda^*$ are eigenvalues and requiring the stability of the equilibrium point $\underline{x} = 0$ forces the λ's to be imaginary numbers.*

---

[10]In order to simplify some formulae in the subsequent proof, we assume an order of the canonical coordinates such that the canonical matrix is slightly different from the usual.

Proof. The complex conjugate of an eigenvalue is itself an eigenvalue for the reality of $\Omega\,H$. Moreover,

$$(\Omega\,H - \lambda 1_{2n})^t = (-H\,\Omega - \lambda 1_{2n}) = \Omega\Omega(H\,\Omega + \lambda 1_{2n}) = \Omega(\Omega\,H + \lambda 1_{2n})\Omega,$$

from which

$$\det(\Omega\,H - \lambda 1_{2n}) = \det(\Omega\,H - \lambda 1_{2n})^t = \det(\Omega\,H + \lambda 1_{2n}),$$

which proves that the opposite of an eigenvalue is again an eigenvalue. In order to have stability, no eigenvalue must have positive real part, and in the present case this forces the λ's to be imaginary numbers.　　　**QED**

We can now prove the following theorem.

THEOREM 2.33 *If the origin is an elliptic point of the quadratic Hamiltonian $H = \frac{1}{2}\underline{x}^t\,H\underline{x}$, then there exists a linear invertible transformation S*

$$\underline{x} = S\underline{X}, \quad \underline{X}^t = (Q_1, P_1, \ldots, Q_n, P_n)$$

which is real, canonical and reduces the Hamiltonian to the sum of n harmonic oscillators with positive or negative frequency.

Proof. After the transformation the Hamilton equations becomes

$$\dot{\underline{X}} = \Omega'\,H'\underline{X}, \quad \Omega' = S^{-1}\Omega(S^{-1})^t, \quad H' = S^t\,HS.$$

If S is symplectic, i.e., by definition $S^t\Omega S = \Omega$, also S^{-1} and S^t are symplectic and $\Omega' = \Omega$. In order to construct such a matrix S, solve the eigenvalue problem

$$\Omega\,H\underline{w} = \lambda\underline{w}, \quad \underline{w} \in \mathbb{C}^{2n}, \ \lambda = i\omega \text{ with } \omega \in \mathbb{R}, \tag{2.3.18}$$

and form the $2n \times 2n$ matrix $T = (\underline{w}_1, \underline{w}_1^*, \underline{w}_2, \underline{w}_2^*, \ldots, \underline{w}_n, \underline{w}_n^*)$ which implies

$$T^{-1}\Omega\,HT = \mathrm{diag}(i\omega_1, -i\omega_1, i\omega_2, -i\omega_2, \ldots, i\omega_n, -i\omega_n), \ \omega_k > 0. \tag{2.3.19}$$

Notice that also TD, where D is an arbitrary diagonal matrix, satisfies the same relation (2.3.19) since diagonal matrices commute; this agrees with the fact that every eigenvector is defined up to an arbitrary multiplicative constant. In order to pass to real form, put

$$\underline{w}_k \overset{\mathrm{def}}{=} \underline{u}_k + i\underline{v}_k, \quad \underline{u}_k, \underline{v}_k \in \mathbb{R}^{2n},$$

and define the real matrix $R = (\underline{u}_1, \underline{v}_1, \underline{u}_2, \underline{v}_2, \ldots, \underline{u}_n, \underline{v}_n)$, which satisfies

$$R^{-1}\Omega\,HR = \mathrm{diag}\left[\begin{pmatrix} 0 & \omega_1 \\ -\omega_1 & 0 \end{pmatrix}, \ldots, \begin{pmatrix} 0 & \omega_n \\ -\omega_n & 0 \end{pmatrix}\right] \tag{2.3.20}$$

as a consequence of (2.3.18).

Let us find how **R** differs from being symplectic. To this end we calculate the generic entry of the matrix $\mathbf{T}^t \mathbf{\Omega} \mathbf{T}$, i.e., the skew product $\underline{w}^t \mathbf{\Omega} \underline{w}'$ between two generic eigenvectors. From the relations

$$\underline{w}^t \mathbf{H} \underline{w}' = -\underline{w}^t \mathbf{\Omega}\mathbf{\Omega} \mathbf{H} \underline{w}' = -\lambda' \underline{w}^t \mathbf{\Omega} \underline{w}',$$

$$\underline{w}^t \mathbf{H} \underline{w}' = \underline{w}^t \mathbf{H} \mathbf{\Omega}^t \mathbf{\Omega} \underline{w}' = (\mathbf{\Omega} \mathbf{H} \underline{w})^t \mathbf{\Omega} \underline{w}' = \lambda \underline{w}^t \mathbf{\Omega} \underline{w}',$$

we get $(\lambda + \lambda') \underline{w}^t \mathbf{\Omega} \underline{w}' = 0$. Therefore, $\underline{w}^t \mathbf{\Omega} \underline{w}' = 0$ for $\lambda + \lambda' \neq 0$, from which $\underline{w}^t \mathbf{\Omega} \underline{w}' \neq 0$ for $\lambda + \lambda' = 0$ must hold for the independence of the eigenvectors. Taking into account that the eigenvalues are distinct imaginary numbers, we get that the sole non vanishing case is

$$\underline{w}_k^t \mathbf{\Omega} \underline{w}_k^* = (\underline{u}_k^t + i\underline{v}_k^t)\mathbf{\Omega}(\underline{u}_k - i\underline{v}_k) = -2i\underline{u}_k^t \mathbf{\Omega} \underline{v}_k \neq 0, \quad \forall k.$$

Then, in order to get a symplectic matrix, all we have to do is normalize the eigenvectors. Defining

$$d_k = \frac{1}{\sqrt{\underline{u}_k^t \mathbf{\Omega} \underline{v}_k}}, \quad \mathbf{D} = \text{diag}\left[\begin{pmatrix} d_1 & 0 \\ 0 & d_1 \end{pmatrix}, \ldots, \begin{pmatrix} d_n & 0 \\ 0 & d_n \end{pmatrix}\right],$$

the matrix **RD** results symplectic,

$$\mathbf{D}^t \mathbf{R}^t \mathbf{\Omega} \mathbf{R} \mathbf{D} = \mathbf{\Omega},$$

but in general not real. To make it real, preserving at the same time the symplectic property, define the matrix

$$\mathbf{I} = \text{diag}(\mathbf{a}_1, \mathbf{a}_2, \ldots, \mathbf{a}_n) \quad \text{with}$$

$$\mathbf{a}_k = \begin{pmatrix} 1 & 0 \\ 0 & 1 \end{pmatrix} \quad \text{if } \underline{u}_k^t \mathbf{\Omega} \underline{v}_k > 0,$$

$$\mathbf{a}_k = \begin{pmatrix} i & 0 \\ 0 & -i \end{pmatrix} \quad \text{if } \underline{u}_k^t \mathbf{\Omega} \underline{v}_k < 0,$$

so that the matrix $\mathbf{S} = \mathbf{RDI}$ results real and symplectic. With regard to the equations of motion and recalling (2.3.20), we see that

$$\underline{\dot{X}} = \mathbf{S}^{-1} \mathbf{\Omega} \mathbf{H} \mathbf{S} \underline{X} \quad \text{with}$$

$$\mathbf{S}^{-1} \mathbf{\Omega} \mathbf{H} \mathbf{S} = \text{diag}\left[\begin{pmatrix} 0 & \pm\omega_1 \\ \mp\omega_1 & 0 \end{pmatrix}, \ldots, \begin{pmatrix} 0 & \pm\omega_n \\ \mp\omega_n & 0 \end{pmatrix}\right]$$

holds, where one must take the upper or lower sign when $\underline{u}_k^t \mathbf{\Omega} \underline{v}_k$ is positive or negative, respectively. Therefore, the new form of the Hamiltonian is

$$H = \frac{1}{2} \sum_{k=1}^n \pm \omega_k (P_k^2 + Q_k^2) = \sum_{k=1}^n \pm \omega_k I_k,$$

showing that the Hamiltonian is the sum of n harmonic oscillators with positive or negative frequency.   **QED**

It is important to stress that the signs in the Hamiltonian expression are far from being innocuous when the problem is perturbed. If the Hamiltonian is a positive or negative definite function, a small perturbation does not destroy the stability of the equilibrium; however, if the signs are different, the hypersurface $H = $ const is not compact, and a small perturbation can push the point to infinity.

2.A   Appendix: The Problem of two Fixed Centers

An interesting example of an integrable system is the Euler problem, regarding the motion of a point under the Newtonian action of two fixed masses. The relative Hamiltonian result separable using the elliptic coordinates in the plane:

$$x = \sqrt{(\xi^2 - 1)(1 - \eta^2)}, \quad y = \xi\eta + 1,$$
$$\xi \geq 1, \quad -1 \leq \eta \leq 1.$$

The lines $\xi = $ constant are ellipses with foci at $F_1 \equiv (0,0)$ and $F_2 \equiv (0,2)$, and the lines $\eta = $ constant are confocal hyperbolas.

Taking a unitary mass in F_1 and a mass $m < 1$ in F_2, the Hamiltonian of the Euler problem is

$$H = \frac{1}{2}\left(\frac{\xi^2 - 1}{\xi^2 - \eta^2}p_\xi^2 + \frac{1 - \eta^2}{\xi^2 - \eta^2}p_\eta^2\right) - \frac{1}{\xi + \eta} - \frac{m}{\xi - \eta}.$$

To separate the variables, let us multiply by $\xi^2 - \eta^2$ and, after a reordering of the terms, we find the second first integral y besides the total energy

$$H\xi^2 - \frac{1}{2}(\xi^2 - 1)p_\xi^2 + (1 + m)\xi = -y,$$
$$H\eta^2 + \frac{1}{2}(1 - \eta^2)p_\eta^2 + (1 - m)\eta = -y.$$

A direct calculation shows that

$$y = E_y - \frac{1}{2}G^2 + m\frac{y - 2}{r_2} - H.$$

We seek the complete integral of the Hamilton-Jacobi Equation in the form

$$S(\xi, \eta, H, y, t) = S_\xi(\xi, H, y) + S_\eta(\eta, H, y) - Ht,$$

from which

$$\left(\frac{\partial S_\xi}{\partial \xi}\right)^2 = 2\frac{H\xi^2 + (1+m)\xi + y}{\xi^2 - 1},$$

$$\left(\frac{\partial S_\eta}{\partial \eta}\right)^2 = 2\frac{H\eta^2 + (1-m)\eta + y}{\eta^2 - 1},$$

and the problem is reduced to the quadratures.

The function S generates the canonical transformation

$$\left(\begin{array}{cc} p_\xi & p_\eta \\ \xi & \eta \end{array}\right) \mapsto \left(\begin{array}{cc} H & y \\ \beta_1 & \beta_2 \end{array}\right).$$

Since the transformed Hamiltonian vanishes identically, the new coordinates β_1, β_2 are also first integrals. In particular, from $\frac{\partial S}{\partial y} = $ constant we find

$$\int \frac{d\xi}{\sqrt{(\xi^2 - 1)[H\xi^2 + (1+m)\xi + y]}}$$

$$+ \int \frac{d\eta}{\sqrt{(\eta^2 - 1)[H\eta^2 + (1-m)\eta + y]}} = \text{constant},$$

from which

$$d\tau = \frac{d\xi}{\sqrt{(\xi^2 - 1)[H\xi^2 + (1+m)\xi + y]}},$$

$$d\tau = -\frac{d\eta}{\sqrt{(\eta^2 - 1)[H\eta^2 + (1-m)\eta + y]}},$$

(2.A.1)

τ being a parameter. If we succeed in integrating the two right-hand sides, we find, after inversion, the parametric expressions $\xi(\tau)$ and $\eta(\tau)$ of the orbit. The properties of the orbits can be studied with an accurate qualitative analysis of the position of the roots ξ_1, ξ_2 and η_1, η_2 of the polynomials $H\xi^2 + (1+m) + y$ and $H\eta^2 + (1-m) + y$ with respect to the other two roots -1 and $+1$ of the two denominators: see Cordani (2003) for details. See also Mathúna (2008) for a complete treatment with explicit analytic solutions.

In the generic case of distinct roots, the program EULER, written in the MAPLE language, performs the two integrations analytically, implementing the general method described in Whittaker & Watson (1952, Section 22.7 of Chapter XXII) and Bowman (1961, Chapter IX). Given the initial position and velocity, the program displays the resulting orbit.

It turns out that for negative energy there exist four types of bound orbits: see Figure 2.4.

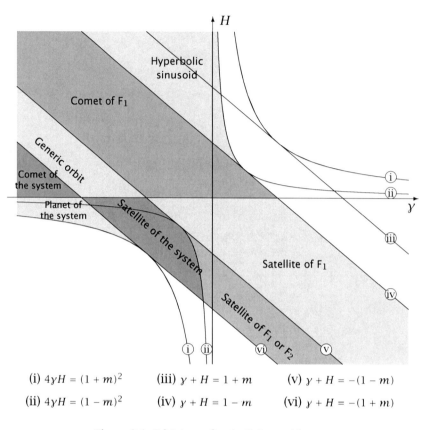

$$(i)\ 4yH = (1+m)^2 \qquad (iii)\ y+H = 1+m \qquad (v)\ y+H = -(1-m)$$

$$(ii)\ 4yH = (1-m)^2 \qquad (iv)\ y+H = 1-m \qquad (vi)\ y+H = -(1+m)$$

Figure 2.4: Orbit types for the Euler problem.

(i) When $\xi_1 \le \xi \le \xi_2$, with $\xi_1, \xi_2 > 1$, while $-1 \le \eta \le 1$, the point P turns around F_1 and F_2, touching alternatively the two ellipses $\xi = \xi_1$ and $\xi = \xi_2$ and filling an "elliptic channel." We say that P is a planet of the system.

(ii) When $1 \le \xi \le \xi_2$, where ξ_2 is the only root > 1, while $-1 \le \eta \le 1$, the point P moves inside the ellipse $\xi = \xi_2$, crossing the y-axis between F_1 and F_2. We say that P is a satellite of the system.

(iii) When ξ_2 is the only root > 1, while $-1 \le \eta \le \eta_1$, where η_1 is the only root strictly contained in $(-1, 1)$, the point P turns around F_1 filling the area delimited by the ellipse $\xi = \xi_2$ and the hyperbola $\eta = \eta_1$.We say that P is a satellite of F_1.

(iv) When $1 \le \xi \le \xi_2$, where ξ_2 is the only root > 1, while $-1 \le \eta \le \eta_1$ or $\eta_2 \le \eta \le 1$, where $\eta_1 < \eta_2$ are the two roots strictly contained in

$(-1, 1)$, the point P moves inside the area, delimited by the ellipse $\xi = \xi_2$ and the hyperbola $\eta = \eta_1$, which contains F_1. Alternatively, it moves inside the area, delimited by the ellipse $\xi = \xi_2$ and the hyperbola $\eta = \eta_2$, which contains F_2. The two motions are selected by the initial position. We say that P is a satellite of F_1 or F_2.

The reader can usefully compare the *exact* motion given by EULER with the numerical output of KEPLER.

CHAPTER 3

Perturbation Theory

A torus is a large convex moulding,
usually at the base of a column.

Given the Hamiltonian H_0 of a completely integrable system, the *perturbed problem* is described by the Hamiltonian $H = H_0 + \varepsilon H_p$, where H_p is a function whose numerical value is of the same order of H_0, and $\varepsilon << 1$. The perturbed problem thus differs slightly from the unperturbed one, but unfortunately the same is not true for the solution: a small perturbation can give rise to *secular* effects, i.e., to a slow but progressive wandering from the unperturbed, and known for infinite time, solution.

In this chapter we will describe the methods of perturbation theory. In the first section of the chapter only *formal* series expansion are considered, while in the second section the subtle problem of their convergence is considered, and the key points of the proof of the KAM theorem on perpetual stability are given. In the third section Nekhoroshev theorem on stability for an exponentially long time is stated, along with a sketch of the proof. In the fourth section we sum up the geography of the phase space. In the fifth section the results of the two theorems are extended to the case of elliptic equilibrium points. In the two final appendices the Diophantine inequality and the route to chaos through the homoclinic tangle mechanism are briefly studied.

3.1   Formal Expansions

The basic idea of perturbation theory, i.e., considering a perturbation in some sense "small" and thus proceeding to series expansions, goes back to the early ages of classical mechanics. Its actual form, however, is due to Poincaré (1892–1893–1899), and may be summarized in this way: find a canonical transformation, which differs from the identity in a quantity of order ε, such that the transformed Hamiltonian is integrable up to the second order terms. One may obviously iterate the procedure, pushing the perturbation to third order, and so on. If the process converges (but this is the key point), to higher orders one obtains better and better approximations.

In the sequel we adopt this point of view, but with a technical improvement that simplifies some calculations: as suggested in Hori (1966) and Deprit (1969), use of the Lie series is made, avoiding those inversion problems which occur typically for methods involving the generating function of a canonical transformation.

3.1.1   Lie Series and Formal Canonical Transformations

Let us consider a one-parameter family of diffeomorphisms Φ_τ, $\tau \in \mathbb{R}$, defined in some analytic manifold N. Such a family is said to be a *flow* when

(i) for $\tau = 0$ one has $\Phi_0 = $ identity,

(ii) $\Phi_{\tau'} \circ \Phi_{\tau''} = \Phi_{\tau'+\tau''}$.

If the flow is defined for every τ, it is a one-parameter group of diffeomorphisms. Otherwise, Φ_τ represents a *local* flow. Let $x = x_1, \ldots, x_n$ be local coordinates on N, and $x_\tau = \Phi_\tau x$ the coordinates of the transformed point. To every flow Φ_τ corresponds the vector field (a sort of velocity field)

$$X(x) = \frac{d}{d\tau}(\Phi_\tau x)\Big|_{\tau=0},$$

called the *generator* of the flow.

One may reverse the argument, starting from a vector field X and defining the corresponding flow Φ_τ as that which satisfies

$$\frac{d}{d\tau}(\Phi_\tau x) = X(\Phi_\tau x),$$

i.e., as the solution of the dynamical system

$$\frac{dx}{d\tau} = X(x). \tag{3.1.1}$$

Let $f : N \to \mathbb{R}$ be any function of the state of the system, which we shall also call a *dynamical variable*. Then the evolution of f under the flow generated by the vector field X is naturally defined as the function $f(\Phi_\tau x)$. If f is differentiable, then its derivative with respect to τ is

$$\frac{df}{d\tau} = \sum_i X^i \frac{\partial f}{\partial x^i}.$$

This suggests defining the operator

$$\mathcal{L}_X f = \sum_i X^i \frac{\partial f}{\partial x^i},$$

which maps the space of the analytic functions on N into itself. $\mathcal{L}_X f$ is called the *Lie derivative* of f along the flow generated by X. It is plainly a linear operator, which satisfies the Leibniz rule

$$\mathcal{L}_X(fg) = f\mathcal{L}_X g + g\mathcal{L}_X f. \tag{3.1.2}$$

Given two vector fields X and Y, the commutator $[\mathcal{L}_X, \mathcal{L}_Y]$ is again a derivative with respect to a vector field, since

$$[\mathcal{L}_X, \mathcal{L}_Y] = \mathcal{L}_{[X,Y]},$$

where, by definition,

$$[X, Y]^j = \sum_{h=1}^{n} \left(X^h \frac{\partial Y^j}{\partial x^h} - Y^h \frac{\partial X^j}{\partial x^h} \right) \tag{3.1.3}$$

is the *Lie bracket*. In fact, the second derivatives cancel out.

As one can easily verify, the commutator between vector fields possesses the following properties:

(i) it is linear, i.e., $[X, \alpha Y + \beta Z] = \alpha[X, Y] + \beta[X, Z]$;

(ii) it is skew symmetric, i.e., $[X, Y] + [Y, X] = 0$;

(iii) it satisfies the Jacobi identity, i.e.,

$$[X, [Y, Z]] + [Y, [Z, X]] + [Z, [X, Y]] = 0$$

for every vector field. This shows that the vector fields form a Lie algebra with respect to the commutator.

The Lie derivative allows us to construct the general solution of the dynamical system (3.1.1) explicitly, though at a formal level. Let us look for a solution which satisfies the initial conditions $x(0) = x_0$ and of the form

$$x(\tau) = x_0 + \tau x^{(1)} + \tau^2 x^{(2)} + \cdots,$$

namely a power expansion with coefficients

$$x^{(s)} = \frac{1}{s!}\frac{d^s}{d\tau^s}(\Phi_\tau x_0)\Big|_{\tau=o}.$$

On the other hand, since Φ_τ is the flow generated by the vector field X, it is immediately seen that

$$\frac{d}{d\tau}(\Phi_\tau x_0)\Big|_{\tau=o} = X|_{x=x_0},\dots,\frac{d^s}{d\tau^s}(\Phi_\tau x_0)\Big|_{\tau=o} = \mathcal{L}_X^{s-1}X\Big|_{x=x_0},\dots,$$

where $\mathcal{L}_X X = (\mathcal{L}_X X_1,\dots,\mathcal{L}_X X_n)$. Remarking that $\mathcal{L}_X x = X$, the solution can be given in the explicit form

$$x(\tau) = x_0 + \tau\mathcal{L}_X x|_{x=x_0} + \frac{\tau^2}{2!}\mathcal{L}_X^2 x\Big|_{x=x_0} + \frac{\tau^3}{3!}\mathcal{L}_X^3 x\Big|_{x=x_0} + \cdots$$

$$\overset{\text{def}}{=} \exp(\tau\mathcal{L}_X)x_0.$$

The exponential Lie operator $\exp(\tau\mathcal{L}_X)$ we have just defined is clearly linear and provides us with a flow. In fact, for $\tau = 0$ it is the identity, and moreover

$$\exp(t\mathcal{L}_X) \circ \exp(\tau\mathcal{L}_X) = \sum_{r\geq 0}\frac{t^r}{r!}\mathcal{L}_X^r \sum_{s\geq 0}\frac{\tau^s}{s!}\mathcal{L}_X^s$$

$$= \sum_{k\geq 0}\frac{1}{k!}\mathcal{L}_X^k \sum_{j=0}^{k}\binom{k}{j}t^j\tau^{k-j} = \sum_{k\geq 0}\frac{(t+\tau)^k}{k!}\mathcal{L}_X^k = \exp\left((t+\tau)\mathcal{L}_X\right).$$

A basic feature of the exponential Lie operator is the *exchange property*, which will be used in the reduction to normal form of a perturbed Hamiltonian. Put

$$x_\tau = \exp(\tau\mathcal{L}_X)x = x + \tau X + \cdots,$$

and consider it as a coordinate change: $x \mapsto x_\tau$. Given some $f(x)$, define $f_\tau(x) = f(x_\tau)$. Then

$$\frac{df_\tau(x)}{d\tau}\Big|_{\tau=0} = \frac{df(x_\tau)}{d\tau}\Big|_{\tau=0} = \sum_{k=1}^{n}\frac{\partial f(x_\tau)}{\partial x_\tau^k}\frac{dx_\tau^k}{d\tau}\Big|_{\tau=0}$$

$$= \sum_{k=1}^{n}X^k\frac{\partial f(x)}{\partial x^k} = \mathcal{L}_X f(x),$$

from which

$$f(\exp(\tau\mathcal{L}_X)x) = f(x) + \tau\frac{df(x_\tau)}{d\tau}\Big|_{\tau=0} + \frac{1}{2!}\tau^2\frac{d^2 f(x_\tau)}{d\tau^2}\Big|_{\tau=0} + \cdots$$

$$= f(x) + \tau\mathcal{L}_X f(x) + \frac{1}{2!}\tau^2\mathcal{L}_X^2 f(x) + \cdots = \exp(\tau\mathcal{L}_X)f(x),$$

showing that the exponential Lie operator can be moved in front of the function, so that the transformed function can be calculated without substitution of variables. In particular, by taking $f : \mathbb{R} \to \mathbb{R}$ and $X = 1$ one finds the elementary Taylor series development.

Until now the manifold N has been supposed to be generic. From now on N will be a $2n$-dimensional symplectic manifold, and the vector field X_χ the symplectic gradient of some Hamiltonian $\chi : N \to \mathbb{R}$,

$$(X_\chi)^k = \sum_{h=1}^{2n} \Omega^{kh} \frac{\partial \chi}{\partial x^h},$$

(Ω^{kh}) being the inverse of the matrix associated to a symplectic 2-form. To simplify the notation we put $\mathcal{L}_{X_\chi} = \mathcal{L}_\chi$. The Poisson bracket between two dynamical variables $F, G : N \to \mathbb{R}$ can thus be written

$$\{F, G\} \overset{\text{def}}{=} \sum_{h,k=1}^{2n} \frac{\partial G}{\partial x^h} \Omega^{hk} \frac{\partial F}{\partial x^k} = \mathcal{L}_F G = -\mathcal{L}_G F,$$

for the skew symmetry of the symplectic 2-form.

The closure property of the 2-form ensures the validity of the classical Jacobian formula

$$\{H, \{F, G\}\} + \{F, \{G, H\}\} + \{G, \{H, F\}\} = 0,$$

which can be written as

$$\mathcal{L}_H \{F, G\} = \{F, \mathcal{L}_H G\} + \{\mathcal{L}_H F, G\},$$

similar to the Leibniz formula (3.1.2).

As far as we are concerned, the key property is: the transformation $x \mapsto x_\tau = \exp(\tau \mathcal{L}_\chi)x$ is canonical for every τ. It is an obvious consequence of the fact that x_τ is the solution at "time" τ of the Hamilton equations with Hamiltonian χ.

With these tools at hand we can study the perturbations of Hamiltonian systems.

3.1.2 Homological Equation and Its Formal Solution

Consider the perturbed Hamiltonian system

$$H(x) = H_0(x) + \varepsilon H_p(x), \quad x \in N,$$

where N is a symplectic manifold. We want to find a canonical transformation

$$x \mapsto x' = \exp(\varepsilon \mathcal{L}_\chi)x = x + \varepsilon \mathcal{L}_\chi x + \mathcal{O}(\varepsilon^2)$$

with generator χ, transforming H into a completely integrable Hamiltonian up to second order terms. Even this somewhat "minimal" requirement can be satisfied in general only under certain conditions.

Recalling the exchange property, we can write

$$H(x') = \exp(\varepsilon \mathcal{L}_\chi)H = H + \varepsilon \mathcal{L}_\chi H + \cdots$$
$$= H_0 + \varepsilon(H_p + \mathcal{L}_\chi H_0) + \mathcal{O}(\varepsilon^2).$$

We must find the unknown generator χ in such a way that the first order term $H_p + \mathcal{L}_\chi H_0$ becomes integrable. To this end, let us assume that the local coordinates x's are action-angle variables

$$x = I_1, \ldots, I_n, \varphi_1, \ldots, \varphi_n.$$

Hence the unperturbed Hamiltonian H_0 will be a function of the action variables only. Without loss of generality, we suppose that $H_0 = H_0(I_1, \ldots, I_d)$, $d \le n$. In this case, H_0 is said to be $(n - d)$-fold *totally degenerate*. Later on, we will also consider the *nondegenerate* Hamiltonian H_0, i.e., such that $\det\left(\frac{\partial^2 H_0}{\partial I_h \partial I_k}\right) \ne 0$. If $d = n$, the unperturbed Hamiltonian is said to be *totally nondegenerate*. Notice that a Hamiltonian can be totally nondegenerate but degenerate: for example, $H_0 = I_1 + I_2^2$.

Let us suppose first that the unperturbed system is totally nondegenerate, and put

$$\omega_i(I) = \frac{\partial H_0}{\partial I_i}, \quad i = 1, \ldots, n.$$

The vector with components $\omega_i(I)$ is called the *frequency vector*. This name is due to the following fact. The Hamilton equations of the unperturbed system

$$\dot{I}_h = 0, \quad \dot{\varphi}_h = \omega_h(I),$$

are easily integrated

$$I_h(t) = I_h^0, \quad \varphi_h(t) = \omega_h(I^0)t + \varphi_h^0,$$

with I_h^0 and φ_h^0 integration constants. The unperturbed angles evolve with constant angular velocity, whence the name.

Write the first order perturbing term as

$$H_p - \mathcal{L}_{H_0}\chi = H_p - \omega \cdot \partial \chi.$$

Obviously H_p is also a function of the angles; otherwise, the perturbed system would be integrable. It is tempting to try to find the unknown χ such that the disturbing term goes to zero, but this is not possible: in fact the term $\omega \cdot \partial \chi$ has vanishing mean value with respect to the angles, while

H_p is a generic function. Hereafter an overbar will denote the averaging of a function over the angles:

$$\overline{f} = \frac{1}{(2\pi)^n} \int_{\mathbb{T}^n} f(\varphi_1, \ldots, \varphi_n) \, d\varphi_1 \ldots d\varphi_n,$$

where $\mathbb{T}^n = \overbrace{S^1 \times \cdots \times S^1}^{n \text{ times}}$ is the n-dimensional torus. Expanding into a Fourier series, we get

$$\chi(I, \varphi) = \overline{\chi}(I) + \sum_{k \in \mathbb{Z}^n - \{0\}} \chi_k(I) e^{ik \cdot \varphi},$$

$$H_p(I, \varphi) = \overline{H}_p(I) + \sum_{k \in \mathbb{Z}^n - \{0\}} H_{pk}(I) e^{ik \cdot \varphi},$$

where

$$\chi_k(I) = \frac{1}{(2\pi)^n} \int_{\mathbb{T}^n} \chi(I, \varphi) e^{-ik \cdot \varphi} \, d\varphi_1 \ldots d\varphi_n,$$

$$H_{pk}(I) = \frac{1}{(2\pi)^n} \int_{\mathbb{T}^n} H_p(I, \varphi) e^{-ik \cdot \varphi} \, d\varphi_1 \ldots d\varphi_n,$$

and $k \in \mathbb{Z}^n - \{0\}$ is an n-dimensional vector with integer components $(k_1, \ldots, k_n)$ not all zero. Choosing

$$\chi_k(I) = \frac{H_{pk}(I)}{ik \cdot \omega},$$

plainly annihilates the disturbing term except for its mean value part, leaving us with the transformed Hamiltonian

$$H(I, \varphi) = H_0(I) + \varepsilon \overline{H}_p(I) + \mathcal{O}(\varepsilon^2). \tag{3.1.4}$$

We call *resonant* those frequency vectors ω for which there exists some $k \in \mathbb{Z}^n - \{0\}$ such that $k \cdot \omega = 0$. Excluding the resonant frequency vectors, we are thus able to solve the equation

$$H_p - \overline{H}_p = \omega \cdot \partial \chi \tag{3.1.5}$$

and to push the unwanted dependence on the angles up to second order terms. Equation (3.1.5) was called by Poincaré the *homological equation*.

The procedure can be obviously iterated. Applying to (3.1.4) the operator $\exp(\varepsilon^2 \mathcal{L}_{\chi_2})$ and solving a second homological equation, the unwanted dependence on the angles is pushed to third order, and so forth. To give an idea of how the iterative process works, we report explicitly the first four terms, putting, for simplicity, $\mathcal{L}_{\chi_j} = \mathcal{L}_j$. Then

$$\prod_{j=4}^{1} \exp(\varepsilon^j \mathcal{L}_j) H = H_0 + \varepsilon \left(\mathcal{L}_1 H_0 + H_p \right) + \varepsilon^2 \left(\mathcal{L}_2 H_0 + \frac{1}{2} \mathcal{L}_1^2 H_0 + \mathcal{L}_1 H_p \right)$$

(3.1.6)

$$+ \varepsilon^3 \left(\mathcal{L}_3 H_0 + \frac{1}{2} \mathcal{L}_1^2 H_p + \mathcal{L}_2 H_p + \mathcal{L}_2 \mathcal{L}_1 H_0 + \frac{1}{6} \mathcal{L}_1^3 H_0 \right)$$

$$+ \varepsilon^4 \left(\begin{array}{c} \mathcal{L}_4 H_0 + \frac{1}{2} \mathcal{L}_2 \mathcal{L}_1^2 H_0 + \mathcal{L}_2 \mathcal{L}_1 H_p + \mathcal{L}_3 \mathcal{L}_1 H_0 \\ + \mathcal{L}_3 H_p + \frac{1}{2} \mathcal{L}_2^2 H_0 + \frac{1}{24} \mathcal{L}_1^4 H_0 + \frac{1}{6} \mathcal{L}_1^3 H_p \end{array} \right) + \mathcal{O}(\varepsilon^5).$$

As is easily seen, the term of order r is of the type

$$\varepsilon^r \left(H_r(\chi_{r-1}, \chi_{r-2}, \ldots, \chi_1, H_0, H_p) - \omega \cdot \partial \chi_r \right),$$

with H_r some function. Having solved the homological equations of order $\leq$ $r-1$, the respective generators are all known, and the homological equation of order r

$$H_r - \overline{H}_r = \omega \cdot \partial \chi_r$$

can, at least formally, be solved, leaving us with the term $\varepsilon^r \overline{H}_r(I)$. The transformed Hamiltonian is said to be in *normal form* to order r, and the process we have described is referred to as the *normalization procedure*.

We have proved that, at least for a totally nondegenerate H_0 and excluding the resonant frequency vectors, a perturbed but truncated at some finite order problem can be made completely integrable. Let us now consider an $(n - d)$-fold totally degenerate system. In this case, only the first d components $\omega_a(I)$ of the frequency vector are different from zero, so that only the first d angles (the *fast angles*) evolve for the unperturbed Hamiltonian. If the perturbation is switched on, the other angles (called the *slow angles*) evolve too, but very slowly due to the smallness of the perturbation:

$$\dot{I}_j = -\varepsilon \frac{\partial H_p}{\partial \varphi_j}, \quad \dot{\varphi}_a = \omega_a + \varepsilon \frac{\partial H_p}{\partial I_a}, \quad \dot{\varphi}_h = \varepsilon \frac{\partial H_p}{\partial I_h},$$

$$j = 1, \ldots, n, \quad a = 1, \ldots, d, \quad h = d + 1, \ldots, n.$$

The normalization process can clearly be performed only for the fast angles. These are eliminated from the perturbation up to some finite order, while the slow angles survive in the normalized Hamiltonian; loosely speaking, there is no fast angular evolution on which to average. This means that the normalized Hamiltonian is not integrable, and one must devise a further strategy if he wants to continue in an analytical way.

In every case, i.e., either total degeneracy of some order or total nondegeneracy, a perturbed Hamiltonian is said to be in normal form at order r

if

$$H = H_0 + \sum_{j=1}^{r} \varepsilon^j \overline{H}_j + \mathcal{O}(\varepsilon^{r+1}), \quad \{H_0, \overline{H}_j\} = 0 \quad \forall j \leq r.$$

3.2 Perpetual Stability and KAM Theorem

In this section (where we consider nondegenerate analytical Hamiltonians, i.e., $\det(\partial^2 H_0 / \partial I_h \partial I_k) \neq 0$) we tackle the crucial question: Does the normalization process converge when $r \to \infty$? Clearly, the delicate points are two:

(i) the convergence of the formal solution of the homological equation;

(ii) the convergence of the ε-development.

Untill the fundamental work of Kolmogorov (1954), two diametrically opposed answers had been given. For the astronomers, interested in the computation of perturbed orbits, the answer was (more or less tacitly) affirmative:[1] the negative one seemed to be a disaster, making the series expansions pointless. On the contrary, for the statistical physicists, interested in the possibility of applying the ergodic theorem, the answer was decisively negative. That of Kolmogorov was, in some sense, a Solomon's verdict: the ultimate fate of a slightly perturbed orbit depends on the initial conditions, so that for the same perturbed Hamiltonian two different orbit types in general coexist, i.e., orbits staying in the neighborhood of an unperturbed one *for infinite time* (we speak of *perpetual stability*) and others departing indefinitely.

To make this statement more precise, let us consider the unperturbed motions. For a completely integrable system, the phase space is foliated with n-dimensional tori. If the normalization process converged, the transformed Hamiltonian would still be integrable, and the tori, though slightly deformed, would continuously fill the phase space. But complete integrability is an exception, and even a very small perturbation would destroy it. The point is that this destruction does not happen abruptly: for small perturbations only a few tori disappear, their number growing with the intensity of the perturbation. But what kind of tori are destroyed first? Since by assumption H_0 is nondegenerate, hence $\det(\partial \omega_h / \partial I_k) \neq 0$, the frequency vectors can substitute the actions in order to label the tori. The frequency vectors can be classified (with a *Diophantine inequality*) on the basis of the more or less reciprocal irrationality of their components. The Kolmogorov theorem then states that those tori whose frequency vectors have components reciprocally close to rationality are the first ones, as ε grows, to be destroyed, while those characterized by a sufficiently strong irrationality are only slightly deformed, giving rise to perpetual stability.

---

[1] With obviously some notable exceptions, for example Poincaré.

The Kolmogorov theorem was stated in Kolmogorov (1954) with just a sketch of the proof. A complete and rigorous proof was subsequently given in Arnold (1963), Moser (1955), Moser (1962), and Moser (1967), but following different lines. The original idea of Kolmogorov was then reconsidered in Benettin, Galgani, Giorgilli & Strelcyn (1984) and completed. The whole argument is now referred to as KAM theory from the acronym of the author names. In the next subsection we prove the Kolmogorov theorem, closely following Benettin *et al.* (1984), to which we refer for the details omitted in the proof. Then, we will briefly outline Arnold's proof. The following two steps are basic in both proofs (compare with points (i) and (ii) mentioned above).

(i) Solve a truncated linearized problem that reduces the system into one that is closer to being completely integrable. This comes down to solving some homological equation(s), and it is in this step that small denominators come into play.

(ii) Use a rapidly converging iteration scheme to get rid of the perturbation term. This iteration scheme is nothing but an adaptation of Newton's (or the quadratic) method. The need for such a rapidly decaying iteration is forced by the requirement to beat the explosive growth of the perturbative terms due to small denominators. The quadratic method consists essentially, at every stage, in taking H_0 *plus all the previous averaged terms* as the new unperturbed Hamiltonian, so that every perturbative term is the square of the preceding one and decreases as ε^{2^r}, instead of ε^r.[2]

Before presenting the various theorems, we must return to the Lie series and to the solution of the homological equation through the Fourier series in order to investigate their convergence: remember, that the "solution" already found is only formal.

Let us fix some notation. For $v \in \mathbb{C}^n$, we put $\|v\| = \max_i |v_i|$. Analogously, the norm of a matrix (M_{ij}) is defined as for an n^2-vector, namely by $\|M\| = \max_{ij} |M_{ij}|$. Moreover we put

$$I = \{I_1, \ldots, I_n\} \in \mathcal{B} \subset \mathbb{R}^n, \quad \varphi = \{\varphi_1, \ldots, \varphi_n\} \in \mathbb{T}^n,$$

$\mathcal{B}$ being an open ball in $\mathbb{R}^n$. Being interested in the analytic case, we will consider complex extensions of subsets of $\mathbb{R}^{2n}$ and of real analytic functions

---

[2]In effect the quadratic method is very useful in making the proof easier but it is not strictly necessary. Direct proofs of the convergences of the perturbative series have been produced on the basis of a suitable grouping of terms and consequent cancellation. See Chierchia & Falcolini (1994), Chierchia & Falcolini (1996), and Giorgilli & Locatelli (1997). These methods are conceptually more natural, enlightening the deep mechanism which leads to the convergence while, in some sense, the quadratic method resorts to brute force.

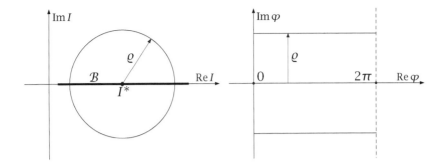

Figure 3.1: The domain $\mathcal{D}_\varrho$ for $n = 1$.

defined there. Having fixed $I^* \in \mathcal{B}$ and a positive $\varrho < 1$ so small that the real closed ball of radius ϱ centered at I^* is contained in $\mathcal{B}$, a central role will be played by the subset $\mathcal{D}_\varrho$ of $\mathbb{C}^{2n}$ defined by

$$\mathcal{D}_\varrho = \{(I, \varphi) \in \mathbb{C}^{2n} : \|I - I^*\| \le \varrho, \|\operatorname{Im}\varphi\| \le \varrho\},$$

where $\operatorname{Im}\varphi = (\operatorname{Im}\varphi_1, \ldots, \operatorname{Im}\varphi_n)$: see Figure 3.1. Clearly, the domain $\mathcal{D}_\varrho$ is the complex extension of the real n-dimensional torus $I = I^*$. For what concern functions, let $\mathcal{A}_\varrho$ be the set of all complex analytic functions in the interior of $\mathcal{D}_\varrho$ that are real for real values of the variables. Their norm is defined by

$$\|f\|_\varrho = \sup |f(z)|, \ z \in \mathcal{D}_\varrho, \ f \in \mathcal{A}_\varrho.$$

In the case of vector-valued functions $f = (f_1, \ldots, f_n)$ with values in $\mathbb{C}^n$, we also write $f \in \mathcal{A}_\varrho$ if $f_h \in \mathcal{A}_\varrho$, $h = 1, \ldots, n$ and define

$$\|f\|_\varrho = \max_h \|f_h\|_\varrho.$$

Analogously, if $\mathbf{D}$ is an $(n \times n)$ matrix whose elements D_{hk} belong to $\mathcal{A}_\varrho$, we set

$$\|\mathbf{D}\|_\varrho = \max_{hi} \|D_{hi}\|_\varrho.$$

3.2.1 Cauchy Inequality

Being interested in giving estimates, we will make use of the Cauchy inequality. Given $f \in \mathcal{A}_\varrho$, a positive $\delta < \varrho$, and nonnegative integers j_i, h_i ($i = 1, \ldots, n$), we prove that

$$\left\| \frac{\partial^{j_1 + \cdots + j_n + h_1 + \cdots + h_n}}{\partial I_1^{j_1} \ldots \partial I_n^{j_n} \partial \varphi_1^{h_1} \ldots \partial \varphi_n^{h_n}} f(I, \varphi) \right\|_{\varrho - \delta} \le \frac{j_1! \ldots j_n! h_1! \ldots h_n!}{\delta^{j_1 + \cdots + j_n + h_1 + \cdots + h_n}} \|f\|_\varrho. \quad (3.2.1)$$

Let us first comment on this formula. In order to check the convergence of some series, we would like the right-hand side to be small, but δ is the only parameter at our disposal: choosing it large we restrict the domain on the left-hand side so that, iterating the application of the Cauchy inequality, the domain measure risks to vanish. We are between the two fires of conflicting requirements, and the delicate point is to prove convergence while keeping at the same time the domain measure finite.

For the proof of the Cauchy inequality, consider an open disk $\Delta_\varrho(z_0)$ centered at the point z_0 of the complex plane $\mathbb{C}$. Consider a function f analytic and bounded on the disk $\Delta_\varrho(z_0)$. The *supremum norm* $\|f\|_\varrho$ of f in the domain $\Delta_\varrho(z_0)$ is defined as

$$\|f\|_\varrho = \sup_{z \in \Delta_\varrho(z_0)} |f(z)| .$$

First we prove the Cauchy inequality for the derivative f' at the origin:

$$|f'(z_0)| \le \frac{1}{\varrho} \|f\|_\varrho .$$

More generally, for the s-th derivative $f^{(s)}$ one has the estimate

$$\left| f^{(s)}(z_0) \right| \le \frac{s!}{\varrho^s} \|f\|_\varrho .$$

Notice that this estimate cannot be improved in general. In fact, let $z_0 = 0$, $\varrho = 1$, and consider the function $f(z) = z^s$: then $\|f\|_1 = 1$, so that the Cauchy inequality gives $\left| f^{(s)}(0) \right| \le s!$. The proof of the inequalities above is an easy consequence of the Cauchy integral formula

$$f^{(s)}(z) = \frac{s!}{2\pi i} \oint \frac{f(\zeta)}{(\zeta - z)^{s+1}} \, d\zeta.$$

Indeed, let $z = z_0$ and let the integration path be the circle of radius ϱ with center at z_0. If $\zeta - z_0 = \varrho e^{i\theta}$, then $d\zeta = i(\zeta - z_0) \, d\theta$ and

$$\left| f^{(s)}(z_0) \right| \le \frac{s!}{2\pi} \oint \left| \frac{f(\zeta)}{(\zeta - z_0)^{s+1}} \right| \varrho \, d\theta,$$

so that the claim follows by replacing $|\zeta - z_0| = \varrho$ and $|f(\zeta)| \le \|f\|_\varrho$.

In the case of n variables the extension is straightforward. Let the domain $\Delta_\varrho(z_0)$ be the polydisk of radius ϱ centered at $z_0 \in \mathbb{C}^n$, namely

$$\Delta_\varrho(z_0) = \{z \in \mathbb{C}^n : |z - z_0| \le \varrho\},$$

where $|z| = \max_j \left| z_j \right|$. This is nothing but the Cartesian product of complex disks of radius ϱ in the complex plane. Define the supremum norm of an analytic function f as above. Then

$$\left| \frac{\partial f}{\partial z_j}(z_0) \right| \le \frac{1}{\varrho} \|f\|_\varrho, \quad 1 \le j \le n.$$

The estimates can be extended to more general domains. Indeed, let G be a subset of $\mathbb{C}^n$, and consider the domain

$$G_\varrho = \bigcup_{z \in G} \Delta_\varrho(z),$$

that is, the domain G_ϱ is the union of closed polydisks of radius ϱ around every point of G.

The supremum norm of a complex function $f : G_\varrho \to \mathbb{C}$ is defined in a natural way as

$$\|f\|_\varrho = \sup_{z \in G_\rho} |f(z)|.$$

PROPOSITION 3.1 *Let f be analytic in the interior of G_ϱ, and let $\|f\|_\varrho$ be finite. Then for any positive $\delta \le \varrho$ one has*

$$\left\| \frac{\partial f}{\partial z_j} \right\|_{\varrho-\delta} \le \frac{1}{\delta} \|f\|_\varrho.$$

Proof. Let us consider a point $z \in G_{\varrho-\delta}$, i.e., the union of polydisk of radius $\varrho - \delta$ centered at every point of G. Remark that the polydisk $\Delta_\delta(z)$ is a subset of G_ϱ, so that f is analytic and bounded on the polydisk; moreover, we have the estimate $|f(z')| \le \|f\|_\varrho$ for all $z' \in \Delta_\delta(z)$. By the Cauchy estimate, we immediately get

$$\left| \frac{\partial f}{\partial z_j}(z) \right| \le \frac{1}{\delta} \|f\|_\varrho, \quad 1 \le j \le n.$$

Since this is true for every point $z \in G_{\varrho-\delta}$, we obtain the result.   **QED**

Taking for G the n-dimensional torus $I = I^*$ and consequently $G_\varrho = \mathcal{D}_\varrho$, the generalization (3.2.1) follows.

3.2.2   Convergence of Lie Series

First we note that, given two functions $\chi, f \in \mathcal{A}_\varrho$ with

$$\|\partial \chi\|_\rho \overset{\text{def}}{=} \max \left(\left\| \frac{\partial \chi}{\partial I} \right\|_\varrho, \left\| \frac{\partial \chi}{\partial \varphi} \right\|_\varrho \right), \tag{3.2.2}$$

and any positive $\delta < \varrho$, the Cauchy inequality and a count of the terms in Poisson brackets give

$$\left\|\mathcal{L}_\chi f\right\|_{\varrho-\delta} \leq 2n \frac{\|\partial \chi\|_\rho}{\delta} \|f\|_\varrho ,$$

$$\left\|\mathcal{L}_\chi^2 f\right\|_{\varrho-\delta} \leq 4n(2n+1) \left(\frac{\|\partial \chi\|_\rho}{\delta}\right)^2 \|f\|_\varrho . \tag{3.2.3}$$

Let χ be the generator of a (for a moment formal) canonical transformation

$$Q : J, \psi \rightarrow I, \varphi.$$

The existence of such a canonical transformation Q with associated operator $\exp \mathcal{L}_\chi$, and some relevant estimates are then given by the following proposition.

PROPOSITION 3.2 *Assume* $\|\partial \chi\|_\rho < \frac{\delta}{2}$. *Then, for all initial data* $(J, \psi) \in \mathcal{D}_{\varrho-\delta}$, *the solution of the canonical system*

$$\frac{dI}{d\tau} = -\frac{\partial \chi}{\partial \varphi}, \quad \frac{d\varphi}{d\tau} = \frac{\partial \chi}{\partial I}$$

at $\tau = 1$ *exists in* $\mathcal{D}_\varrho$, *thus defining a canonical transformation* $Q : \mathcal{D}_{\varrho-\delta} \rightarrow \mathcal{D}_\varrho$, $Q \in \mathcal{A}_{\varrho-\delta}$. *Then, the operator* $\exp \mathcal{L}_\chi : \mathcal{A}_\varrho \rightarrow \mathcal{A}_{\varrho-\delta}$, *with* $\exp \mathcal{L}_\chi f = f \circ Q$, *is well defined. Moreover, one has the estimates*

$$\|Q - identity\|_{\varrho-\delta} \leq \|\partial \chi\|_\rho \tag{3.2.4}$$

and

$$\|\exp \mathcal{L}_\chi f\|_{\varrho-\delta} \leq \|f\|_\varrho ,$$

$$\|\exp \mathcal{L}_\chi f - f\|_{\varrho-\delta} \leq 4n \frac{\|\partial \chi\|_\rho}{\delta} \|f\|_\varrho , \tag{3.2.5}$$

$$\|\exp \mathcal{L}_\chi f - f - \mathcal{L}_\chi f\|_{\varrho-\delta} \leq 16n(2n+1) \left(\frac{\|\partial \chi\|_\rho}{\delta}\right)^2 \|f\|_\varrho .$$

Sketch of the proof. The standard existence and uniqueness theorem guarantees that the mapping

$$Q^\tau(J, \psi) = (I(\tau), \varphi(\tau))$$

and, in particular, the mapping Q^1 is defined. This mapping is obviously canonical, being the "time-one" solution of a canonical system. Estimate (3.2.4) is an immediate consequence of the mean value theorem. The first of (3.2.5) is trivial. The second and the third inequalities follow from the Taylor formulae for f of first and second order respectively, namely

$$\exp \mathcal{L}_X f - f = \left.\frac{df}{d\tau}\right|_{\tau'} = \left.\mathcal{L}_X f\right|_{\tau'},$$

$$\exp \mathcal{L}_X f - f - \frac{df}{d\tau} = \left.\frac{1}{2}\frac{d^2 f}{d\tau^2}\right|_{\tau''} = \left.\frac{1}{2}\mathcal{L}_X^2 f\right|_{\tau''},$$

with $0 < \tau', \tau'' < 1$, and from (3.2.3).   **QED**

Having verified that the Lie series expansions are not only formal, we investigate the solution of the homological equation, which is an unavoidable passage in every perturbative theory.

3.2.3 Homological Equation and Its Solution

In this subsection we only consider functions of the angles, so that the actions will be considered as parameters, and thus disregarded.

We know that the homological equation

$$\sum_i \lambda_i \frac{\partial F(\varphi)}{\partial \varphi_i} = G(\varphi)$$

is formally solvable if $\overline{G} = 0$. Here $\lambda = (\lambda_1, \ldots, \lambda_n) \in \mathbb{R}^n$ is a known *fixed arbitrary* frequency vector that we suppose to be nonresonant. The formal solution is

$$F(\varphi) = \sum_{k \in \mathbb{Z}^n - \{0\}} \frac{g_k}{ik \cdot \lambda} e^{ik \cdot \varphi}, \qquad (3.2.6)$$

where g_k are the Fourier coefficients of G. At first sight, it seems that this sum cannot converge in general, i.e., for arbitrary frequency vectors $(\lambda_1, \ldots, \lambda_n)$, even if nonresonant. In fact, the set in $\mathbb{R}^n$ for which the denominator vanishes is dense (since the rational numbers are dense on the real line and every nonempty open set in $\mathbb{R}$ contains at least one rational number), and for hope of convergence we must exclude from the sum not only the resonant term, but also those for which $k \cdot \lambda$ is very small. This is the celebrated problem of the "small denominators." The situation may appear hopeless but, fortunately, a classical result in Diophantine theory (see Appendix 3.A) guarantees that $k \cdot \lambda$ can be bounded from below without being left with an empty set. In more precise terms, the inequality

$$|k \cdot \lambda| \geq \frac{\gamma}{|k|^n} \quad \forall k \in \mathbb{Z}^n - \{0\}, \quad |k| \overset{\text{def}}{=} \sum_i |k_i|, \qquad (3.2.7)$$

for some positive y is satisfied by a set Ω_y of real vectors λ of large relative measure, the complement of this set having Lebesgue measure $\mathcal{O}(y)$. Inequality (3.2.7) is referred to as the *Diophantine condition* or *inequality*.

This is a key point. In fact, since G is by hypothesis analytic, we prove that its Fourier coefficients g_k decay exponentially with $|k|$, while $\frac{1}{|k\cdot\lambda|}$ grows at most as a power, owing to the Diophantine condition. This allows us to prove the convergence of (3.2.6).

It is easy to estimate the exponential decay of the Fourier coefficients of an analytical function $G(\varphi)$, $\varphi \in S^1$. The set $|\operatorname{Im}\varphi| < \varrho$ can be represented as a thin circular strip surrounding the unit circle in the complex plane of the variable $z = e^{i\varphi}$. Then

$$g_k = \frac{1}{2\pi} \int_0^{2\pi} G(\varphi)e^{-ik\varphi}\,d\varphi = \frac{1}{2\pi} \int_0^{2\pi} G(\varphi \pm i\varrho)e^{\pm k\varrho}e^{-ik\varphi}\,d\varphi, \quad k \in \mathbb{Z}$$

since, for the analyticity of G, it is possible to shift the integration path from the unit to the inner or outer circle of the strip. Taking the minus sign for $k > 0$ and the plus sign for $k < 0$, we obtain

$$|g_k| \leq \|G\|_\varrho\, e^{-|k|\varrho}.$$

The argument is straightforwardly extended to the n-dimensional case, with $k \in \mathbb{Z}^n$ and $|k| \stackrel{\text{def}}{=} \sum_i |k_i|$,

We will also need this proposition, which is the inverse of the previous statement.

PROPOSITION 3.3 *Suppose that for some positive constants C and ϱ with $\varrho \leq 1$ and for each $k \in \mathbb{Z}^n$ one has the sequence of constants $|f_k|$ such that $|f_k| \leq Ce^{-|k|\varrho}$. Define the function $F(\varphi) = \sum_k f_k e^{ik\cdot\varphi}$. Then, for any positive $\delta < \varrho$ one has $F \in \mathcal{A}_{\varrho-\delta}$ and*

$$\|F\|_{\varrho-\delta} \leq C\left(\frac{4}{\delta}\right)^n.$$

Proof. Let $\|\operatorname{Im}\varphi\| \leq \varrho - \delta$. Then

$$\|F\|_{\varrho-\delta} = \sup\left|\sum_{k\in\mathbb{Z}^n} f_k e^{ik\cdot\varphi}\right| \leq C\sum_{k\in\mathbb{Z}^n} e^{-|k|\varrho}e^{|k|(\varrho-\delta)}$$

$$= C\sum_{k\in\mathbb{Z}^n} e^{-|k|\delta} = C\left(\sum_{k\in\mathbb{Z}} e^{-|k|\delta}\right)^n$$

$$< 2^n C\left(\sum_{k=0}^{\infty} e^{-\delta k}\right)^n = 2^n C\frac{1}{(1-e^{-\delta})^n} < C\left(\frac{4}{\delta}\right)^n,$$

because $\frac{1}{1-e^{-\delta}} < \frac{2}{\delta}$ for any positive $\delta < 1$.   **QED**

We can now prove the following

PROPOSITION 3.4 *Consider the homological equation*

$$\sum_i \lambda_i \frac{\partial F(\varphi)}{\partial \varphi_i} = G(\varphi),$$

and assume $\lambda \in \Omega_\gamma$ *and* $G \in \mathcal{A}_\varrho$ *for some positive* γ *and* ϱ *with* $\overline{G} = 0$. *Then, for any positive* $\delta < \varrho$, *the homological equation admits a unique solution* $F \in \mathcal{A}_{\varrho-\delta}$ *with* $\overline{F} = 0$. *For this solution one has the basic estimates*

$$\|F\|_{\varrho-\delta} \leq \frac{\sigma_n}{\gamma \delta^{2n}} \|G\|_\varrho, \quad \left\|\frac{\partial F}{\partial \varphi}\right\|_{\varrho-\delta} \leq \frac{\sigma_n}{\gamma \delta^{2n+1}} \|G\|_\varrho,$$

where $\sigma_n = 2^{4n+1} \left(\frac{n+1}{e}\right)^{n+1}$.

Proof. The formal solution of the homological equation is given by (3.2.6), where now we assume that the frequency vector λ satisfies the Diophantine condition (3.2.7). By $G \in \mathcal{A}_\varrho$ and exponential decaying of Fourier coefficients, one has

$$|f_k| \leq \frac{|k|^n}{\gamma} \|G\|_\varrho e^{-|k|\varrho}. \tag{3.2.8}$$

For any strictly positive K, s, δ, by putting $\frac{K\delta}{s} = x$ one easily verifies that

$$K^s \leq \left(\frac{s}{e\delta}\right)^s e^{K\delta},$$

from which, putting $K = |k|$ and $s = n$, we get

$$|f_k| \leq C e^{-|k|(\varrho-\delta)}, \text{ with } C = \frac{1}{\gamma}\left(\frac{n}{e\delta}\right)^n \|G\|_\varrho.$$

For any $\delta < \varrho$ and with $\varrho - \delta$ in place of ϱ, Proposition 3.3 ensures that

$$F(\varphi) \stackrel{\text{def}}{=} \sum_{k \neq 0} f_k e^{ik\cdot\varphi} \in \mathcal{A}_{\varrho-2\delta},$$

$$\|F\|_{\varrho-2\delta} \leq \frac{1}{\gamma}\left(\frac{4n}{e\delta^2}\right)^n \|G\|_\varrho.$$

Taking $\frac{\delta}{2}$ instead of δ, the statement of the proposition and the first estimate are proved, because $\sigma_n > \left(\frac{16n}{e}\right)^n$.

Let us prove the latter estimate. From (3.2.6), the Fourier coefficients of $\frac{\partial F}{\partial \varphi_j}$ result in the form $h_{k_j} = \frac{k_j g_k}{\lambda \cdot k}$, $j = 1, \ldots, n$, so that from the Diophantine inequality one gets

$$\left| h_{k_j} \right| \leq \frac{|k|^{n+1}}{\gamma} \|G\|_\varrho \, e^{-|k|\varrho}.$$

But, with $K = |k|$ and $s = n + 1$, this gives

$$\left| h_{k_j} \right| \leq C e^{-|k|(\varrho - \delta)}, \text{ with } C = \frac{1}{\gamma} \left(\frac{n+1}{e\delta} \right)^{n+1} \|G\|_\varrho,$$

so that from Proposition 3.3, with $\varrho - \delta$ in place of ϱ, one obtains

$$\left\| \frac{\partial F}{\partial \varphi} \right\|_{\varrho - 2\delta} \leq \frac{4^n}{\gamma \delta^{2n+1}} \left(\frac{n+1}{e} \right)^{n+1} \|G\|_\varrho.$$

By taking again $\frac{\delta}{2}$ instead of δ, one therefore obtains also the second estimate.   **QED**

REMARK 3.1 The above estimates are not optimal. With deeper arguments optimal estimates can be given with δ^{2n} and δ^{2n+1} replaced by δ^n and δ^{n+1} respectively, and σ_n replaced by $\frac{2^{4n+1} n!}{2^n - 1}$. See Rüssmann (1975).

Let us comment on this result. The set Ω_γ is a Cantor set (see Appendix 3.A) of large measure, which is both good and bad news: unfortunately, the bad outweighs the good. Indeed, we can now push the perturbation to second order almost everywhere but, because of the complete loss of differentiability, this Pyrrhic victory on the small denominators prevents us from taking the further steps, which would push the perturbation to higher orders. In order to escape this essential difficulty, two different strategies have been proposed by Kolmogorov and by Arnold, respectively.

3.2.4   KAM Theorem (According to Kolmogorov)

An important thing we have learned is that the goal of classical perturbation theory to completely eliminate the angles in the transformed Hamiltonian is hopeless. Even a very small perturbation destroys the foliation of the phase space in invariant tori, and the n first integrals in involution of the unperturbed problem no longer exist.

The basic idea of Kolmogorov is to focus attention on the conservation problem of the tori, instead of on the existence problem of the first integrals in involution, and to investigate if *all* tori are destroyed, or if some survive for infinite times. To this end, he *fixes* the frequency vector, and states that, if this vector satisfies the Diophantine condition and *the perturbation is sufficiently small*, the corresponding torus is only slightly deformed by the perturbation, but not destroyed. Considering only *one* fixed torus allows us clearly to avoid the difficulty of the lack of differentiability.

To state and prove the Kolmogorov theorem, we start with a rearrangement of the perturbed Hamiltonian $H(I, \varphi)$, which we suppose already expressed as a function of the action-angle variables. Having fixed the values of the action variables, which we can safely put to zero, by a Taylor expansion in I one can write the Hamiltonian H in the form

$$H(I, \varphi) = a + A(\varphi) + [\lambda + B(\varphi)] \cdot I + \frac{1}{2} I \cdot \mathbf{D}(\varphi) I + R(I, \varphi).$$

Here λ is the frequency vector of the unperturbed Hamiltonian, $a \in \mathbb{R}$ is a constant that is uniquely defined by the condition $\overline{A} = 0$, while A, B_i, D_{ij} and $R \in \mathcal{A}_\varrho$, with R of the order $\|I\|^3$. One has clearly

$$a = \overline{H}(0), \qquad A(\varphi) = H(0, \varphi) - a,$$

$$B_i = \frac{\partial H}{\partial I_i}(0, \varphi) - \lambda_i, \quad D_{ij}(\varphi) = \frac{\partial^2 H}{\partial I_i \partial I_j}(0, \varphi).$$

Let us suppose for a moment that $A(\varphi) + B(\varphi) \cdot I = 0$. Then the Hamilton equations would be

$$\dot{\varphi}_i = \lambda_i + \sum_j D_{ij} I_j + \frac{\partial R}{\partial I_i},$$

$$\dot{I}_i = -\frac{\partial}{\partial \varphi_i} \left[\frac{1}{2} I \cdot \mathbf{D} I + R \right],$$

which admit the *particular* solution

$$I_i(t) = 0, \quad \varphi_i(t) = \lambda_i t + \varphi_i^0,$$

as one sees using the fact that R is of order $\|I\|^3$. In other words, if one is not guaranteed to have a foliation of the whole phase space into invariant tori, however, by a simple inspection one can still ascertains the invariance of *one* torus supporting quasi-periodic motions with angular frequency λ_i.

The Kolmogorov theorem just states that the disturbing term $A(\varphi) + B(\varphi) \cdot I$ can be removed with a canonical transformation $Q : J, \psi \to I, \varphi$. In the new canonical variables J, ψ the torus $J_i = 0$ is thus invariant for the Hamiltonian flow induced by the transformed Hamiltonian $H \circ Q$. In terms of the original variables I, φ this torus is described by the parametric equations $(I, \varphi) = Q(0, \psi)$, and is invariant for the Hamiltonian flow induced by H. This torus is a small perturbation of the torus $I_i = 0$, which is by assumption invariant for the unperturbed Hamiltonian, supporting quasi-periodic motions with the same angular frequencies λ_i.

We can now state the main theorem of this chapter.

THEOREM 3.5 (KOLMOGOROV) *Consider the Hamiltonian*

$$H(I, \varphi) = H^0(I, \varphi) + H^p(I, \varphi)$$

defined in the domain $\mathcal{D}_\varrho$ *by*

$$H^0(I, \varphi) = a + \lambda \cdot I + \frac{1}{2}I \cdot \mathbf{D}(\varphi)I + R(I, \varphi),$$
$$H^p(I, \varphi) = A(\varphi) + B(\varphi) \cdot I$$

with $\|H\|_\varrho < 1$ *and R of order* $\|I\|^3$ *. Let us suppose that*

(i) $\lambda \in \Omega_y$: *the unperturbed frequency vector satisfies the Diophantine condition,*

(ii) $\det \mathbf{D} \neq 0$,

then there exist positive numbers η *and* ϱ' *with* $\varrho' < \varrho$ *such that, if* H^p *is small enough to have*

(iii) $max\left(\|A\|_\varrho, \|B\|_\varrho\right) \leq \eta$,

one can construct a canonical analytical transformation

$$Q : \mathcal{D}_{\varrho'} \to \mathcal{D}_\varrho, \ Q \in \mathcal{A}_{\varrho'},$$

which brings the Hamiltonian H into the form

$$H'(J, \psi) \equiv (H \circ Q)(J, \psi) = a' + \lambda \cdot J + R'(J, \psi),$$

where $R' \in \mathcal{A}_{\varrho'}$ *is, as a function of J, of order* $\|J\|^2$ *. This canonical transformation is near the identity, in the sense that* $\|Q - \text{identity}\|_{\varrho'} \to 0$ *as* $\|H^p\|_\varrho \to 0$.

Notice that $D_{ij} \in \mathcal{A}_\varrho$ and (ii) are equivalent to the existence of a positive m such that

(ii') $\|\mathbf{D}v\|_\varrho \leq \frac{1}{m}\|v\|, \quad m\|v\| \leq \|\overline{\mathbf{D}}v\|, \quad \forall v \in \mathbb{C}^n$.

The scheme of the proof of the Kolmogorov theorem is as follows. One performs a sequence of canonical analytical transformations such that the disturbing term H^p at step r decreases with r, its norm $\left\|H_r^p\right\|_{\varrho_r}$ being essentially of the order of $\left\|H_{r-1}^p\right\|_{\varrho_{r-1}}^2$ (Newton's quadratic method), while the other parameters ϱ_r and m_r are kept controlled and strictly positive. The convergence of the scheme with $\left\|H_r^p\right\|_{\varrho_r} \to 0, \ \varrho_r \to \varrho_\infty > 0$, and $m_r \to m_\infty > 0$ as $r \to \infty$ is then established.

Let us now describe a step of the iterative process, which amounts to implement the first of the two basic points on page 92. We stress that iterative Lemma 3.6 below and in particular expression (3.2.17) *are the mainstay of the whole theorem.*

If a canonical change of variables with generating function χ is performed, one obtains in place of H the new Hamiltonian H' which we can decompose as $H' = H'^0 + H'^p$, fully analogously to the previous decomposition $H = H^0 + H^p$. Precisely, one has

$$H'^0 = a' + \lambda \cdot J + \frac{1}{2} J \cdot \mathbf{D}'(\psi) J + R'(J, \psi),$$
$$H'^p = A'(\psi) + B'(\psi) \cdot J$$

with $R'(J, \psi)$ of order $\|J\|^3$. In the spirit of the perturbation theory, one thinks that both H'^p and the generating function χ are of first order and chooses χ in order to eliminate the unwanted terms of the same order in the new Hamiltonian H'. To this end one first writes the identity

$$H' = H^0 + H^p + \mathcal{L}_\chi H^0 + [\mathcal{L}_\chi H^p + \exp \mathcal{L}_\chi H - H - \mathcal{L}_\chi H], \qquad (3.2.9)$$

where all terms which must to be considered of the second order have been collected into the bracket [...], in agreement with the estimate given in the third of (3.2.5). Then one tries to choose χ in such a way that the first order terms in H', namely $H^p + \mathcal{L}_\chi H^0$, do not contribute to H'^p. This is obtained by imposing

$$H^p + \mathcal{L}_\chi H^0 = c + \mathcal{O}\left(\|I\|^2\right),$$

where c is a constant.

Following Kolmogorov (1954), we show that this condition is met by a generating function χ of the form

$$\chi = \xi \cdot \varphi + X(\varphi) + Y(\varphi) \cdot I,$$

where the constant $\xi \in \mathbb{R}^n$ and the functions $X(\varphi), Y_i(\varphi)$ are to be suitably determined. The generating function is the sum of two terms: the term $Y(\varphi) \cdot I$ acts in the "angle direction" and is needed to straighten out the flow up to order $\mathcal{O}(\varepsilon^2)$, while the other term $\xi \cdot \varphi + X(\varphi)$ acts in the "action direction" and is needed to keep the frequency of the torus fixed. A straightforward calculation gives

$$H^p + \mathcal{L}_\chi H^0 = -\sum_i \xi_i \lambda_i + A(\varphi) - \sum_i \lambda_i \frac{\partial X}{\partial \varphi_i}$$
$$+ \sum_h \left[B_h(\varphi) - \sum_i D_{ih}(\varphi) \left(\xi_i + \frac{\partial X}{\partial \varphi_i} \right) - \sum_i \lambda_i \frac{\partial Y_h}{\partial \varphi_i} \right] I_h + \mathcal{O}\left(\|I\|^2\right).$$

Then one must impose

$$\sum_i \lambda_i \frac{\partial X}{\partial \varphi_i} = A(\varphi), \tag{3.2.10a}$$

$$\sum_i \lambda_i \frac{\partial Y_h}{\partial \varphi_i} = B_h(\varphi) - \sum_i D_{ih}(\varphi) \left(\xi_i + \frac{\partial X}{\partial \varphi_i} \right), \tag{3.2.10b}$$

which are two homological equations. By Proposition 3.4, Equation (3.2.10a) can be solved for the unknown X because $\overline{A} = 0$. Then one has to determine the unknown constants ξ_i in such a way that the mean value of the right-hand side of (3.2.10b) vanishes. This leads to a linear system of equations for ξ_i, which in matrix notation can be written as

$$\overline{\mathbf{D}}\xi = \overline{B - \mathbf{D} \frac{\partial X}{\partial \varphi}}. \tag{3.2.11}$$

This equation can be solved since, by hypothesis, $\det \overline{\mathbf{D}} \neq 0$. Also Equation (3.2.10b) in the unknown Y can therefore be solved.

The existence of the wanted generating function χ is thus guaranteed, and one must now prove the existence of the corresponding canonical transformation Q. Therefore, in order to apply Proposition 3.2, we exhibit the following basic estimate involving the generating function χ: Equations (3.2.10a), (3.2.10b), and (3.2.11) in the unknown $\xi_i, X(\varphi)$, and $Y_i(\varphi)$, which define the generating function χ, can be solved with $X, Y_i \in \mathcal{A}_{\tilde{\varrho}}, \tilde{\varrho} = \varrho - 2\delta$ for any positive $\delta < \frac{\varrho}{2}$, and, for the quantity $\|\partial\chi\|_{\tilde{\rho}}$ defined by (3.2.2), one gets

$$\|\partial\chi\|_{\tilde{\rho}} \leq (4n+1)\sigma_n^2 \frac{\eta}{\gamma^2 m^3 \delta^{2(2n+1)}}. \tag{3.2.12}$$

This inequality, concerning a function of the solutions of the two homological Equations (3.2.10a), (3.2.10b), can be verified with a twofold application of Proposition 3.4.

Let us choose two real numbers ϱ_* and m_* (bounding parameters) such that $0 < \varrho_* < \varrho$, $0 < m_* < m$. Define moreover the constant $\Lambda_n = 2(4n+1)^2\sigma_n^2$. Assume that η (which is a measure of the perturbing term) is so small that

$$m - n\Lambda_n \frac{\eta}{\varrho_*^2 \gamma^2 m^3 \delta^{4n+3}} > m_*. \tag{3.2.13}$$

Remarking that the inequality (3.2.12) can be written as

$$\frac{\|\partial\chi\|_{\tilde{\rho}}}{\delta} < \frac{1}{2n(4n+1)} (m - m_*)\varrho_*^2$$

owing to (3.2.13) and since $\varrho, m < 1$, one sees that $\|\partial\chi\|_{\tilde{\rho}} < \frac{\delta}{2}$ holds, so that Proposition 3.2 can be applied. Thus χ generates a canonical transformation with domain $\mathcal{D}_{\varrho'}$, where $\varrho' = \tilde{\varrho} - \delta = \varrho - 3\delta$.

In order to iterate the process, as described in point (ii) on page 92, we also need to estimate the new η' and m' entering the decomposition of the Hamiltonian in the Kolmogorov theorem. For what concerns A', by its definition, using expression (3.2.9) for H' and recalling that, by virtue of the choice of χ, only the last term contributes to A', one gets

$$\|A'\|_{\varrho'} \leq 2\|\mathcal{L}_\chi H^p + H \circ Q - H - \mathcal{L}_\chi H\|_{\varrho'}.$$

From this expression, after some calculations one obtains

$$\|A'\|_{\varrho'} \leq \Lambda_n^2 \frac{\eta^2}{\gamma^4 m^6 \delta^{8n+6}},$$

and analogously

$$\|B'\|_{\varrho'} \leq \frac{1}{2}\Lambda_n^2 \frac{\eta^2}{\varrho_* \gamma^4 m^6 \delta^{8n+6}},$$

so that

$$\max\left(\|A'\|_{\varrho'}, \|B'\|_{\varrho'}\right) \leq \eta',$$

with $\varrho' = \varrho - 3\delta$ and $\eta' = \Lambda_n^2 \frac{\eta^2}{\varrho_* \gamma^4 m^6 \delta^{8n+6}}$. Let us come to the estimate for m'. From the definition of $\mathbf{D}$ and $\mathbf{D}'$, the Cauchy inequality, and (3.2.14), one gets

$$\|(\mathbf{D}' - \mathbf{D})v\|_{\varrho'} \leq n\Lambda_n \frac{\eta}{\varrho_*^2 \gamma^2 m^3 \delta^{4n+3}} \|v\|,$$

from which it is easy to obtain conditions (ii') with $\mathbf{D}'$ in place of $\mathbf{D}$ and $m' = m - n\Lambda_n \frac{\eta}{\varrho_*^2 \gamma^2 m^3 \delta^{4n+3}}$ in place of m.

We will also need the inequality

$$\|f \circ Q - f\|_{\varrho'} < \frac{1}{2}\sqrt{\varrho_* \eta'}\|f\|_{\varrho}. \tag{3.2.14}$$

From the second of (3.2.5), with $\tilde{\varrho}$ in place of ϱ, one has

$$\|\exp \mathcal{L}_\chi f - f\|_{\tilde{\varrho}-\delta} \leq 4n\frac{\|\partial\chi\|_{\tilde{\varrho}}}{\delta}\|f\|_{\tilde{\varrho}},$$

from which (3.2.14) follows using (3.2.12) and

$$\|f\|_{\tilde{\varrho}} \leq \|f\|_{\varrho}, \qquad \frac{2n}{4n+1} < \frac{1}{2}.$$

Summing up, we can state the following

LEMMA 3.6 (ITERATIVE) *For given positive numbers $\gamma, \varrho, m, \eta < 1$ and $0 < \varrho_* < \varrho$, $0 < m_* < m$, consider the Hamiltonian $H(I, \varphi)$ as in the Kolmogorov theorem and satisfying (i), (ii), (iii). For any positive δ, so small that $\varrho - 3\delta > \varrho_*$, let us assume that η is so small that*

$$m - n\Lambda_n \frac{\eta}{\varrho_*^2 \gamma^2 m^3 \delta^{4n+3}} > m_*$$

where

$$\Lambda_n = 2(4n + 1)^2 \sigma_n^2,$$

with σ_n defined in Proposition 3.4. Then one can find an analytical canonical transformation $Q : \mathcal{D}_{\varrho-3\delta} \to \mathcal{D}_\varrho$, $Q \in \mathcal{A}_{\varrho-3\delta}$, such that the transformed Hamiltonian $H' = H \circ Q$ can be decomposed in a way analogous to H with corresponding primed quantities $a', A', B', \mathbf{D}'$, and R', but with the same λ, and satisfies analogous conditions with positive parameters $\varrho', m', \eta' < 1$ given by

$$\varrho' = \varrho - 3\delta > \varrho_*, \tag{3.2.15}$$

$$m' = m - n\Lambda_n \frac{\eta}{\varrho_*^2 y^2 m^3 \delta^{4n+3}} > m_*, \tag{3.2.16}$$

$$\eta' = \eta^2 \frac{\Lambda_n^2}{\varrho_* y^4 m^6 \delta^{8n+6}}, \tag{3.2.17}$$

with $\|H'\|_{\varrho'} \le 1$. Moreover, for any $f \in \mathcal{A}_\varrho$, (3.2.14) holds.

Having proved the iterative lemma, we are in possession of all the tools needed for the proof of the Kolmogorov theorem, and we can complete the latter of the two basic steps mentioned on page 92. The crucial point is the quadratic dependence on the old perturbing parameter η in the relation (3.2.17), and this quadratic dependence clearly implements Newton's method; the factor multiplying the quadratic term in the right-hand side member takes into account the small denominators, as suggested by the presence of y.

Proof (Kolmogorov theorem). One has to repeatedly apply the iterative lemma in order to eliminate the perturbation $H^p(\varphi, I) = A(\varphi) + B(\varphi) \cdot I$, verifying that the parameters ϱ_r and m_r do not vanish, as the iterative order $r \to \infty$: this is the *consistency problem*. We must therefore find numerical sequences for δ_r, m_r, η_r which satisfy the relations (3.2.15)-(3.2.17) and ensure that $\eta_r \to 0$.

For this, let us start by choosing the sequence

$$\eta_r = \frac{\eta_0}{2^{2\tau r}}, \quad \tau = 4n + 3, \quad \eta_0 > 0. \tag{3.2.18}$$

Since $\eta_r \to 0$ for $r \to \infty$, and

$$\max\left(\|A\|_{\varrho_r}, \|B\|_{\varrho_r}\right) \le \eta_r,$$

then $H_r^p \to 0$. Let us fix arbitrarily the initial values $\varrho_0, m_0 > 0$ and the "bounding parameters" $0 < \varrho_* < \varrho_0$, $0 < m_* < m_0$, leaving η_0 for the moment unspecified, whose value will be found accordingly with the consistency problem. From (3.2.16), (3.2.17), and (3.2.15) we have

$$m_{r+1} = m_r - \frac{n}{\sqrt{\varrho_*^3}} \sqrt{\eta_{r+1}}, \tag{3.2.19}$$

$$\delta_r^{2\tau} = \frac{\Lambda_n^2}{\gamma^4 m_r^6 \varrho_*} \frac{\eta_r^2}{\eta_{r+1}}, \tag{3.2.20}$$

$$\varrho_{r+1} = \varrho_r - 3\delta_r. \tag{3.2.21}$$

The relation (3.2.19) gives m_r at the generic order r, (3.2.20) gives δ_r, and (3.2.21) gives ϱ_r, all as some functions of η_0. Clearly, the successions $\{m_r\}$ and $\{\varrho_r\}$ are decreasing: we want to show that for every ϱ_* and m_* it is possible to find η_0 in a way that satisfies the consistency problem, i.e.,

$$\varrho_r > \varrho_*, \quad m_r > m_* \quad \forall r.$$

From (3.2.19) we obtain

$$m_\infty = m_0 - \frac{n}{\sqrt{\varrho_*^3}} \sum_{r=0}^{\infty} \sqrt{\eta_{r+1}}.$$

Since $\sum_{r=0}^{\infty} \frac{1}{2^{\tau(r+1)}} = \frac{1}{2^\tau - 1}$, by imposing $m_\infty > m_*$ we obtain that η_0 must satisfy the inequality

$$\eta_0 < \frac{\varrho_*^3 (m_0 - m_*)^2}{n^2} (2^\tau - 1)^2. \tag{3.2.22}$$

Relation (3.2.21) gives $\varrho_\infty = \varrho_0 - \sum_{r=0}^{\infty} 3\delta_r$. Taking into account that, for the chosen sequence (3.2.18), the relation

$$\left(\frac{\eta_r^2}{\eta_{r+1}} \right)^{\frac{1}{2\tau}} = \frac{\eta_0^{\frac{1}{2\tau}}}{2^{r-1}}$$

holds, we obtain

$$\sum_{r=0}^{\infty} \left(\frac{\eta_r^2}{\eta_{r+1}} \right)^{\frac{1}{2\tau}} = 4\eta_0^{\frac{1}{2\tau}}.$$

Let us impose $\varrho_\infty > \varrho_*$. Since we have already guaranteed that $m_r > m_*$, taking (3.2.20) into account we obtain that η_0 must satisfy the inequality

$$\eta_0 < \frac{1}{\Lambda_n^2} \varrho_* \gamma^4 m_*^6 \left(\frac{\varrho_0 - \varrho_*}{12} \right)^{2\tau}. \tag{3.2.23}$$

Clearly one must satisfy the more restrictive of the two conditions (3.2.22) or (3.2.23).

Finally, we come to the ultimate task: the convergence of the sequence of canonical transformations. Starting from the initial perturbed Hamiltonian H defined in $\mathcal{D}_{\varrho_0}$ with $\|H\|_{\varrho_0} \leq 1$, characterized by positive parameters $\gamma, \varrho_0, m_0 < 1$, consider the quantity η_0, satisfying the more restrictive of the two preceding conditions, and assume

$$\max \left(\|A\|_{\varrho_0}, \|B\|_{\varrho_0} \right) < \eta_0.$$

Then one can recursively apply the iterative lemma, defining at every step $r \geq 1$ a canonical transformation $Q_r : \mathcal{D}_{\varrho_r} \to \mathcal{D}_{\varrho_{r-1}}$, with the corresponding operator $\exp \mathcal{L}_{\chi_r} : \mathcal{A}_{\varrho_{r-1}} \to \mathcal{A}_{\varrho_r}$. Furthermore, from (3.2.14), one has the estimate

$$\|\exp \mathcal{L}_{\chi_r} f - f\|_{\varrho_r} < \sqrt{\varrho_* \eta_r} \|f\|_{\varrho_{r-1}}. \tag{3.2.24}$$

We can now define the composite canonical transformation $\hat{Q}_r : \mathcal{D}_{\varrho_r} \to \mathcal{D}_{\varrho_0}$ by $\hat{Q}_r = Q_1 \circ \cdots \circ Q_r$ and the corresponding composite operator $\widehat{U}_r : \mathcal{A}_{\varrho_0} \to \mathcal{A}_{\varrho_r}$, defined by $\widehat{U}_r f = f \circ \hat{Q}_r$, or equivalently by $\widehat{U}_r = \exp \mathcal{L}_{\chi_r} \widehat{U}_{r-1}$, with $\widehat{U}_0 =$ identity. Clearly, in order to prove the convergence of the sequence $\{\hat{Q}_r\}$ of canonical transformations restricted to $\mathcal{D}_{\varrho_\infty}$, it is sufficient to prove the convergence of the corresponding sequence $\{\widehat{U}_r\}$ of operators for every $f \in \mathcal{A}_{\varrho_0}$. This in turn is seen by remarking that, from (3.2.24) and the first of (3.2.5), one has

$$\left\| \left(\widehat{U}_{r+1} - \widehat{U}_r \right) f \right\|_{\varrho_{r+1}} = \left\| \exp \mathcal{L}_{\chi_{r+1}} \left(\widehat{U}_r f \right) - \widehat{U}_r f \right\|_{\varrho_{r+1}}$$

$$\leq \sqrt{\varrho_* \eta_{r+1}} \left\| \widehat{U}_r f \right\|_{\varrho_r} \leq \sqrt{\varrho_* \eta_{r+1}} \|f\|_{\varrho_0}$$

and also, applying the triangular inequality, for any $j \geq 1$

$$\left\| \left(\widehat{U}_{r+j} - \widehat{U}_r \right) f \right\|_{\varrho_{r+j}}$$

$$= \left\| \left(\widehat{U}_{r+j} - \widehat{U}_{r+j-1} \right) f + \left(\widehat{U}_{r+j-1} - \widehat{U}_{r+j-2} \right) f + \ldots \right\|_{\varrho_{r+j}}$$

$$\leq \|f\|_{\varrho_0} \sum_{s=r}^{r+j-1} \sqrt{\varrho_* \eta_{s+1}}.$$

Thus, as the series

$$\sum_{r=1}^{\infty} \sqrt{\varrho_* \eta_r} = \sum_{r=1}^{\infty} \frac{\sqrt{\varrho_* \eta_0}}{2^{\tau r}} = \frac{\sqrt{\varrho_* \eta_0}}{2^{\tau} - 1}$$

converges, one deduces that the sequence $\widehat{U}_r f$ converges uniformly for any $f \in \mathcal{A}_{\varrho_0}$. Therefore, by the Weierstrass theorem one then has

$$\lim_{r \to \infty} \widehat{U}_r f = \widehat{U}_\infty f \in \mathcal{A}_{\varrho_\infty}.$$

In particular, for the Hamiltonian $H_\infty = \lim\limits_{r \to \infty} \widehat{\mathscr{A}}_r H$, one has $H_\infty = H_\infty^0 + H_\infty^p$, where, by construction, $H_\infty^p(I, \varphi) = A_\infty(\varphi) + B_\infty(\varphi) \cdot I = 0$. Finally, the mapping $\widehat{Q}_\infty = \lim\limits_{r \to \infty} \widehat{Q}_r$ turns out to be canonical again by virtue of the Weierstrass theorem, as a uniform limit of canonical mappings.   **QED**

Let us review and comment on the hypotheses of the Kolmogorov theorem, and the role that they play in the proof.

(i) The unperturbed frequencies $\lambda_1, \lambda_2, \ldots, \lambda_n$ must satisfy the Diophantine condition (3.2.7). This condition is necessary to ensure that the two homological Equations (3.2.10a), (3.2.10b) are solvable owing to Proposition 3.4, so that an iterative step can be performed. This condition means that the corresponding unperturbed torus does not support periodic orbits or even those too close to periodicity; it is physically intuitive that otherwise the effects of the perturbation are cumulative, and the torus will be destroyed sooner or later.

(ii) The unperturbed Hamiltonian must be nondegenerate. This condition is necessary to ensure the invertibility of the relation between actions and frequencies, so that at each iterative step one can tune the actions to keep the frequencies fixed.

(iii) The perturbation must be sufficiently small, and the threshold η_0 is given by the most restrictive of the two conditions (3.2.22), (3.2.23), usually the latter. In this case one sees that $\eta_0 = \mathcal{O}(\gamma^4)$, which is a sensible result: lowering γ weakens the Diophantine condition, so that more tori would be conserved, and this requires a smaller perturbation. The $2n$-dimensional Lebesgue measure of the set of tori, whose existence is guaranteed by the Kolmogorov theorem, is positive and the measure of its complement tends to zero as the size of the perturbation tends to zero. Thinking topologically, one would call the invariant tori exceptional, as the complement of an open dense set: see Appendix 3.A. However, since they have a large Lebesgue measure, they are more the rule than the exception, and a very slightly perturbed Hamiltonian system behaves practically as integrable.

One can now immediately deduce from the above theorem the following corollary, more suited for the applications and which is known as the Kolmogorov-Arnold-Moser (KAM) theorem.

COROLLARY 3.7 (KAM THEOREM) *Given the total perturbed analytic Hamiltonian*

$$H(I, \varphi) = H_0(I) + \varepsilon H_p(I, \varphi),$$

with a nondegenerate unperturbed Hamiltonian, i.e., $\det\left(\frac{\partial^2 H_0}{\partial I_h \partial I_j}\right) \neq 0$, *for every set* I_h^* *of the actions such that the unperturbed frequencies* $\omega_h(I^*) = \frac{\partial H_0}{\partial I_h}(I^*)$ *satisfy the Diophantine condition, the tori* $I_h^* =$ *constant survive, though slightly deformed, to sufficiently small perturbations.*

Proof. We may safely suppose $I^* = 0$, after a possible translation in the actions. From the mere definitions one has that

$$\lambda_i = \omega_i(0), \quad a = H_0(0) + \varepsilon \overline{H}_p(0),$$

$$A(\varphi) = \varepsilon[H_p(0,\varphi) - \overline{H}_p(0)], \quad B_h(\varphi) = \varepsilon \frac{\partial H_p}{\partial I_h}(0,\varphi),$$

$$(\mathbf{D}(0,\varphi))_{hi} = \frac{\partial^2 H_0}{\partial I_h \partial I_i}(0) + \varepsilon \frac{\partial^2 H_p}{\partial I_h \partial I_i}(0,\varphi).$$

From the expressions in the second line it is obvious that the condition on the smallness of the perturbative parameter ε implies the condition (iii) of the Kolmogorov theorem, hence that ε and η have basically the same meaning. From the expression in the third line and the nondegeneracy of H_0, condition (ii) follows. Therefore, for the Kolmogorov theorem there exists a canonical transformation sending the perturbed Hamiltonian to a form for which the conservation of the torus $I^* = 0$ is evident.   **QED**

3.2.5   KAM Theorem (According to Arnold)

Completing the program dictated by the first of the two basic steps mentioned on page 92 and pushing the perturbation to second order entails the disastrous consequence of a complete loss of differentiability, and thus the impossibility to undertake any further perturbative step. In order to escape the difficulty, Arnold (1963) proposed a strategy which is different from that of Kolmogorov. Its starting point is the introduction of an *ultraviolet cut-off* $K \in \mathbb{R}$ of an analytical function F, i.e., the splitting of F as the sum of an *infrared* and an *ultraviolet* part:

$$F = F^{\leq K} + F^{>K}$$

with

$$F^{\leq K} = \sum_{|k| \leq K} f_k e^{ik \cdot \varphi}, \quad F^{>K} = \sum_{|k| > K} f_k e^{ik \cdot \varphi}.$$

Taking into account the exponential decay of the Fourier components of an analytical function, it is easy to estimate the ultraviolet part.

PROPOSITION 3.8 *If F is analytical on $\mathcal{D}_\varrho$, then for $0 < \delta < \varrho$ there exists a constant $C(n,\delta)$ such that*

$$\left\|F^{>K}\right\|_{\varrho-\delta} \leq C(n,\delta)\,\|F\|_\varrho\, e^{-\frac{1}{2}\delta K}.$$

Proof. Because of the analyticity of F in $\mathcal{D}_\varrho$,

$$|f_k| \leq \|F\|_\varrho \, e^{-|k|\varrho}$$

holds, from which

$$\left\|F^{>K}\right\|_{\varrho-\delta} = \sup \left| \sum_{|k|>K} f_k e^{ik \cdot [\varphi + i(\varrho-\delta)]} \right|$$

$$\leq \sum_{|k|>K} |f_k| \, e^{|k|(\varrho-\delta)} \leq \|F\|_\varrho \sum_{|k|>K} e^{-|k|\delta}$$

$$\leq \|F\|_\varrho \, e^{-\frac{1}{2}\delta K} \left(\sum_{j \in \mathbb{Z}} e^{-\frac{1}{2}|j|\delta} \right)^n = \|F\|_\varrho \, e^{-\frac{1}{2}\delta K} \left(\frac{1 + e^{-\delta/2}}{1 - e^{-\delta/2}} \right)^n,$$

which is the result. **QED**

This result can be exploited to make the ultraviolet part small. Taking for example $K = \frac{2}{\delta} r \log \varepsilon^{-1}$ we get

$$\left\|F^{>K}\right\|_{\varrho-\delta} \leq C \|F\|_\varrho \, \varepsilon^r,$$

while, with $K = 1/\varepsilon$, the ultraviolet part becomes exponentially small:

$$\left\|F^{>K}\right\|_{\varrho-\delta} \leq C \|F\|_\varrho \, e^{-\frac{1}{2}\frac{\delta}{\varepsilon}}.$$

Arnold's idea is to work at every perturbative step considering only the infrared part, which has a finite number of Fourier components, and to stay far enough from the relative resonances, to avoid any problem of convergence. At the first step, in order to push the perturbation at order ε^2, one takes $K = \frac{2}{\delta} \log \varepsilon^{-1}$ and rules out the relative ultraviolet part, which has a term ε in front and gives a contribution of second order. Adopting the quadratic method and taking at the r step the ultraviolet cut-off $2^r K$, the iteration leads to a sequence of Hamiltonians closer and closer to integrable but in shrinking domains, which in the limit reduce to the points of a Cantor set (see Figure 3.2 on page 116). One can therefore obtain the KAM theorem for $r \to \infty$.

We remark that the Arnold strategy is more efficient compared with that of Kolmogorov, leading to a relation of the type $\varepsilon = \mathcal{O}(\gamma^2)$ instead of $\varepsilon = \mathcal{O}(\gamma^4)$. The difference is due to the double resort of Kolmogorov to the homological equation; see (3.2.10a) and (3.2.10b).

An equivalent but probably more expressive version of the theorem has been formulated by Chierchia & Gallavotti (1982) and Pöschel (1982): given a quasi-integrable Hamiltonian $H(I, \varphi)$, there exists a canonical transformation $C : I', \varphi' \to I, \varphi$, a Cantor subset $\mathcal{D}$ of the action space, and a completely integrable Hamiltonian $K(I)$ such that

$$H(C(I', \varphi')) \overset{\text{def}}{=} H'(I', \varphi') \overset{\mathcal{D}}{=} K(I');$$

the symbol $\overset{\mathcal{D}}{=}$ means that the equality holds only for $I' \in \mathcal{D}$, also extending to all derivatives. Moreover, the frequency map $\omega(I') = \partial K/\partial I'$ transforms $\mathcal{D}$ into the set of Diophantine frequencies, with $\gamma = \mathcal{O}(\sqrt{\varepsilon})$.

3.2.6 Isoenergetic KAM Theorem

Sometimes the nondegeneracy condition $\det(\partial^2 H_0/\partial I_h \partial I_k) \neq 0$ is violated, but

$$\det \begin{pmatrix} \frac{\partial^2 H_0}{\partial I_h \partial I_k} & \frac{\partial H_0}{\partial I_h} \\ \frac{\partial H_0}{\partial I_k} & 0 \end{pmatrix} \neq 0 \tag{3.2.25}$$

holds. As we shall prove, the condition guarantees that the $(2n - 1)$-dimensional manifold of constant energy is filled with conserved tori. The inequality (3.2.25) is called the *isoenergetic nondegeneracy condition* and is independent of the usual nondegeneracy condition. There exist Hamiltonians which satisfy only one of the two conditions; for example, $H_0 = I_1^2 + I_2$ is degenerate but not isoenergetic degenerate, while $H_0 = \log(I_1/I_2)$ is isoenergetic degenerate but not degenerate.

The condition (3.2.25) for isoenergetic nondegeneracy means that on a level surface of the energy the ratios of, e.g., the first $n - 1$ frequencies with the last one are functionally independent. In coordinate-free language, we say that the map

$$M_h \to (\omega_1 : \omega_2 : \cdots : \omega_n) \in P_{\mathbb{R}}^{n-1}, \quad \text{with } M_h : H_0(I) = h, \tag{3.2.26}$$

between a level surface of the energy on the action space and the $(n - 1)$-dimensional projective space of the frequency ratio, is locally invertible.

The frequency ratios come into play in reducing a Hamiltonian system with the energy integral, which amounts to dividing out the kernel of the restriction to the surface of the energy of the canonical 2-form. Indeed, let us solve $H(q, p) = h$ with respect to, e.g., p_n:

$$p_n + \mathcal{P}(Q, P, \tau) = 0, \quad Q = q^1, \ldots, q^{n-1}, \ P = p_1, \ldots, p_{n-1}, \ \tau = q^n.$$

The dynamics takes place on a $(2n - 1)$-dimensional hypersurface. Let i be the inclusion map. The pull-backs of the canonical 1-form and 2-form of the phase space to this hypersurface are

$$i^*\Theta = P dQ - \mathcal{P}(Q, P, \tau) d\tau,$$
$$i^*\Omega = dP \wedge dQ - d\mathcal{P}(Q, P, \tau) \wedge d\tau.$$

The matrix of the 2-form $i^*\Omega$ is

$$\begin{pmatrix} \mathbf{0}_{n-1} & -\mathbf{1}_{n-1} & -\partial_Q P \\ \mathbf{1}_{n-1} & \mathbf{0}_{n-1} & -\partial_P P \\ \partial_Q P & \partial_P P & 0 \end{pmatrix}$$

and has rank $2n-2$, and thus has a 1-dimensional kernel. The lines tangent to the kernel are called the *characteristic lines* of the 1-form, and, as is easy to check, satisfy the equations

$$\frac{dQ}{d\tau} = \frac{\partial P}{\partial P}, \quad \frac{dP}{d\tau} = -\frac{\partial P}{\partial Q}.$$

One of the angles, i.e., $q^n = \tau$, is thus taken as the new "time," at least locally, and the $n-1$ frequencies of the reduced system become just the ratios of the map (3.2.26). The isoenegetic nondegeneracy condition arises by imposing the usual nondegeneracy condition on the reduced system.

In order to prove condition (3.2.25), let us consider the composition of the two maps

$$\begin{pmatrix} I_1 \\ \vdots \\ I_n \\ \lambda \end{pmatrix} \overset{(i)}{\mapsto} \begin{pmatrix} x_1(I) \\ \vdots \\ x_{n-1}(I) \\ h(I) \\ \lambda \end{pmatrix} \overset{(ii)}{\mapsto} \begin{pmatrix} \lambda w_1(x,h) \\ \vdots \\ \lambda w_n(x,h) \\ H_0(h) \end{pmatrix},$$

where $x_1,\ldots,x_{n-1}$ are local coordinates on M_h and $\lambda \neq 0$. Since $H_0(h)$ is the identity, for every fixed value of h the map (ii) is equivalent to (3.2.26). Taking into account that (i) is invertible by construction, we require that the Jacobian matrix of the map composition $(I,\lambda) \mapsto (\lambda w(I), H_0(I))$ have maximal rank. A direct calculation shows that its determinant differs from the determinant in the isoenergetic nondegeneracy condition for a multiplicative factor λ^{n-1}, hence condition (3.2.25) follows.

For the two-degrees-of-freedom system the isoenergetic condition ensures the perpetual stability for *all* the orbits, not only for those satisfying the Diophantine condition. Indeed, the 2-dimensional tori form an insurmountable barrier which forbids the dynamics to invade the whole 3-dimensional manifold of constant energy.

3.3 Exponentially Long Stability and Nekhoroshev Theorem

The KAM theorem is a mathematical result of paramount importance, clarifying the equivocal and contradictory statements claimed by astronomers and statistical physicists. Nevertheless, it is almost useless in practical applications. The set of the stable tori is bizarre and counterintuitive, though of large measure; as a consequence, it is impossible to say if some initial conditions, never exactly known, give rise to a stable motion. Moreover, KAM theory says nothing about what happens to the destroyed tori: this is sometimes erroneously interpreted as dictating that *all* the destroyed tori are replaced by chaotic motion.

Nekhoroshev observed that this unpleasant situation is basically due to the fact that we are seeking results which are valid for an infinite lapse of time. On the contrary, if we are content with finite times, but possibly exceeding the lifetime of the universe, we can recover statements uniformly valid on the whole action space. To this end, we must also study what happens inside the resonances, introducing a *resonant normal form*. The result is that if the unperturbed Hamiltonian satisfies a *steepness* (or, more simply, a *quasi-convexity*) condition, the actions oscillate but require a time exponentially long in the inverse of the perturbative parameter to evolve away. The original proof is given in Nekhoroshev (1977) and Nekhoroshev (1979).

The proof consists of two steps, an *analytical* and a *geometrical* part. With the first step one brings the Hamiltonian to a normal form in a local domain and up to some order r, in such a way that the non-normalized remainder is exponentially small, as suggested by Proposition 3.8; from this, a local stability result follows. The exponential dependence of time on the inverse of the perturbative parameter follows from a suitable choice of the order of normalization. The second step consists in constructing a covering of the action space with local domains, or *resonant blocks*, characterized by a given set of resonances of some order, where the Hamiltonian acquires a local normal form as above. Then one proves that the dynamical evolution is confined inside a block for an exponentially long time, from which the final result follows.

The basic "elementary" idea of the first step consists in taking the order r to be a suitable function of ε, growing to infinity when ε goes to zero. This is in fact the heart of the exponential estimates in the analytic part of the theorem. The essence of the proof is to show that the normalized term of order r grows "only" as some power of r^r; surprisingly, such an apparently terrible growth gives rise to the desired exponential estimates. Indeed, it is clear from (3.1.6) that at every r perturbative step and working far enough from all resonances up to order rK, the worst added small denominator will be

$$\min_{0<|k|\leq rK} (|k \cdot \omega|) = \frac{\gamma}{(rK)^n},$$

thanks to Diophantine inequality. The r perturbative term $R(r)$ will not exceed $\sim \varepsilon^r \left(\frac{r^n K^n}{\gamma}\right)^r$ with $\gamma \simeq \sqrt{\varepsilon}$. Take $K = \varepsilon^{-1/4nr}$ and seek the r_{opt} value minimizing $R(r)$. A straightforward computation gives

$$r_{\text{opt}} = \frac{1}{e}\varepsilon^{-\frac{1}{4n}}, \quad R(r_{\text{opt}}) = e^{-\frac{n}{e}\frac{1}{\varepsilon^{1/4n}}},$$

showing that R can be made exponentially small in the perturbative parameter ε. Incidentally, we remark that the exponent $1/4n$ is not the optimal one, because of the heuristic and non-rigorous procedure we followed: the probable optimal value has been found by Lochak (1992).

We have thus reached a first result: in a sufficiently nonresonant domain, $\mathcal{W}$ say, of the action space the motion is stable for an exponentially long time.

To understand how this domain is structured, let us first define the *Arnold web*: in the frequency space it is given by frequencies satisfying the resonance relations $\omega \cdot k = 0$, along with a neighborhood decreasing with the square root of the perturbation and exponentially with the order $\sum_{j=1}^{n} |k_j|$ of the resonance itself. The Arnold web is therefore the union of the neighborhoods of all the hyperplanes of codimension one through the origin and with rational slope. Assume for simplicity $n = 3$. In Figure 3.2 a section with the plane $\omega_3 = 1$ in the 3-dimensional frequency space is shown, thus with equation $k_1\omega_1 + k_2\omega_2 + k_3 = 0$. The "skeleton" is formed by the lines whose slope and intersection with the axes take rational values, "fleshed out" with the resonance strips. The Arnold web is *connected, open and dense* in the action space with, however, a relatively small measure vanishing with the square root of the perturbative parameter; see also Appendix 3.A, where one shows that the Diophantine condition (3.2.7) is chosen just in order to have a small measure of the Arnold web. On a 2-dimensional energy surface of the action space an image of Figure 3.2 appears, distorted under the diffeomorphism given by the local inverse of the frequency map. The smallness property can also be verified directly, taking into account that the width of a resonance strip is proportional to $\sqrt{\varepsilon}\exp(-\sigma |k|)$, as will be made clear below with the pendulum model: arguing as in the proof of Proposition 3.3, one sees that the sum over all the resonances converges to a finite value $\simeq \mathcal{O}(\sqrt{\varepsilon})$.

The sufficiently nonresonant domain $\mathcal{W}$ of the action space is therefore the complement of a "truncated" Arnold web, for which only the resonances up to the order determined by r_{opt} are considered. The domain $\mathcal{W}$ contains infinitely many, but at the same time very tiny resonances, which prevents the perpetual stability but also ensures the stability for exponentially long times. Clearly, when the order increases, the domain $\mathcal{W}$ is more and more fragmented and tends to the Cantor set of the conserved KAM tori. Notice that what can be displayed with the numerical computations (as for example in Figure 8.11 on page 249) is just a truncated Arnold web, since enlightening the entire web would require an infinite integration time.

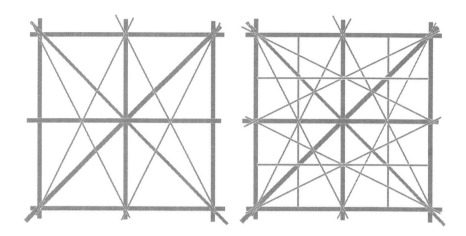

Figure 3.2: The Arnold web in frequency space, "truncated" at a low order corresponding to an ultraviolet cut-off K (left) and at a larger order (right) with a cut-off $K' > K$. When the order K increases, the complement of the Arnold web is more and more fragmented and tends to the Cantor set of the conserved KAM tori.

In order to describe the second step, let us first study what happens inside the resonances of multiplicity 1, thus characterized by a *single* resonance relation $k^{(\text{res})} \cdot \omega = 0$. Translate the actions so that the resonant torus is parametrized by $I = 0$, then perform a linear canonical transformation such that the first transformed angle is $\varphi_r = k^{(\text{res})} \cdot \varphi$, from which $\omega_r = 0$ follows; let $I_r, I'_2 \ldots, I'_n$ and $\varphi'_2, \ldots, \varphi'_n$ be the other transformed action-angle variables. Normalize the Hamiltonian to first order, eliminating all angles but the resonant φ_r, then develop the unperturbed Hamiltonian in a Taylor series and the perturbative part in a Fourier series. Up to a constant, we obtain a Hamiltonian of the type

$$H(I, \varphi) = \frac{1}{2} A I_r^2 - \varepsilon c \cos \varphi_r + \mathcal{O}(\varepsilon^2) + \ldots, \tag{3.3.1}$$

where the dots indicate the sum of harmonics of higher order and c is the first harmonic of the normalized perturbative part. Bearing in mind the definition of φ_r, we get that c coincides with the $k^{(\text{res})}$ harmonic of the perturbation Hamiltonian:

$$c = \frac{1}{(2\pi)^n} \int_{\mathbb{T}^n} H_p(I_r, \varphi_r, I', \varphi') e^{-i\varphi_r} d\varphi_r \, d\varphi'_2 \ldots d\varphi'_n$$

$$= \frac{1}{(2\pi)^n} \int_{\mathbb{T}^n} H_p(I, \varphi) e^{-ik^{(\text{res})} \cdot \varphi} d\varphi_1 \ldots d\varphi_n = H_{p\,k^{(\text{res})}}.$$

If H_p is analytic, the Fourier coefficients are exponentially small; hence c is exponentially small with $\left| k^{(\text{res})} \right|$. Clearly (3.3.1) describes the dynamics of

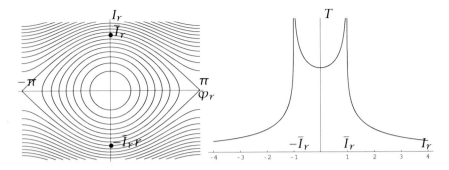

Figure 3.3: Energy level lines and period of a pendulum.

a pendulum with a perturbation of order ε^2, and the excluded harmonics do not qualitatively change the motion.

The left panel of Figure 3.3 shows the well-known dynamics of the pendulum. $I_r = 0$, $\varphi_r = 0$ is the stable equilibrium point, while $I_r = 0$, $\varphi_r = \pm\pi$ is unstable. The two curves connecting the unstable point to itself are named *separatrices*, because they separate the plane in three regions with different dynamical properties. Above and below the separatrices the angle φ_r circulates clockwise and counterclockwise, while inside the separatrices it librates. The frequency of libration is $\sqrt{A\varepsilon c}$ for motions closest to the stable equilibrium point, and thus small as the square root of the perturbation and exponentially small with the order $\left|k^{(\text{res})}\right|$ of the resonance. This frequency decreases to zero when approaching the separatrices and on a separatrix an infinite time occurs to travel from the unstable point to itself; see the right panel of Figure 3.3, where the period of the librating and circulating motion is drawn. The width of the resonant zone is $2\bar{I}_r = 4\sqrt{\varepsilon c/A}$, thus small as the square root of the perturbation and exponentially small with the order $\left|k^{(\text{res})}\right|$ of the resonance: this entails an increasing difficulty in visualizing resonances of high order.

In (3.3.1), invoking the elliptic functions we can replace I_r and φ_r with the action-angle variables of a pendulum, leaving $I'_2, \dots, I'_n$ and $\varphi'_2, \dots, \varphi'_n$ unchanged. These new action-angle variables are called *resonant*, and allow us to again normalize the Hamiltonian (3.3.1) to the optimal order, getting so a remainder exponentially small. Neglecting this remainder, we are left with an integrable Hamiltonian which, if the perturbation Hamiltonian satisfies the quasi-convexity condition, does not move away the actions, as we shall prove in a moment. Therefore, *a progressive drift of the actions can be caused only by the remainder, but requires an exponentially long time.*

Let us show how the quasi-convexity condition comes into play. Consider the generic case of a resonance of multiplicity $d \le n - 1$, thus characterized by d relations $k^{(s)} \cdot \omega = 0$, $s = 1, \dots, d$, with $\left|k^{(s)}\right| \le K$ (ultraviolet cut-off). Let $\mathcal{L}$ be a *resonant lattice*, that is, the subgroup of $\mathbb{Z}^n$ of dimension $d \leq n - 1$ generated by the vectors $k^{(s)}$. The associated *resonant manifold* $M_\mathcal{L}$, of dimension $n - d$, is defined as

$$M_\mathcal{L} = \{I : v \cdot \omega(I) = 0, \quad \forall v \in \mathcal{L}\}.$$

As already remarked, when K grows the associated resonant manifolds become thicker, while the complementary set is more and more fragmented. In a $\sqrt{\varepsilon}$-neighborhood of $M_\mathcal{L}$, but far from other resonances, the best thing one can do is to eliminate the nonresonant angles, giving to the Hamiltonian the resonant normal form

$$H(I, \varphi) = H_0(I) + \varepsilon g^{(r)} + \varepsilon^{r+1} R(I, \varphi), \quad g^{(r)}(I, \varphi) = \sum_{k \in \mathcal{L}} g_k^{(r)}(I) e^{ik \cdot \varphi}.$$

Let us neglect the remainder $R(I, \varphi)$. The normal part is no longer integrable for $d \geq 2$, but we can still investigate how it moves the actions. From the Hamilton equations it is clear that $\dot{I}$ is a linear combination of vectors $k \in \mathcal{L}$:

$$\dot{I} = \varepsilon \sum_{k \in \mathcal{L}} C_k(I, \varphi) k, \quad C_k(I, \varphi) = i g_k^{(r)}(I) e^{ik \cdot \varphi},$$

so that the dynamics produced by the normal form is flattened on the plane $\Pi_\mathcal{L}(I^*)$ parallel to $\mathcal{L}$ through the initial point I^*. This plane is called the *plane of fast drift,* and a transverse motion can be produced only by the remainder. Nevertheless, without further assumptions on the unperturbed Hamiltonian H_0, the motion on $\Pi_\mathcal{L}$ could be unbounded. In order to provide a confinement on the plane of fast drift, we suppose H_0 to be *quasi-convex.* More precisely, H_0 is said to be quasi-convex if

$$\omega(I) \cdot v = 0 \quad \text{and} \quad v \cdot H_0''(I) v = 0 \;\; \forall I \Rightarrow v = 0, \qquad (3.3.2)$$

where $H_0''(I) = \frac{\partial^2 H_0}{\partial I \partial I}$. Quasi-convexity is clearly a generalization of the convexity (concavity) property, i.e.,

$$v \cdot H_0''(I) v > 0 \; (< 0) \quad \forall v \neq 0.$$

A convex function is, for example, $H_0 = I_1^2 + I_2^2 + I_3^2$ while a quasi-convex but not convex function is $H_0 = I_1^2 + I_2^2 + I_3$. Quasi-convexity has two relevant consequences (see Figure 3.4).

(i) The d-dimensional plane $\Pi_\mathcal{L}$ of fast drift and the $(n - d)$-dimensional resonant manifold $M_\mathcal{L}$ are transversal in the intersection point I^*, that is, $\dim(T_{I^*} M_\mathcal{L} \oplus \Pi_\mathcal{L}) = n$. To prove it, take any vector $v \in \Pi_\mathcal{L}$ and define the scalar function $f(I) = \omega(I) \cdot v$ which clearly vanishes on $M_\mathcal{L}$; thus the gradient $\nabla f(I) = H_0''(I) v$ is orthogonal to $T_{I^*} M_\mathcal{L}$ for

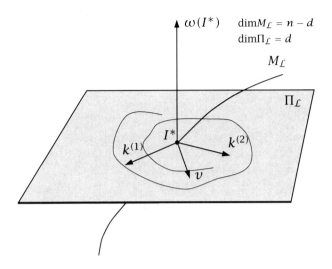

Figure 3.4: The plane $\Pi_{\mathcal{L}}$ of fast drift and the resonant manifold $M_{\mathcal{L}}$. If H_0 is (quasi-)convex, $\Pi_{\mathcal{L}}$ and $M_{\mathcal{L}}$ are transversal and the dynamics produced by the normal form is confined in a neighborhood of the intersection point I^*.

every $v \in \Pi_{\mathcal{L}}$. If there is no transversality, by definition a vector u which belongs at the same time to $T_{I^*}M_{\mathcal{L}}$ and $\Pi_{\mathcal{L}}$ does exist, that is,

$$\exists u \neq 0 : u \cdot H_0''(I^*)u = 0 \quad \text{and} \quad \omega(I^*) \cdot u = 0, \qquad (3.3.3)$$

and this is against the quasi-convexity. Therefore, H_0 quasi-convex implies transversality.

(ii) For a small vector $v \in \Pi_{\mathcal{L}}$, consider the series development

$$H_0(I^* + v) = H_0(I^*) + \omega(I^*) \cdot v + \frac{1}{2}v \cdot H_0''(I^*)v + \dots;$$

the linear term vanishes while the quadratic term has a definite sign for the quasi-convexity hypothesis Therefore, energy conservation provides the required confinement.

In the original proof of Nekhoroshev, instead of the quasi-convexity a weaker property called *steepness* is considered, which does not imply full transversality but only, in some sense, tangency of finite order. An example of a nonsteep Hamiltonian is $H_0 = \frac{1}{2}(I_1^2 - I_2^2)$, since the lines $I_1 \pm I_2 = 0$ are simultaneously the "plane" of fast drift and the resonant manifold.

We can lastly enunciate the following

THEOREM 3.9 (NEKHOROSHEV) *Given the total perturbed analytic Hamiltonian*

$$H(I, \varphi) = H_0(I) + \varepsilon H_p(I, \varphi),$$

with $H_0(I)$ *nondegenerate and quasi-convex (or, more generally, steep), if ε is sufficiently small one has*

$$\|I(t) - I(0)\| \le L\varepsilon^b \quad \text{for } |t| \le T \exp\left(\frac{1}{\varepsilon^a}\right)$$

uniformly for all *orbits.*

L and T are some dimensional constants which can be estimated, while a and b are stability exponents, whose optimal value in the convex case is very likely $a = b = \frac{1}{2n}$; see Lochak (1992). This value can be improved *inside* the resonances, where it becomes $a = b = \frac{1}{2(n-d)}$: loosely speaking, every resonance relation behaves like a first integral, and reduces of a unity the number of freedom degrees. We come to the surprising conclusion that resonances are not only a source of dynamical instability, through destruction of the KAM tori and their relative perpetual stability, but also a source of stability for a finite but exponentially long time, whose length grows with the multiplicity d of the resonance in which the orbit is "trapped."

3.4   Geography of the Phase Space

We draw some final summarizing comments on the geography of the phase space. The orbits of an n-dimensional perturbed Hamiltonian system are basically of two types: regular and chaotic.

Their main characteristics are the following.

(i) *Regular orbits.* They in turn may be classified as orbits on:

   a) *KAM tori*, i.e., those obtained by slightly deforming the unperturbed tori satisfying the Diophantine inequality, as dictated by the KAM theorem. They are also called *strongly nonresonant tori*. The dynamical evolution of the action-angle variables is quasi-periodic and the trajectories densely fill the respective tori; see Figure 3.5 (left).

   b) *Regular resonant tori.* The unperturbed tori not satisfying the Diophantine inequality break down under the perturbation, and the adapted (or *normal resonant*) Hamiltonian turns out to be that of a perturbed pendulum for resonances of single multiplicity. Again, from the KAM theorem one expects the existence of motions on regular resonant tori, which are obtained by deforming the tori of the unperturbed, thus integrable, pendulum. The relative orbits wind around a series of nested "small tubes" which, in turn, wind themselves around a KAM torus but without touching it: the chaotic orbits take place between KAM and regular

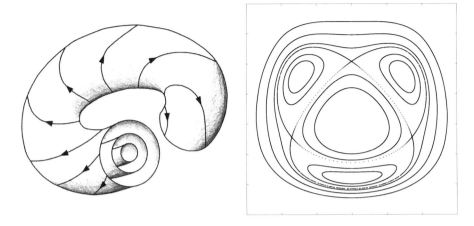

Figure 3.5: Left: nested tori for a two-degree-of-freedom system. Right: a section in the resonant case (any resemblance to extraterrestrial beings is entirely coincidental).

resonant tori, which are nested around the resonant trajectory of the unperturbed case. Moreover, the primary resonances can develop secondary resonances (a tube around a tube around a KAM torus) reproducing the same pattern, and so forth. To give an idea of how KAM and regular resonant tori are arranged for a two-degree-of-freedom system, see Figure 3.5 (right), where a section is considered. The three smaller islands represent two resonant tori, one nested into the other, and the three centers are the traces of the resonant orbit. Notice the separatrices between KAM and regular resonant tori; magnifying the picture, the chaos around the separatrices and in particular the hyperbolic points would be made visible. Further examples of the geography of KAM and regular resonant tori can be generated by the reader himself with the program KEPLER; see Chapter 7, item (i) on page 222. For three, or more, degrees of freedom, the tori cannot be visualized any longer, so we resort to a tool (the *Frequency Modulation Indicator*, described later) which allows us to display the resonance distribution; for an example, see Figure 3.6.

(ii) *Chaotic orbits*, whose dynamical evolution is very sensitive to the initial conditions. They may be further classified into three types.

a) *Stochastic layer orbits*, starting in the thin stochastic layer surrounding the separatrices between KAM and regular resonant tori. Loosely speaking, these orbits "hesitate" among clockwise or counterclockwise circulation and libration, giving rise to chaotic

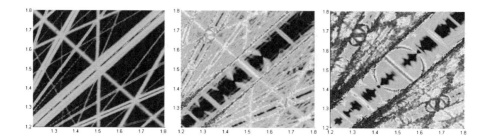

Figure 3.6: The progressive destruction of the Arnold web when the perturbative parameter increases. Left: $\varepsilon = 0.01$. Center: $\varepsilon = 0.06$. Right: $\varepsilon = 0.1$.

dynamics. A rigorous analysis of the chaotic motion requires us to introduce some concepts which revolve around homoclinic tangle, Smale horseshoe, and symbolic dynamics: see Appendix 3.B for a very short introduction. Notice that stochastic layer orbits can also be found *inside* the resonance zones, in the much thiner stochastic layer surrounding the secondary resonances. One can proceed further, considering the stochastic layer surrounding the resonances of third level, et cetera.

b) *Double (or multiple) resonant orbits*, starting at the crossing between two, or more, single resonances. Here the normal averaged Hamiltonian depends on two or more angles and therefore is not integrable. These orbits, though chaotic, are trapped for very long times in the zone of the crossing.

c) *Chirikov orbits*, starting in a zone where the resonances overlap. This happens when the perturbative parameter is larger than a threshold value; see Figure 1.1 or Figure 3.6, where the progressive destruction of the Arnold web is shown. The enlargement of the resonance strips entails their overlapping and consequently a widespread chaos diffusion.

For a generic perturbed Hamiltonian system, when the perturbative parameter grows, the nature of the phase space changes, covering, in ascending order, three different situations.

(i) *KAM*: practically all points are regular, all unperturbed tori are conserved, and the dynamics is basically controlled by the KAM theorem. More exactly, the measure of the destroyed tori is negligible, so in practice all orbits appear regular and the Nekhoroshev theorem, though applicable, is unimportant.

(ii) *Nekhoroshev*: the measure of the destroyed tori forming the Arnold web is small but not negligible. In the Nekhoroshev case the dynamics is still controlled almost everywhere by the KAM theorem apart from the Arnold web, where it is just controlled by the Nekhoroshev theorem. A point of a stochastic layer orbit (i.e., exactly on the border of a resonance) can in principle travel along the whole Arnold web, reaching the neighborhood of every point in phase space but in a very long time, which grows exponentially with the inverse of the perturbative parameter. This phenomenon, whose existence is not in general proven, is known as *Arnold diffusion* (Arnold 1964).

(iii) *Chirikov*: the global measure of the resonances does not leave any place for invariant tori, and the dynamics is no longer controlled by the KAM and Nekhoroshev theorems but by the Chirikov (1979) overlapping criterion. The resonances overlap, and the motion can jump from one resonance to another, giving rise to large-scale diffusion with a time scale much shorter with respect to the Arnold diffusion, so that the system is fully chaotic. We stress the difference: while the Arnold diffusion is *along* the resonances, the Chirikov diffusion is *across* the resonances.

3.5 Elliptic Equilibrium Points

Given an n-dimensional system admitting an elliptic equilibrium point, let us develop its Hamiltonian in a Taylor series about this point, which we safely suppose placed in the origin $q = 0 = p$. The first derivatives at the origin are vanishing by definition of equilibrium, while the quadratic part can be reduced to the Hamiltonian of n independent harmonic oscillators by means of a canonical transformation (Theorem 2.33 on page 76). The Hamiltonian is written as

$$H = \frac{1}{2} \sum_{k=1}^{n} \omega_k (q_k^2 + p_k^2) + H_3(q, p) + H_4(q, p) + \ldots, \quad \omega \in \mathbb{R}^n, \quad (3.5.1)$$

where $H_s(q, p)$ is a homogeneous polynomial of degree s.

It is only natural to consider the completely integrable quadratic part as the unperturbed Hamiltonian H_0 and the sum of the other polynomials as the perturbation, the distance from the origin playing the role of perturbative parameter. However, passing to action-angle coordinates shows that $H_0 = \sum \omega_k I_k$ is degenerate, so that KAM and Nekhoroshev theorems cannot be applied directly. The idea to escape this difficulty consists in normalizing the first few terms of the perturbation, adding them to H_0: in general, this will "remove" the degeneration.

The normalization of the polynomial perturbation is accomplished by the following theorem, whose name originates from historical reasons but whose proof directly follows the general scheme of Subsection 3.1.2.

THEOREM 3.10 (BIRKHOFF) *Suppose that the frequencies ω in the Hamiltonian (3.5.1) do not satisfy any resonance relation of order less than or equal to K. Then there exists a canonical transformation in a neighborhood of the equilibrium point such that the Hamiltonian is reduced to a Birkhoff normal form, i.e., to a polynomial in the action variables of degree $[K/2]$, up to terms in q, p of degree $K + 1$.*

Proof. Switch to action-angle variables. Writing the trigonometric functions in exponential form:

$$q_k = \frac{1}{2}\sqrt{2I_k}(e^{i\varphi_k} + e^{-i\varphi_k}), \quad p_k = \frac{1}{2i}\sqrt{2I_k}(e^{i\varphi_k} - e^{-i\varphi_k}),$$

shows that the angle average $\overline{M}$ of the generic monomial

$$M = q_1^{h_1} p_1^{h_2} \ldots q_n^{h_{2n-1}} p_n^{h_{2n}}$$

is nonnull if and only if *all* the exponents $h_1, \ldots, h_{2n}$ are even; in this case

$$\overline{M} = cI_1^{(h_1+h_2)/2} \ldots I_n^{(h_{2n-1}+h_{2n})/2},$$

with c a numerical factor. As in the general case, the nonresonance condition is necessary in order to solve the homological equation at every perturbative step.   **QED**

For $K = 4$ the perturbed Hamiltonian can be written as

$$H = \underline{\omega}^t \underline{I} + \frac{1}{2}\underline{I}^t \mathbf{A}\underline{I} + H_5(q, p) + \ldots,$$

$\mathbf{A}$ being a symmetric $n \times n$ real matrix. If

$$\det \mathbf{A} \neq 0 \quad \text{and/or} \quad \det \begin{pmatrix} \mathbf{A} & \underline{\omega} \\ \underline{\omega}^t & 0 \end{pmatrix} \neq 0,$$

the system is nondegenerate and/or isoenergetically nondegenerate in a small neighborhood of the equilibrium point, and the KAM theorem can be applied.

Moreover, if the quasi-convexity or quasi-concavity property (3.3.2) holds (in the present case, if the matrix $\mathbf{A}$ has nonnegative or nonpositive eigenvalues), the Nekhoroshev theorem can also be applied but not directly. The technical difficulty comes from the singularity of the action-angle variables at the origin, which prevents the straightforward application of the Nekhoroshev theorem. Working with action-angle coordinates as in Lochak (1995), one has to exclude a small cusp-shaped region around each hyperplane $q_k^2 + p_k^2 = 0$. Hence, the "stability region" that one finds, although arriving arbitrarily close to the equilibrium point, does not include any neighborhood of it. One is thus forced to work in Cartesian q, p coordinates, as in Fassò, Guzzo & Benettin (1998), to completely prove the claim.

Notice that the problem had already been studied in the previous papers of Giorgilli (1988) and Giorgilli, Delshams, Fontich, Galgani & Simó (1989), but assuming the very strong (and difficult to check in practice) condition that the frequencies of the linear term in the actions satisfy a Diophantine condition; within this assumption, the proof can take as unperturbed Hamiltonian only the linear term.

3.A   Appendix: Results from Diophantine Theory

We will prove some results from Diophantine theory. A frequency vector $\lambda \in \mathbb{R}^n$ is said to be *resonant* if there exists a $k \in \mathbb{Z}^n$, $k \neq 0$ such that $\lambda \cdot k = 0$. A frequency vector is said to be *strongly nonresonant* if one can find a positive function ψ such that

$$|k \cdot \lambda| \geq \psi(|k|) \quad \forall k \in \mathbb{Z}^n, \ k \neq 0, \quad |k| = \sum_i |k_i|.$$

Given an open bounded subset $\mathcal{D} \subset \mathbb{R}^n$, the question is whether one can determine ψ in such a way that the subset of the strongly nonresonant frequencies in $\mathcal{D}$, namely the set

$$\Omega = \{\lambda \in \mathcal{D} : |k \cdot \lambda| \geq \psi(|k|)\} \quad \forall k \in \mathbb{Z}^n, \ k \neq 0,$$

has a large measure in $\mathcal{D}$.

A simple procedure to determine such a ψ is presented in Giorgilli (1989). Pick a nonzero $k \in \mathbb{Z}^n$, and consider the set

$$\widetilde{\Omega}_k = \{\lambda \in \mathcal{D} : |k \cdot \lambda| < \psi(|k|)\},$$

which is in fact the set of the λ's that are close to resonance with k. Consider the plane in $\mathbb{R}^n$ through the origin orthogonal to k, i.e., the plane which contains all the resonant frequency vectors, and the set of points whose distance from the plane is less than $\frac{\psi(|k|)}{\sqrt{k_1^2 + \ldots + k_n^2}}$: this set of points clearly contains the set $\widetilde{\Omega}_k$.

LEMMA 3.11 *Take $x \in \mathbb{R}^n$, $\|x\| = \sqrt{x_1^2 + x_2^2 + \ldots + x_n^2}$ the Euclidean norm of x and $|x| = \sum_j |x|$. Then*

$$|x| \leq \sqrt{n} \, \|x\|.$$

Proof. Given two positive numbers a and b, from $(a - b)^2 \geq 0$ it follows that $ab \leq (a^2 + b^2)/2$, hence

$$|x|^2 = \sum_{j,k} |x_j||x_k| \leq \frac{1}{2} \sum_{j,k} (x_j^2 + x_k^k) = n \, \|x\|^2,$$

from which the lemma follows.   **QED**

The measure $\mu\left(\tilde{\Omega}_k\right)$ is bounded by

$$\mu\left(\tilde{\Omega}_k\right) \le \sqrt{n}C\frac{\psi(|k|)}{|k|},$$

where C is a constant that depends only on the domain $\mathcal{D}$. Then the measure of the complement $\tilde{\Omega}$ of Ω in $\mathcal{D}$ cannot exceed

$$\mu\left(\tilde{\Omega}\right) = \mu\left(\bigcup_{k\ne 0}\tilde{\Omega}_k\right) \le \sum_{k\ne 0}\mu\left(\tilde{\Omega}_k\right) \le \sqrt{n}C\sum_{k\ne 0}\frac{\psi(|k|)}{|k|}.$$

Writing now

$$\sum_{k\ne 0}\frac{\psi(|k|)}{|k|} = \sum_{s>0}\sum_{|k|=s}\frac{\psi(s)}{s},$$

and using the fact that the number of vectors $k \in \mathbb{Z}^n$ that satisfies $|k| = s$ does not exceed $2^n s^{n-1}$, as one verifies with a recursive procedure, we finally get

$$\mu\left(\bigcup_{k\ne 0}\tilde{\Omega}_k\right) \le 2^n\sqrt{n}C\sum_{s>0}s^{n-2}\psi(s).$$

Then it is enough to choose $\psi(s) = \frac{\gamma}{s^\tau}$ with suitable constants $\gamma > 0$ and $\tau > n-1$ in order to get that the complement of Ω in $\mathcal{D}$ has a measure that is small with γ. Such a result, although obtained with rough estimates, is optimal with relation to the value of τ. Indeed, for $\tau < n-1$ the set Ω is empty, while for $\tau = n-1$ the set Ω is nonempty, but has zero measure: see for example Rüssmann (1975).

The set Ω is therefore the complement of the "strange" set $\tilde{\Omega}$ which is of a small measure, though open and dense.[3] Still more strange appears the set Ω itself, which, as a Cantor[4] set of nonnull measure, is characterized by the following properties:

(i) it is a closed set, as the complement of an open set;

(ii) it is *nowhere dense*, by property (i) and as the complement of a dense set; hence, it has empty interior, i.e., it does not contain open subsets. This prevents the complete integrability of the perturbed Hamiltonian;

---

[3] The set $\tilde{\Omega}$ is open as the union of open sets; it is dense since the set of the resonant frequencies is already dense in $\mathbb{R}^n$, just as the rational numbers are dense in $\mathbb{R}$.

[4] The prototype of a Cantor set is obtained by removing the open middle third $(1/3, 2/3)$ from the closed interval $[0, 1]$, then removing the two open middle thirds $(1/9, 2/9)$ and $(7/9, 8/9)$, and so on, iterating the process *ad infinitum*. This set however has null measure, in contrast with Ω. The key to obtaining Cantor sets with positive measure is to remove less and less as one proceeds.

(iii) none of its points is an isolated point; i.e., the neighborhood of any point of the set contains at least another point of the set. In practice, it is therefore impossible to tell if a given point, whose coordinates are never exactly known, belongs to Ω, but the small measure of its complement makes this probable for a randomly chosen point.

If $n \geq 3$ the complement $\tilde{\Omega}$ of Ω, i.e., the "bad set" where the tori are destroyed by the perturbation, is not only open and dense but also connected. It is usually called an *Arnold web* and is the image, under the map $\omega \mapsto I$, of a small neighborhood of measure $\mathcal{O}(\sqrt{\varepsilon})$ of the set of the hyperplanes $\omega \cdot k = 0$, passing through the origin of the frequency space and having rational slope.

3.B   Appendix: Homoclinic Tangle and Chaos

Let us examine what happens when we slightly perturb an integrable Hamiltonian system possessing an unstable, or hyperbolic, equilibrium point.

To be concrete, we consider the standard map S of Figure 1.1, where the hyperbolic point P is placed at 0 identified with 2π. In the 0 point two distinct lines intersect, each one characterized by being invariant under the action of the map but exhibiting two contrasting behaviors: while every point of the first line is pushed toward the equilibrium point, those of the latter are moved away. The two lines are called the *stable* and *unstable manifold*, respectively. The situation is reminiscent of the pendulum, where however the stable and unstable manifolds join smoothly together, forming a separatrix (see Figure 3.3). But this is just the characteristic of the integrable systems, while in general we expect that the two lines intersect in a point, different from the hyperbolic one, called *homoclinic*.[5] Notice that the two lines are forced to intersect from the fact that the map is symplectic, thus area preserving, which forbids spiral trajectories. Moreover, a line cannot intersect itself, since there would be a point without a unique inverse.

The fundamental feature of a homoclinic point is that if there exists a single such point on a stable and an unstable invariant manifold corresponding to a particular hyperbolic fixed point, then there exist an infinite number of homoclinic points on the same invariant manifolds. We prove the statement by induction. Our base case is the initial homoclinic point. Thus, assume that there exist n homoclinic points for these invariant manifolds. Let M_s be the stable manifold and M_u be the unstable manifold, and let H be the homoclinic point farthest from the fixed point along the unstable manifold. Since M_s and M_u are invariant manifolds, $S(\mathrm{H}) \in M_s, M_u$. Since $S(\mathrm{H}) \in M_s, M_u$, $S(\mathrm{H})$ is either a homoclinic point or the fixed point.

---

[5]When the two stable and unstable manifolds start from two different hyperbolic points, the intersection point is called *heteroclinic*.

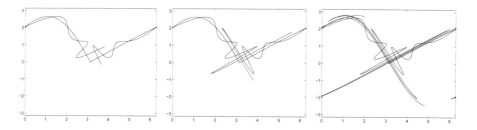

Figure 3.7: The homoclinic tangle, i.e., the web or trellis which Poincaré (1892–1893–1899, Vol. 3, page 389) described but did not attempt to draw.

Since S takes a point on the unstable manifold away from the fixed point, $S(H)$ cannot be the fixed point. Thus, it is a homoclinic point. For the same reason, it is not one of the n homoclinic points that we already have. Thus, there exist $n + 1$ homoclinic points. Therefore, the number of homoclinic points on the corresponding invariant manifolds is infinite.

The infinite number of homoclinic points entails that they crowd more and more densely in the neighborhood of the fixed point on the stable and unstable manifolds. Consequently, the lobes formed by the two orbits are more and more stretched out in order to preserve the areas, and inevitably they cross mutually, forming a very intricate pattern, called *homoclinic tangle* (see Figure 3.7), where the hyperbolic point is in π and the intersection point of the stable and unstable manifolds is in $0(2\pi)$.

The dynamics near the homoclinic tangle can be described as follows (see Figure 3.8). Pick some distance d and a homoclinc point H. Since H is on the stable manifold, it will map (along M_s) within d of the fixed point P after some number of iterations of S. Call this number n_s. Since H is on the unstable manifold, it will map (along M_u) within d of P after some number of iterations of S^{-1}. Call this number n_u. Consider a box D around the origin that extends a distance d (small) along each manifold. If we iterate S^{-1} on this box n_s times, P will remain in the box, but the box will stretch along the stable manifold to cover H. Thus, $H \in S^{-n_s}(D)$. If we iterate S on this box n_u times, P will remain in the box, but the box will stretch along the unstable manifold to cover H. Thus, $H \in S^{n_u}(D)$. Then consider the box $S^{-n_s}(D)$. If we let $S^{n_s+n_u}$ act on this box, we will end up with the box $S^{n_u}(D)$. Since the two boxes must cross over each other, we have just found a map whose behavior is reproduced qualitatively by the *Smale horseshoe map* $\mathcal{H}$.

This map takes a square Q, stretches it vertically by a factor > 2 and contracts it horizontally by a factor $< 1/2$. Then it bends the resulting rectangle in the shape of just a horseshoe, and superimposes it with the initial square, leaving out a central vertical strip. The inverse map $\mathcal{H}^{-1}$ instead stretches horizontally and contracts vertically, leaving out a central horizontal strip after the bend. As one easily verifies, $\mathcal{H}^{\infty}(Q)$ is exactly

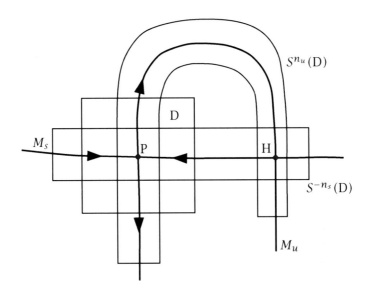

Figure 3.8: The horseshoe in the neighborhood of a hyperbolic fixed point.

the Cantor set of footnote 4 on page 126 formed by vertical lines, while $\mathcal{H}^{-\infty}(Q)$ is formed similarly by horizontal lines. The invariant set $\Lambda = \mathcal{H}^{\infty}(Q) \cap \mathcal{H}^{-\infty}(Q)$ is evidently the set of points in the initial square that never leave it under forward or backward iteration, so that we can consider the "dynamics" generated by the action of $\mathcal{H}$ on Λ.

To study this dynamics, it is convenient to introduce a third mathematical object, the *symbolic dynamics.* To every point of Λ we associate univocally a bi-infinite sequence of numbers

$$\{\ldots, a_{-3}, a_{-2}, a_{-1} | a_0, a_1, a_2, a_3, \ldots\} \text{ with } a_k = 0 \text{ or } 1,$$

a sort of address which instructs $\mathcal{H}$ on how to locate the point. It is also possible to define a distance between two sequences, comparing the finite subsequences centered around the vertical bar, thus generating a topology on the set of the sequences. It turns out that the action of $\mathcal{H}$ on Λ corresponds to shift of one place a sequence. In this way we reduce the study of the horseshoe map to that of shifts on symbol sequences. In other words, the horseshoe map, the map near a hyperbolic point, and the shift map are, in essence, only different ways of describing the same phenomenon, i.e., the chaos, and we have the possibility to choose the most suited method.

Using the symbolic dynamics, all in all it is simple to prove some basic features of the chaotical dynamics, which can just be considered as the "fingerprints" of the chaos:

(i) Sensitive dependence on initial condition (butterfly effect);

(ii) Existence of at least one dense orbit;

(iii) Density of the set of periodic orbits.

We stress that these few lines give only an extremely shallow idea of the argument. For example, we have not even touched on the dissipative systems, where the *strange attractors*, i.e., attractors of fractal type, arise. For a serious study see Smale (1967), Guckenheimer & Holmes (1983), Wiggins (1990), and Ott (1993).

CHAPTER 4

Numerical Tools I: ODE Integration

*The high speed computing machines make it possible
to enjoy the advantage of intricate methods.*
— P.C. HAMMER & J.W. HOLLINGSWORTH (1955)

In the following chapters we will study some concrete quasi-integrable Hamiltonian systems. Their analytical approximate normal form will be deduced and compared with the "true" motion, obtained from numerical integration. Moreover, the geography of the resonances will be detected thanks to the tools described in the next chapter, which again require a numerical integration.

In the present chapter we give the rudiments of the numerical integration of the Ordinary Differential Equations (ODE). The chapter must be viewed as a didactical service; indeed no generality is claimed, rather only the methods which come into play in the attached MATLAB programs are considered. In particular, the integration of stiff problems[1] will not be touched, while the geometric methods, usually ignored in basic courses, will receive some attention. For a more complete and serious study of the numerical integration of ODE, see Hairer, Norsett & Wanner (1993), Hairer & Wanner (1996), Hairer, Lubich & Wanner (2002), Butcher (2003), Iserles (1996), Quarteroni, Sacco & Saleri (2000), and Press, Teukolski, Wetterling & Flannery (1992).

---

[1] Stiffness occurs in a problem where there are two or more very different time scales on which the dependent variable is changing.

4.1 Cauchy Problem

In the present chapter we are concerned with the *initial value problem* or *Cauchy problem*:

$$\begin{cases} \dfrac{d}{dt}x(t) = f(t,x), \ \forall t \in \Delta \subset \mathbb{R}, \\ \qquad\quad x(t_0) = x_0 \in \mathbb{R}. \end{cases} \qquad (4.1.1)$$

Our problem has as input the function $f : \Delta \times \mathbb{R} \to \mathbb{R}$ and the initial value x_0, and must return as output a function $x : \Delta \to \mathbb{R}$ satisfying the initial condition. In the applications, x and f will be in general $\mathbb{R}^n$-valued, but that is irrelevant for the discussion in the present chapter.

The following theorem is fundamental.

THEOREM 4.1 *Let $f(t,x)$ be continuous in both the arguments and satisfy a Lipschitz condition in x. Then the solution of the Cauchy problem exists and is unique.*

We recall that a function $g(x)$ satisfies a Lipschitz condition in $[a,b]$ if a constant L exists such that

$$|g(x_1) - g(x_2)| \le L|x_1 - x_2|, \ \forall x \in [a,b].$$

Clearly, differentiability with bounded derivative is stronger than the Lipschitz condition and includes it.

With regard to numerical integration, we start by illustrating the Euler method, which is very simple and well suited for introducing basic ideas and definitions. Notice, however, that it is *never* used in practice because of the large error that is accumulated as the process goes on, but it is still important to study because the error analysis is easier to understand. Its poor performance is due to the fact that it is a *one-stage* and a *one-step* method.

In order to get better methods, two different strategies can be adopted. The Runge–Kutta and the extrapolation methods maintain the structure of the one-step method, thus the numerical result of every step depends only on the previous one, but they increase their accuracy at the price of an increase of functional evaluations at each time step. In contrast, the multi-step methods require only one functional evaluation at each step, and their accuracy can be increased at the expense of increasing the number of steps.

A further drawback of the Euler method is that it is neither symplectic nor conservative: the various geometric methods try to satisfy just one of the two properties.

4.2 Euler Method

The intuitive idea behind the method is very simple. In the Cauchy problem, consider $x(t)$ as the coordinate of a particle moving with velocity $f(t,x)$; then, the simple principle is that, in a short period of time, so short that there has not been time for the velocity to change significantly, the change in position will be approximately equal to the change in time multiplied by velocity. More formally, let us consider a succession of points, or *nodes*, $t_k = t_0 + kh$, $k = 0, 1, 2, \ldots$ where h is the *stepsize*. We want to approximate the values of x in the nodes t_k; therefore x_k will denote the approximate value of $x(t_k)$. To find these values, let us assume that we know the value at t_k and approximate the derivative with the incremental ratio. We get

$$t_{k+1} = t_k + h, \quad x_{k+1} = x_k + hf(t_k, x_k), \qquad (4.2.1)$$

which is the (*explicit* or *forward*) Euler method. All the fancy Runge–Kutta or multi-step methods that we shall discuss in the sequel are nothing but a generalization of this basic scheme.

Let us define the *local truncation error*

$$\tau(t, x(t); h) = \frac{x(t+h) - x(t)}{h} - f(t, x),$$

which is a measure of how much the exact solution does not satisfy the numerical scheme (4.2.1). In other words, $h\tau(t, x(t); h)$ is the residual arising when we pretend that the exact solution satisfies the numerical scheme. Then

$$\frac{d}{dt}x(t) = f(t, x) \Rightarrow \lim_{h \to 0} \tau(t, x(t); h) = 0,$$

which expresses the *consistency* of the numerical method (4.2.1) with respect to the Cauchy problem (4.1.1). In general, a numerical method is said to be consistent if its local truncation error is infinitesimal with respect to h. Moreover, a method has order p if, $\forall t \in \Delta$, the solution $x(t)$ of the Cauchy problem (4.1.1) fulfills the condition

$$\tau(h) = \mathcal{O}(h^p) \quad \text{for } h \to 0.$$

Using Taylor expansions, it is easy to prove that the forward Euler method has order one.

However, the request that a numerical method be consistent is insufficient, since the errors could accumulate catastrophically at every step. Rather, we require that a method be *convergent* of order p, i.e.,

$$|x(t_k) - x_k| = \mathcal{O}(h^p), \quad \forall k.$$

It turns out that a numerical method is convergent of order p if it is consistent of order p and, moreover, if it is *stable*.

In order to define the notion of stability, let us consider the following Cauchy model problem:

$$\begin{cases} \dfrac{d}{dt} x(t) = \lambda x(t), \ \lambda < 0, \\ \qquad x(0) = 1, \end{cases} \qquad (4.2.2)$$

whose solution is $x(t) = e^{\lambda t}$, thus $\lim\limits_{t \to \infty} x(t) = 0$. We say that a numerical scheme is *(absolutely) stable* if it gives an approximation such that

$$x_k \to 0 \ \text{ for } t_k \to \infty.$$

Notice that the request that a numerical method be stable arises, before anything else, from the need of keeping under control the (unavoidable) errors introduced by the finite arithmetic of the computer. Indeed, if the numerical method were not stable, the rounding errors would make the computed solution completely useless.

It easy to check the conditions for the stability of the forward Euler problem. The model problem (4.2.2) gives

$$\begin{cases} x_{k+1} - x_k = \lambda h x_k, \\ \qquad x_0 = 1, \end{cases}$$

whose solution is

$$x_k = (1 + \lambda h)^k x_0 = (1 + \lambda h)^k.$$

The stability condition is satisfied if and only if

$$|1 + \lambda h| < 1 \Leftrightarrow h < \frac{2}{|\lambda|},$$

therefore the method is said to be *conditionally* stable, i.e., stable only if the step is sufficiently small.

An important variant is the *implicit* or *backward* Euler method:

$$t_{k+1} = t_k + h, \ \ x_{k+1} = x_k + h f(t_{k+1}, x_{k+1}), \qquad (4.2.3)$$

where the slope of the curve is evaluated at the final point of the integration interval, instead of at the initial one as in the explicit methods. This requires an additional computational effort, because one must also solve an implicit algebraic equation in order to determine the x_{k+1} value, which may be justified when a particularly stable method is required.[2] Indeed, the stability analysis leads to the condition

$$\frac{1}{(1 - \lambda h)^k} < 1,$$

---

[2] For example, when the problem is stiff.

which is satisfied for every value of h: the method is *unconditionally* stable.

To get an intuitive idea of the different behaviors of the two methods, let us consider a system of two equations describing the planar motion of a mass around a stable point. Energy conservation forces the mass to move on a closed trajectory, for example, a circle. Applying the explicit Euler method, one obtains a trajectory which moves the mass away from the equilibrium point, while with the implicit method the mass "spirals" toward it. Thus the first method adds energy and destabilizes the system, while the latter subtracts energy and makes it more stable. Obviously, an intermediate behavior would be preferable: this and other considerations will lead us later to introduce the *symplectic* Euler method.

4.3 Runge-Kutta Methods

The Euler method (4.2.1) can be derived from the Taylor series expansion, stopping at the second term. By retaining more terms, we can generate higher order single-step methods. For example, retaining one additional term in the Taylor series

$$x(t + h) = x(t) + h\frac{dx(t)}{dt} + \frac{h^2}{2}\frac{d^2x(t)}{dt^2} + \frac{h^3}{6}\frac{d^3x(t)}{dt^3} + \cdots ,$$

gives the second order method

$$x_{k+1} = x_k + h\dot{x}_k + \frac{h^2}{2}\ddot{x}_k. \tag{4.3.1}$$

This approach requires the computation of higher derivatives of $x(t)$, which can be obtained by differentiating the first equation of (4.1.1) using the chain rule:

$$\dot{x}_k = f(t_k, x_k) \overset{\text{def}}{=} f_k,$$
$$\ddot{x}_k = f_t(t_k, x_k) + f_x(t_k, x_k)f(t_k, x_k) \overset{\text{def}}{=} f_{t,k} + f_{x,k}f_k, \tag{4.3.2}$$

where the subscripts indicate partial derivatives with respect to the given variable. As the order increases, the expressions for the derivatives rapidly become too complicated to be practical to compute; moreover, every single Cauchy problem would require an explicit and dedicated treatment, while one wants a code which works in the generic case. *Runge-Kutta methods* are single-step methods similar in motivation to Taylor series methods, but they do not require computation of higher derivatives. Instead, Runge-Kutta methods simulate the effect of higher derivatives by evaluating f several times between t_k and t_{k+1}.

We provide an example of this technique for a second order method. Then

$$x_{k+1} = x_k + h(b_1 K_1 + b_2 K_2),$$
$$K_1 = f_k, \quad K_2 = f(t_k + hc, x_k + hcK_1).$$

Expanding K_2 in a Taylor series and truncating at the second order, we get

$$K_2 = f_k + hc(f_{t,k} + K_1 f_{x,k}) + \mathcal{O}(h^2),$$

from which

$$x_{k+1} = x_k + hf_k(b_1 + b_2) + h^2 c b_2(f_{t,k} + f_{x,k} f_k) + \mathcal{O}(h^3).$$

Comparing with (4.3.1)–(4.3.2) and forcing the coefficients in the two expansions to agree up to higher order terms, we obtain that the coefficients must satisfy $b_1 + b_2 = 1$ and $cb_2 = 1/2$. Thus, there are infinitely many Runge-Kutta methods of second order.

In its most general form, a Runge-Kutta method can be written as

$$x_{k+1} = x_k + hF(t_k, x_k, h; f),$$

where F is the increment function defined as follows:

$$F(t_k, x_k, h; f) = \sum_{i=1}^{s} b_i K_i,$$

$$K_i = f(t_k + c_i h, x_k + h \sum_{j=1}^{s} a_{ij} K_j), \quad i = 1, 2, \ldots, s$$

and s denotes the number of *stages* of the method. The coefficients $\{a_{ij}\}$, $\{c_i\}$ and $\{b_i\}$ fully characterize a Runge-Kutta method and are usually collected in a *Butcher array* or *Butcher table*

$$
\begin{array}{c|cccc}
c_1 & a_{11} & a_{12} & \cdots & a_{1s} \\
c_2 & a_{21} & a_{22} & & a_{2s} \\
\vdots & \vdots & & \ddots & \vdots \\
c_s & a_{s1} & a_{s2} & \cdots & a_{ss} \\
\hline
& b_1 & b_2 & \cdots & b_s
\end{array}
\qquad \text{or} \qquad
\begin{array}{c|c}
\underline{c} & \mathbf{A} \\
\hline
& \underline{b}^t
\end{array}.
$$

For example, the best known is the following fourth order and fourth stage method:

$$
\begin{array}{c|cccc}
0 & 0 & 0 & 0 & 0 \\
1/2 & 1/2 & 0 & 0 & 0 \\
1/2 & 0 & 1/2 & 0 & 0 \\
1 & 0 & 0 & 1 & 0 \\
\hline
& 1/6 & 1/3 & 1/3 & 1/6
\end{array}
$$

which is explicit, since $a_{ij} = 0$ for every $j \geq i$.

In general, the method is consistent if and only if $\sum_{i=1}^{s} b_i = 1$. As Butcher proved, the order of an s-stage explicit Runge-Kutta method cannot be greater than s. For the first stages, the relation is given by

order	1	2	3	4	5	6	7	8
$s_{\min}$	1	2	3	4	6	7	9	11

Up to now, we have considered only methods with a *fixed* stepsize, chosen in advance before the integration process. In choosing the stepsize for a numerical solution of an ODE, we want to take large steps to reduce the computational cost but, at the same time, we must also take into account both stability and accuracy, which require small steps. It is desirable to have a method possessing a *stepsize adaptivity*, i.e., a method which is able to modify the stepsize, taking it smaller when the integration error grows and vice versa. It is thus necessary to have an efficient *a posteriori* estimator of the local error available, since the *a priori* local error estimates are too difficult to be used in practice.

At the present time, the best method not requiring extra functional evaluations consists of using simultaneously two different Runge-Kutta methods with s stages, of order p and $q > p$ respectively, which share the same values of the **A** and $\underline{c}$ Butcher parameters. These methods are synthetically represented by the modified table

$$
\begin{array}{c|c}
\underline{c} & \mathbf{A} \\
\hline
& b_p{}^t \\
& b_q{}^t
\end{array}
$$

where the method of order p is identified by the coefficients $\underline{c}$, **A**, and b_p, while that of order q is identified by $\underline{c}$, **A**, and b_q. Taking the difference between the approximate solutions produced by the two methods provides an estimate of the local error for the scheme of lower order. On the other hand, since the coefficients K_i coincide, this difference is given by $h \sum_{i=1}^{s} (b_{i,q} - b_{i,p}) K_i$ and no extra functional evaluation is required. The error is then compared with the tolerance value fixed by the user: the computation relative to the higher order is consequently retained, lengthening in case the stepsize, or is rejected and computed again with a smaller stepsize. Several pairs have been found, for example, that of Dormand-Prince: see the already quoted books for their explicit description.

4.4 Gragg–Bulirsch–Stoer Method

This method exploits the powerful idea of Richardson *extrapolation*. Quoting from Press *et al.* (1992): "The idea is to consider the final answer of a numerical calculation as itself being an analytic function (if a complicated one) of an adjustable parameter like the stepsize h. That analytic function can be probed by performing the calculation with various values of h, none of them being necessarily small enough to yield the accuracy that we desire. When we know enough about the function, we fit it to some analytic form, and then evaluate it at that mythical and golden point $h = 0$. Richardson extrapolation is a method for turning straw into gold!" In practice, a large interval H is spanned by different sequences of finer and finer substeps,

$$h = \frac{H}{n}, \text{ with } n = 2, 4, 6, 8, 12, 16, 24, 32, 48, 64, 96, \ldots [n_j = 2n_{j-2}]$$

$$\text{or } n = 2, 4, 6, 8, 10, \ldots [n_j = 2j],$$

the latter probably being more efficient. The result is extrapolated to an answer that is supposed to correspond to infinitely fine substeps.

The integrations are done by the *modified midpoint method*:

$$x_1 = x_0 + h f_0,$$

$$\vdots$$

$$x_k = x_{k-2} + 2h f_{k-1}, \ k = 2, 3, \ldots, n.$$

The final result is obtained by averaging x_n and the estimate available from Euler method, to obtain

$$x(t_0 + H) = \frac{1}{2}[x_n + (x_{n-1} + h f_n)].$$

The modified midpoint method is a second order method, but with the advantage of requiring (asymptotically for large n) only one evaluation per step h instead of the two required by second order Runge–Kutta. But the principal usefulness of the modified midpoint method derives from the fact that the error, expressed as a power series in h, contains only even powers of the stepsize, as discovered by Gragg, so we can gain two orders at a time.

With regard to the choice of fitting function, Bulirsch and Stoer first recognized the strength of rational functions. Nevertheless, more recent experience suggests that for smooth problems straightforward polynomial extrapolation is slightly more efficient than rational function extrapolation.

4.5 Adams–Bashforth–Moulton Methods

Let us now consider multi-step schemes. Unlike to one-step schemes, here the previously computed points come into play. In an s-step scheme approximation, x_{k+1} is computed from the data

$$(t_k, x_k), \ (t_{k-1}, x_{k-1}), \ldots, (t_{k-s+1}, x_{k-s+1}). \tag{4.5.1}$$

The data are given by previous steps or by a start-up computation.

To derive multi-step schemes, apply the fundamental theorem of integral calculus to the Cauchy problem (4.1.1). Integrating between x_k and x_{k+1} we get

$$x_{k+1} = x_k + \int_{x_k}^{x_{k+1}} f(t, x(t))\, dt.$$

The integral cannot be calculated explicitly, since it depends on the unknown function $x(t)$ itself. However, we know the s values $f_k, f_{k-1}, \dots, f_{k-s+1}$ of the integrand at the points (4.5.1), so it may be replaced by an interpolation polynomial $P(t)$. One distinguishes two cases.

(i) *Adams-Bashforth explicit method.* The polynomial $P(t)$ interpolates

$$f_{k-s+j}, \; j = 1, \dots, s.$$

As P does not depend on x_{k+1}, the method is explicit. Using Lagrangian interpolation, we have for P the representation

$$P(t) = \sum_{j=1}^{s} f_{k-s+j}\, l_{kj}(t), \text{ where } l_{kj}(t) = \prod_{\substack{v=1 \\ v \neq j}}^{s} \frac{t - t_{k-s+v}}{t_{k-s+j} - t_{k-s+v}}.$$

Inserting $P(t)$ into the integral yields

$$x_{k+1} = x_k + h \sum_{j=1}^{s} \beta_{kj} f_{k-s+j}, \text{ where } \beta_{kj} = \frac{1}{h} \int_{x_k}^{x_{k+1}} l_{kj}(t)\, dt.$$

For example, for the first few step numbers we get

$$s = 1: \; x_{k+1} = x_k + h f_k$$
$$s = 2: \; x_{k+1} = x_k + h \left(\frac{3}{2} f_k - \frac{1}{2} f_{k-1} \right) \qquad (4.5.2)$$
$$s = 3: \; x_{k+1} = x_k + h \left(\frac{23}{12} f_k - \frac{4}{3} f_{k-1} + \frac{5}{12} f_{k-2} \right).$$

(ii) *Adams-Moulton implicit method.* As before, $P(t)$ interpolates f_{k-s+j} but now $j = 1, \dots, s, s+1$. As $P(t)$ depends on x_{k+1}, the method is implicit. For example, for the first few step numbers we get

$$s = 1: \; x_{k+1} = x_k + h f_{k+1}$$
$$s = 2: \; x_{k+1} = x_k + h \left(\frac{1}{2} f_{k+1} + \frac{1}{2} f_k \right) \qquad (4.5.3)$$
$$s = 3: \; x_{k+1} = x_k + h \left(\frac{5}{12} f_{k+1} + \frac{2}{3} f_k - \frac{1}{12} f_{k-1} \right).$$

The two schemes can be combined into a *predictor-corrector* method, which first uses a prediction step to approximate the value of the variable, then uses this value to refine the guess: the x_{k+1} value obtained from the predictor method is inserted into the right member of the corrector method, which thus becomes fully explicit. There are actually three separate processes occurring in a predictor-corrector method: the predictor step, which we call P, the evaluation of the derivative f_{k+1} from the latest value of x, which we call E, and the corrector step, which we call C. For example, iterating m times with the corrector would be written $P(EC)^m$. However, the preferred strategy is PECE.

Notice that in (4.5.2) and (4.5.3) the sum of the numerical coefficients in the round brackets is always 1: indeed, a computation using the Taylor series shows that this is the necessary and sufficient condition to have a consistent method. It is possible to show that the s-step Adams-Bashforth method has order s, while the s-step Adams–Moulton method has order $s + 1$.

4.6 Geometric Methods

Thus far we have reviewed numerical methods well suited to integrate a *generic* ODE. Geometric numerical integration deals with numerical integrators that preserve, if any, geometric properties of the ODE flow, and it explains how structure preservation leads to an improved long-time behavior. In particular, we are interested in the symplectic and the time-reversible structures.

Let us suppose that the right-hand side of (4.1.1) is an $\mathbb{R}^2$-valued function (the generalization to $\mathbb{R}^{2n}$ is immediate) such that the Cauchy problem is a Hamiltonian system:

$$\begin{pmatrix} \dot{q} \\ \dot{p} \end{pmatrix} = \Omega \begin{pmatrix} \partial_q H(q,p) \\ \partial_p H(q,p) \end{pmatrix}, \text{ where } \Omega = \begin{pmatrix} 0 & 1 \\ -1 & 0 \end{pmatrix} \qquad (4.6.1)$$

is the (inverse of the) symplectic or canonical matrix. With $x_{k+1} = \Phi_h(x_k)$ we will denote a numerical method with stepsize h. If the numerical method

$$\Phi_h : \begin{pmatrix} q_k \\ p_k \end{pmatrix} \mapsto \begin{pmatrix} q_{k+1} \\ p_{k+1} \end{pmatrix}$$

leaves invariant the symplectic matrix

$$J^t \Omega J = \Omega, \text{ where } J = \partial(q_{k+1}, p_{k+1})/\partial(q_k, p_k),$$

it is said to be a *symplectic method.* It is intuitive that such a method preserves the intrinsic geometric features of the exact flow of (4.6.1): compare with the considerations at the end of the section.

Let us give some basic examples. The explicit Euler method (4.2.1) and the implicit one (4.2.3) are not symplectic. But let us consider the method

$$p_{k+1} = p_k - h\partial_q H(p_{k+1}, q_k),$$
$$q_{k+1} = q_k + h\partial_p H(p_{k+1}, q_k),$$

which treats the p variable by the implicit Euler method and the q variable by the explicit one. Similarly, we also consider

$$p_{k+1} = p_k - h\partial_q H(p_k, q_{k+1}),$$
$$q_{k+1} = q_k + h\partial_p H(p_k, q_{k+1}).$$

One easily proves that both methods are symplectic. To this end, consider for example the former method, the proof for the latter being quite similar. The result of differentiation and implicit differentiation of the left- and right-hand sides with respect to p_k, q_k can be expressed in matrix form as

$$\begin{pmatrix} 1 + h\partial_{qp}H & 0 \\ -h\partial_{pp}H & 1 \end{pmatrix} \begin{pmatrix} \partial(q_{k+1}, p_{k+1}) \\ \partial(q_k, p_k) \end{pmatrix} = \begin{pmatrix} 1 & -h\partial_{qq}H \\ 0 & 1 + h\partial_{qp}H \end{pmatrix},$$

where the partial derivatives are all evaluated at (p_{k+1}, q_k). Inversion of the first matrix in this relation allows us to compute the Jacobian matrix $\mathbf{J}$ and to check the symplecticity condition in a straightforward way.

Besides the symplecticity property, the method exhibits a good qualitative behavior of the total energy of a harmonic oscillator.[3] Notice that with the explicit and the implicit Euler method we obtain, respectively,

$$q_{k+1}^2 + p_{k+1}^2 = (1 + h^2)(q_k^2 + p_k^2),$$
$$q_{k+1}^2 + p_{k+1}^2 = \frac{1}{1 + h^2}(q_k^2 + p_k^2),$$

so that the point in phase space gains or loses energy indefinitely. With, e.g., the first symplectic Euler method we get instead

$$q_{k+1}^2 + p_{k+1}^2 = q_k^2 + p_k^2 + h^2(p_{k+1}^2 - q_k^2),$$

so that, even though not exactly conserved, the energy oscillates about the right value without drifting away.

Given a numerical method Φ_h, we define the *adjoint method*

$$\Phi_h^* \stackrel{\text{def}}{=} \Phi_{-h}^{-1},$$

which is clearly a numerical method of the same order. Whenever a method satisfies

$$\Phi_h^* = \Phi_h \text{ i.e, } \Phi_h^{-1} = \Phi_{-h},$$

---

[3] A numerical method cannot be symplectic and *at the same time* conservative.

it is called a *symmetric method*. It is sensible to think that a symmetric method is well suited for the numerical integration of an ODE whose exact flow is time-reversible.

Exchanging h with $-h$ and x_{k+1} with x_k shows that the adjoint of the explicit Euler method is the implicit Euler method and vice versa. Similarly, this is true for the two symplectic Euler methods. Obviously, none of these methods is symmetric.

Given a method Ψ_h of order one, it is immediate to check that the two compositions

$$\Psi_{h/2} \circ \Psi^*_{h/2} \quad \text{and} \quad \Psi^*_{h/2} \circ \Psi_{h/2} \tag{4.6.2}$$

are symmetric of order two. The symmetry follows from the two properties

$$(\Phi_h \circ \Psi_h)^* = \Psi^*_h \circ \Phi^*_h \quad \text{and} \quad (\Phi^*_h)^* = \Phi_h.$$

The order two is a consequence of the fact that symmetric methods always have an even order.

Take the first of the two symplectic Euler methods in the role of Ψ_h. For the generic Hamiltonian system, the compositions (4.6.2) yield

$$q_{k+1/2} = q_k + \frac{h}{2}\partial_p H(p_k, q_{k+1/2})$$

$$p_{k+1} = p_k - \frac{h}{2}\left(\partial_q H(p_k, q_{k+1/2}) + \partial_q H(p_{k+1}, q_{k+1/2})\right)$$

$$q_{k+1} = q_{k+1/2} + \frac{h}{2}\partial_p H(p_{k+1}, q_{k+1/2})$$

and

$$p_{k+1/2} = p_k - \frac{h}{2}\partial_q H(p_{k+1/2}, q_k)$$

$$q_{k+1} = q_k + \frac{h}{2}\left(\partial_p H(p_{k+1/2}, q_k) + \partial_p H(p_{k+1/2}, q_{k+1})\right)$$

$$p_{k+1} = p_{k+1/2} - \frac{h}{2}\partial_q H(p_{k+1/2}, q_{k+1})$$

respectively. These two schemes are called *Störmer-Verlet* methods. They are therefore symplectic and symmetric methods of order two. Moreover, one can prove that they exactly conserve quadratic first integrals, as the angular momentum.

The question is: How can one construct in general symplectic and/or symmetric methods, if possible of high order? A first answer is given by the *splitting method*, which consists in the splitting of a Hamiltonian in the sum

$$H = H^{[1]} + H^{[2]}$$

such that the *exact* flows $\varphi_t^{[1]}$ and $\varphi_t^{[2]}$, generated by $H^{[1]}$ and $H^{[2]}$, respectively, can be calculated. Then the numerical method

$$\Phi_h = \varphi_h^{[1]} \circ \varphi_h^{[2]}$$

is of order one, as follows from a Taylor expansion. Similarly, one can check that the method

$$\Phi_h = \varphi_{h/2}^{[1]} \circ \varphi_h^{[2]} \circ \varphi_{h/2}^{[1]}$$

is of second order; moreover, it is symmetric.

In practice, the delicate point is the choice of a suitable splitting, since in general it is an art to find it. However, in the particular but very important case of a natural Hamiltonian $H = T(p) + V(q)$, the sum of kinetic and potential energy, the splitting is obvious, and the exact flows generated by

$$H^{[1]}(p) = T(p) \text{ and } H^{[2]}(q) = V(q)$$

are given by

$$\varphi_t^{[1]} = \begin{pmatrix} p \\ q + t\partial_p T(p) \end{pmatrix}, \quad \varphi_t^{[2]} = \begin{pmatrix} p - t\partial_q V(q) \\ q \end{pmatrix}.$$

The $T(p)$ component is called the *shift* term, while $V(q)$ is the *kick* term. The resulting splitting methods of first and second order are then equivalent to the symplectic Euler method and to the Störmer–Verlet method, respectively.

The second order is in general too low for the long-time calculation of celestial mechanics, where a high accuracy is required. To increase the order, and hence the accuracy, while preserving symplecticity and symmetry, one considers more general compositions of a given basic one-step method $\Phi_h(x)$ with different stepsizes:

$$\Psi_h \stackrel{\text{def}}{=} \Phi_{y_s h} \circ \ldots \circ \Phi_{y_2 h} \circ \Phi_{y_1 h},$$

where the sequence of real parameters $y_1, y_2, \ldots, y_s$ must be conveniently determined to get the wanted order. If the basic method Φ_h is symplectic, the composition method Ψ_h is also symplectic. If the basic method is symmetric and the parameters satisfy the relation $y_i = y_{s+1-i}$, then the composition method is symmetric.

The splitting method requires to inspect directly the Hamiltonian and cannot be "blindly" applied to a generic problem. In this sense, the sole alternative is given by the *implicit Runge-Kutta-Gauss* (IRK–Gauss) method.

We know that a Runge-Kutta method is identified by two numerical vectors $\underline{b}$ and $\underline{c}$ of dimension s and a numerical $s \times s$ matrix $\mathbf{A}$. For the IRK–Gauss method these coefficients are chosen as follows:

(i) Choose $c_1, c_2, \ldots, c_s$ as the zeros of $\mathcal{P}_s(2x - 1)$, where $\mathcal{P}_k(x)$ is the Legendre polynomial of k degree:

$$\mathcal{P}_k(x) = \frac{1}{2^k k!} \left[\frac{d^k}{dx^k} (x^2 - 1)^k \right].$$

(ii) Choose $b_1, b_2, \ldots, b_s$ to satisfy the conditions

$$\sum_{i=1}^{s} b_i c_i^{j-1} = \frac{1}{j}, \; j = 1, 2, \ldots, s.$$

(iii) Choose $a_{ij}, \; i, j = 1, 2, \ldots s$ to satisfy the conditions

$$\sum_{j=1}^{s} a_{ij} c_j^{\ell-1} = \frac{1}{\ell} c_i^{\ell}, \; i, \ell = 1, 2, \ldots, s.$$

(iv) The first three conditions ensure that the method has order $2s$. If, moreover,

$$b_i a_{ij} + b_j a_{ji} - b_i b_j = 0, \; i, j = 1, 2, \ldots, s$$

holds true, the method is symplectic.

Lastly, we mention that precise statements on the long-time behavior of geometric integrators can be given thanks to the idea of *backward error analysis* (Hairer *et al.* 2002, Chapter IX). In this approach the numerical solution is the *exact* solution of a new Hamiltonian which is a perturbation of the original one in which the stepsize appears as the small parameter. Clearly, such an approach is closely related to the perturbation theory of Hamiltonian systems, so one can benefit from the existing experience in this field. For example, one proves that the numerical integration error for a method of order p is small on exponentially long time intervals. Thus

$$H(p_n, q_n) - H(p_0, q_0) = \mathcal{O}(h^p) \text{ for } nh \leq T_0 e^{b/h},$$

with T_0 and b some positive constants.

4.7   What Methods Are in the MATLAB Programs?

In the five programs POINCARE, HAMILTON, LAGRANGE, KEPLER, and LA-PLACE, some numerical integrator are invoked.

The first two methods ode113 and ode45 come originally with MATLAB, and the brief descriptions reported below are taken literally from the original Help. The following three methods are a MATLAB implementation of

FORTRAN programs originally due to Hairer and Wanner: see Hairer *et al.* (1993) and Hairer & Wanner (1996); the implementation is due to C. von Ludwig. Tom is due to F. Mazzia and R. Pavani: see Mazzia & Pavani (2007). RungeKutta4 and RungeKutta5 are the classical fixed-step methods and are added only for a didactical purpose.

ode113

Variable order Adams-Bashforth-Moulton solver. It may be more efficient than ode45 at stringent tolerances and when the ODE function is particularly expensive to evaluate (as happens very often with the perturbed Kepler problem). ode113 is a multi-step solver: it normally needs the solutions at several preceding time points to compute the current solution. For the KEPLER program, this solver is in general better than ode45 and is the best function to apply as a "first try" for most problems.

ode45

Based on an explicit Runge-Kutta (4,5) formula, the Dormand-Prince pair, which allows an adaptive step. It is a one-step solver: in computing the solution at a point, it only needs the solution at the immediately preceding time point.

IRK-Gauss

Implicit Runge-Kutta (Gauss) with fixed integration step. Symplectic method of order 4, 6, 8, 12, selected by the user.

Dop853

Explicit Runge-Kutta method of order 8(5,3), based on the method of Dormand-Prince, which allows an adaptive stepsize. For tolerance between 10^{-7} and 10^{-13}.

Odex

Extrapolation method, with variable order at any step and adaptive stepsize, based on Gragg-Bulirsch-Stoer algorithm.

Tom

Symmetric method of order 10 with fixed stepsize. In some situations it is the most precise solver.

RungeKutta4

The classical Runge-Kutta of the fourth order with fixed stepsize. For didactical purpose only.

RungeKutta5

The classical Runge-Kutta of the fifth order with fixed stepsize. For didactical purpose only.

The previous methods are all devoted to the integration of first order ODE, while three other methods for ODE of second order, due to Hairer, are supplied for the KEPLER program: see Hairer *et al.* (2002). Obviously, they can be utilized only with the standard integration method.

gni_irk2

Implicit Runge-Kutta (Gauss) of order 12 with fixed integration stepsize. Symplectic method.

gni_lmm2

Linear multi-step method of order 8, explicit and not symplectic, with fixed integration stepsize.

gni_comp

Composition of a given basic one-step method. Here the Störmer–Verlet method is used as basic integrator, but it may be changed by replacing the file `stverl.m` and invoking eventually a first order integration method. It is a symplectic method, with fixed integration stepsize.

The LAPLACE program utilizes the three numerical integrators which come with the HNBODY program of Rauch & Hamilton (2004). The symplectic method, which is the default and is strongly recommended, adopts a splitting of the type

$$H(p,q) = T(p) + V(q) + D(p,q).$$

The first two terms are the kinetic and potential energy of the unperturbed Kepler problems relative to every planet. The last term is called *drift* and takes into account the interaction between the planets: its choice is clearly the most critical, but the original documentation of the HNBODY program does not give further information on it. The other two numerical methods are of Bulirsch-Stoer and Runge-Kutta type.

CHAPTER 5

Numerical Tools II: Detecting Order, Chaos, and Resonances

Frequency modulation (FM) conveys information over a carrier wave by varying its frequency.

Detecting and studying how order and chaos are distributed in quasi-integrable Hamiltonian systems and, in particular, exploring the geography of the resonances is surely a major task in the dynamical systems area.

We briefly recall the principal results of Chapter 3. The dynamics of an n-dimensional quasi-integrable system is generated by a Hamiltonian of the type

$$H(I, \varphi) = H_0(I) + \varepsilon H_p(I, \varphi),$$

where $I = I_1, \ldots, I_n \in \mathbb{R}^n$, $\varphi = \varphi_1, \ldots, \varphi_n \in \mathbb{T}^n$ (n-dimensional torus) are the action-angle variables, and ε is a "small" perturbative parameter. The unperturbed motion is the product of n uniform circular motions in the n planes spanned by the couples of variables I_k, φ_k, $k = 1, \ldots, n$ with frequencies $\omega_k = \partial H_0 / \partial I_k$; if $\det(\partial^2 H_0 / \partial I_k \partial I_h) \neq 0$, this relation defines a locally invertible *frequency map* $I \to \omega(I)$ between action space and frequency space. For the perturbed motion a key role is played by the *resonant tori*, which are defined by a relation of the type $\omega \cdot k = 0$, $k \in \mathbb{Z}^n$. Indeed, from the KAM theorem one knows that only the tori sufficiently far from a resonance condition will survive, though slightly deformed, to a small perturbation. The set of destroyed tori forms the *Arnold web*, which in the frequency space consists of a small neighborhood of the union of all the hyperplanes through the origin and having a rational slope; in the action space the inverse of the frequency map will give a distorted image of this set. The Arnold web is *connected, open, and dense* with, however, a small measure vanishing with the square root of the perturbative parameter.

A well-known method to detect the resonance distribution is the Poincaré section. It is a time-honored tool in the study of dynamical systems and is effective in showing where order and chaos are located. In particular, for perturbed quasi-integrable Hamiltonian systems with two degrees of freedom, it is able to indicate where the neighborhood of a torus is simply deformed by the perturbation, or is destroyed and replaced by a train of resonance islands along with their surrounding chaos. Moreover, it can reveal resonances at various levels, i.e., resonances inside resonances inside resonances ..., et cetera. However, it suffers from two main limitations: it is useless for systems with three or more degrees of freedom, and it is ineffective in revealing the finest details, such as those very thin resonances which can be detected in principle but only when one knows exactly where to look.

In order to overcome these drawbacks several other methods are available, which can be divided into two groups. The first has its root in the *Lyapunov exponents* and is characterized by the fact that all the relative methods also require the numerical integration of the variational equation. The latter utilizes the *Fourier transform* and its further refinements in order to analyze the frequencies on KAM tori; it only requires the integration of the equations of motion.

5.1 Poincaré Section

In the 4-dimensional phase space of a Hamiltonian system with two degrees of freedom, let us fix the value of the energy and consider the 3-dimensional hypersurface given by $H(q, p)$ = const. If the system is quasi-integrable, the generic orbit winds around a 2-dimensional torus. Recording the intersection point of such an orbit with a 2-dimensional plane transverse to the flow, one will see the various intersection points arranged on a closed curve homeomorphic to a circle. Varying the value of the actions, different tori are selected, and the resulting intersection plane is locally foliated by a family of closed curves nested one into the other. If, instead, the initial conditions belong to a (primary or first level) resonance, the relative orbit winds around a "small tube" which, in turn, winds itself around a KAM torus but without touching it: the chaotic orbits take place between KAM and resonant tori. Moreover, the primary resonances can develop secondary resonances

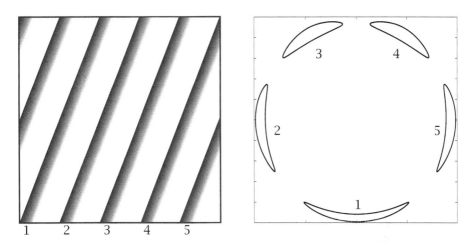

Figure 5.1: A resonant 5:2 regular torus.

(a tube around a tube around a KAM torus) reproducing the same pattern, and so forth.

In Figure 5.1 an example is displayed, regarding a 5:2 resonance. The square in the left picture, with the opposite sides identified, represents the KAM torus broken up by the resonance and replaced by the resonant one. A section along the horizontal or vertical side shows five or two islands, respectively. In the right picture the five islands of the horizontal section are shown. As results from the left picture, during the dynamical evolution they are visited in the order 1-3-5-2-4-1.

With the aid of the program KEPLER, the reader can investigate how the dynamical orbits are arranged in concrete physical cases; see items (i) on page 219 and on page 222. In case of resonance what happens is the following. The two initial frequencies ω_1, ω_2, relative to the two degrees of freedom, collapse into one alone, calculable with a unimodular matrix which transforms the latter new frequency into the vanishing one: $\omega_1, \omega_2 \rightarrow \omega_1', 0$. But a new frequency arises, ω_r say, relative to the traveling time necessary to cover the resonance isles, so that the system again possesses two frequencies. Clearly, the two new frequencies ω_1', ω_r can be in resonance again, more exactly in a resonance of second level, and the reasoning starts again.

5.2 Variational Equation Methods

Once a suitable tangent vector to the orbit has been chosen, all the methods of the present section will examine basically the dynamical evolution of this vector. We touch on the argument only briefly, because the methods will not be further considered or implemented.

5.2.1 The Largest Lyapunov Exponents

Let us consider the system of differential equations

$$\frac{dx}{dt} = f(x), \quad x = x_1, \ldots, x_m \tag{5.2.1}$$

along with the associated variational equations

$$\frac{dv}{dt} = \frac{\partial f}{\partial x} v, \quad v = v_1, \ldots, v_m. \tag{5.2.2}$$

Starting from the fact that chaotic orbits diverge exponentially, the *largest Lyapunov exponent* (LLE) is defined as the limit of $\log \|v(t)\| / t$ when t goes to infinity. Suppose Equation (5.2.1) is Hamiltonian:

$$x = (q_1, \ldots, q_n, p_1, \ldots, p_n),$$
$$f(x) = \left(\frac{\partial H(q, p)}{\partial p}, -\frac{\partial H(q, p)}{\partial q} \right);$$

if the motion is regular, then the LLE is zero, otherwise it is positive.

So far as we are concerned, this tool is affected by two serious limitations.

(i) It is able to discriminate between chaotic and ordered motion, but among regular motions it does not distinguish between KAM tori, i.e., those coming from a slight deformation of the unperturbed ones, and regular resonant tori, i.e., those replacing part of the tori destroyed inside the resonances, where the Hamiltonian can be approximated by a perturbed pendulum. The LLE is thus unable to detect the resonances.

(ii) By definition, the exact computation of the LLE requires an infinite time. In practice, by direct inspection one can reasonably judge after a finite time if the limit is null. This "human" decision is not easily implementable in a computer program, making the automatic analysis of the phase space difficult

5.2.2 The Fast Lyapunov Indicator

In order to reveal the chaos and at the same time to directly detect the resonances, the *fast Lyapunov indicator* (FLI) method has been proposed in Guzzo, Lega & Froeschlé (2002) and Froeschlé, Guzzo & Lega (2000): for every orbit starting from a point of a grid covering the action space, one computes the time occurring to the norm of a suitable tangent vector to reach a threshold value; the computed value is then reported in a graphic as a function of the actions.

The FLI method evidently overcomes point (ii) above. For the first point, the detailed analysis in Guzzo *et al.* (2002) shows that the norm of a tangent vector grows linearly with time for regular orbits, but with different slopes in the two cases of KAM tori and regular resonant motions; clearly, the time occurring to reach the fixed threshold will be different in the two cases. As an example, consider the unperturbed Hamiltonian $H_0 = 1/2 \sum_i I_i^2$, whose frequency map is the identity. In this case, the evolution of the norm of the tangent vector on a KAM torus is approximated by $\|v_I(0)\| t$, and for a regular resonant motion by $\|\Pi_{\Lambda^{\text{orth}}} v_I(0)\| t$. Here, Π_Λ is the Euclidean projection onto the linear space spanned by Λ, which is in general a d-dimensional integer lattice ($d \le n - 1$) defining a resonance through the relation $k \cdot \partial H_0 / \partial I = 0$, $k \in \Lambda$; Λ^{orth} is the linear space orthogonal to Λ. The choice of the initial value of the tangent vector is therefore critical, since if $v_I(0) \in \Lambda^{\text{orth}}$ the FLI is unable to discriminate between the two different motions. It is unclear how much this choice is relevant in the general case. Another critical point is the choice of threshold and integration time, as clearly pointed out and discussed in Cincotta, Giordano & Simó (2003).

5.2.3 Other Methods

The *helicity angles* were introduced by Contopoulos & Voglis (1996): one computes the mean value with respect to the time of the angles which define the orientation of the tangent vector to the orbits. In a chaotic region this value is invariant, while for regular orbits it smoothly changes with the initial conditions. Then the value is computed as usual on a grid of initial conditions. The method is effective, requiring in practice very short time interval of integration to reach a satisfying estimate. Also the *twist angles*, the time derivative of the helicity angles, may be considered: they are invariant for orbits in a chaotic zone, zero on KAM tori, and equal to the libration frequencies for orbits on resonant regular tori.

The *Mean Exponential Growth factor of Nearby Orbits* (MEGNO) method was introduced by Cincotta *et al.* (2003). The method makes use of an integral formula for the computation of the LLE. It is therefore conceptually equivalent but the numerical evaluation requires shorter times.

The *Smaller ALignment Index* (SALI) and the *Generalized ALignment Index* (GALI) methods were introduced by Skokos (2007). One follows the evolution in time of two different tangent vectors v_1 and v_2 and defines SALI as

$$\text{SALI}(t) = \min \left\{ \left\| \frac{v_1}{\|v_1\|} + \frac{v_2}{\|v_2\|} \right\|, \left\| \frac{v_1}{\|v_1\|} - \frac{v_2}{\|v_2\|} \right\| \right\}.$$

Clearly, when the two vectors become collinear, $SALI(t) \rightarrow 0$. Then one shows that $SALI(t) \rightarrow 0$ for chaotic orbits and to a non-null constant for regular orbits. GALI is a generalization whose definition involves $k \geq 2$ vectors and is equivalent to SALI for $k = 2$.

Other methods are *Average Power Law Exponent* (APLE) (see Lukes-Gerakopoulos, Voglis & Efthymiopoulos (2008)) and *Rotation numbers* (see Honjo & Kaneko (2008)).

Note that in Lukes-Gerakopoulos *et al.* (2008) all these methods are quoted as practically equivalent.

5.3   Fourier Transform Methods

While the methods of the previous section are generically applicable to every dynamical system, the Fourier transform methods are specific to the Hamiltonian quasi-integrable systems. This fact is obvious if we remember that the motion of a quasi-integrable system is multiperiodic for most initial conditions.

To make this idea precise, recall that when the motion winds around a KAM torus there exists a canonical transformation from the n original action-angle variables I, φ to new I^∞, φ^∞, such that the new actions are constant and the new angles vary linearly with frequencies ω. For every pair I_h, φ_h let us form the complex variable

$$Z^{(h)}(I^\infty, \varphi^\infty) \overset{\text{def}}{=} I_h(I^\infty, \varphi^\infty) e^{i\varphi_h(I^\infty, \varphi^\infty)}, \quad h = 1, \dots, n \text{ (not summed)},$$

which is defined on a n-dimensional torus parametrized by the angles φ^∞ and can be developed in a Fourier series with the actions I^∞ playing the role of parameters. Its time evolution is therefore

$$Z^{(h)}(t) = \sum_k C_k^{(h)} e^{ik \cdot \omega t}, \quad C_k^{(h)} \in \mathbb{C}, \quad k \in \mathbb{Z}^n - \{0\}, \quad \omega \in \mathbb{R}^n,$$

resembling the old description of the planetary motion through epicycles.

In general, given a function $f(t)$ of the time, its Fourier transform $\hat{f}$ is defined as[1]

$$\hat{f}(\Omega) = \lim_{T \to \infty} \frac{1}{2T} \int_{-T}^{T} f(t) e^{-i\Omega t} dt, \quad \Omega \in \mathbb{R}. \tag{5.3.1}$$

Suppose for a moment that we know the analytical solution for an orbit of the perturbed motion: the Fourier transform $\hat{Z}^{(h)}(\Omega)$ is zero for all values of Ω but $\Omega = k \cdot \omega$, so that $\hat{Z}^{(h)}(k \cdot \omega) = C_k^{(h)}$. In other words, the Fourier transform returns the spectrum, i.e., a list of frequencies along with their

---

[1] Our definition is slightly different from the usual one, where the integration field is the whole real axis and the normalizing factor $2T$ is missing.

amplitudes and phases. *If the motion is regular, these frequencies are all linear combinations, with integer coefficients, of n fundamental ones.* If the computed frequencies show random values instead, the initially selected orbit is chaotic.

In practice, we do not know any analytical solution but only its tabulated values coming from a numerical integration. The subsequent numerical tools of this section also try to recover the spectrum in this case, although with some limitations due to the reduced information in our possession: indeed, the solution is known only at discretized instants and for finite lapses of time.

5.3.1 Fast Fourier Transform (FFT)

Let $f(t)$ be the analytical solution of the equations of motion. With a numerical integration we are able to find its values only at the discrete instants $t_n, n = 0, 1, \ldots, N-1$. Put $f(t_n) = f_n$ and define the sequence of N complex numbers[2]

$$\hat{f}_k = \frac{1}{N} \sum_{n=0}^{N-1} f_n e^{-i\frac{2\pi}{N}nk}, \quad k = 0, 1, \ldots, N - 1. \tag{5.3.2}$$

Clearly, (5.3.2) is the numerical discretization of definition (5.3.1): the total number of points N corresponds to the time interval $2T$, moreover $2\pi n/N = t_n$ and k is the discrete index corresponding to Ω. Every complex number $\hat{f}_k$ gives the amplitude and phase of the "frequency" k, entering the Fourier development of the original $f(t)$.

Implementing (5.3.2) requires the numerical evaluation and sum of N terms for every k, so that the time and complexity of the computation grow as N^2. A well-known factorization method (dating back probably to Gauss and afterward discovered again many times) allows us to drastically reduce this number to $N \log N$: taking into account that typically $N \approx 10^4$ or 10^5, the gain is remarkable. In practice, (5.3.2) is always computed with this algorithm, denoted as *fast Fourier transform* (FFT). For a better numerical performance it requires N to be an integer power of 2.

Applying the FFT to the output of the numerical integration of some equations of motion, we can play with two parameters: the integration step Δt and the finite length $2T$ of the integration interval. The choice of Δt restricts the maximal frequency $\omega_{\max} \sim 1/\Delta t$ entering the sampling, and every $\omega > \omega_{\max}$ escapes the observation. This fact is not very important as far as we are concerned; as it will be evident in the further applications, the chosen integration step will always be widely sufficient and the higher neglected frequencies will be uninteresting. The choice of T is much more

---

[2] Again our definition is slightly different from the usual because of the term $1/N$, taken in order to agree with (5.3.1).

important. The integration in (5.3.1) with a finite window T,

$$\frac{1}{2T} \int_{-T}^{T} e^{i\omega t} e^{-i\Omega t} dt = \frac{\sin\left[(\omega - \Omega)T\right]}{(\omega - \Omega)T} \tag{5.3.3}$$

shows that the exact line centered on $\omega - \Omega$ is instead spread over the whole real line and that the error goes to zero as $1/T$. This is the weak point of the FFT: to improve the result, one must enlarge T and, to get a satisfying precision, one must often take excessively long integration times, during which the instantaneous frequency may change.

5.3.2 Frequency Modified Fourier Transform (FMFT)

The numerical frequency analysis was introduced in Laskar (1990) to detect the chaos in the secular motion of the solar system over 200 Myr. See also Laskar, Froeschlé & Celletti (1992) for a clear exposition of the method. The procedure can be implemented in a computer program: we will use the version called the *frequency modified Fourier transform* (FMFT) presented in Šidlichovský & Nesvorný (1997).

The scenario is the same as for the FFT. Let us assume that we know the tabulated values of a multiperiodic complex function

$$Z(t) = \sum_{k} C_k e^{ik \cdot \omega t}, \quad k \in \mathbb{Z}^n - \{0\}, \quad \omega \in \mathbb{R}^n,$$

over a time span interval $[-T, T]$ and with such a small stepsize that we can very accurately compute its integrals in the interval. The method allows us to compute numerically, but with great precision, the frequencies and complex amplitudes of the truncated approximation

$$Z^{(N)} = \sum_{h=1}^{N} C_h' e^{i\omega_h' t}, \quad N >> n.$$

In the KAM case, one expects that the computed values of the frequencies $\omega_1', \omega_2', \ldots, \omega_N'$ are not random, but are all linear combinations, with integer coefficients, of the n fundamental frequencies $\omega = \omega_1, \ldots, \omega_n$, establishing a map $h \mapsto k$, with $C_h' = C_k$ and $k \cdot \omega = \omega_h'$. We call such a spectrum *regular*.

Let us briefly describe the method. Frequencies and complex amplitudes are computed through an iterative scheme. In order to determine the first frequency ω_1', one searches, with a quadratic interpolation routine, for the maximum amplitude of $\phi(\omega) = \left\langle Z(t), e^{i\omega t} \right\rangle$, where the scalar product $\langle f(t), g(t) \rangle$ is defined by

$$\langle f(t), g(t) \rangle = \frac{1}{2T} \int_{-T}^{T} f(t) \overline{g}(t) \chi(t) \, dt,$$

Figure 5.2: Left: the function $\frac{\sin(2\pi x)}{2\pi x}$. Right: the function $\frac{\sin(2\pi x)}{2\pi x}\frac{1}{1-4x^2}$.

and where $\chi(t)$ is a weight function, that is, a positive and even function with a mean unitary value in $[-T, T]$. Usually, the Hanning filter is used: $\chi(t) = 1 + \cos(\pi t/T)$. Once the first periodic term $e^{i\omega'_1 t}$ is found, its complex amplitude C'_1 is obtained by orthogonal projection of $Z(t)$ onto $e^{i\omega'_1 t}$; then the process is restarted on the remaining part of the function, i.e., $Z_1(t) = Z(t) - C'_1 e^{i\omega'_1 t}$. As the functions $e^{i\omega'_h t}$ are usually not orthogonal, a Gram-Schmidt orthogonalization is also necessary when projecting iteratively on these $e^{i\omega'_h t}$.

In principle, the introduction of the filter χ is not strictly necessary but makes the computations more precise. To understand how, compare the integral (5.3.3) with

$$\frac{1}{2T}\int_{-T}^{T} e^{i\omega t}e^{-i\Omega t}\chi(t)\, dt = \frac{\sin[(\omega - \Omega)T]}{(\omega - \Omega)T}\left[\frac{1}{1 - (\omega - \Omega)^2 \frac{T^2}{\pi^2}}\right].$$

As one sees in Figure 5.2, the presence of the Hanning filter smooths out the lobes external to the central bell.

The FMFT implementation provides two mechanisms for getting a more precise output. These corrections attempt to remove systematic effects on frequencies and amplitudes produced by interaction of different Fourier terms.

The first mechanism is called *linear correction*. Assume that one has two terms in the signal with similar frequencies, $\omega_1 < \omega_2$ say, and one applies FMFT to it. One obtains $\omega_1^* = \omega_1 + \delta\omega_1$ and $\omega_2^* = \omega_2 + \delta\omega_2$ where the error terms $\delta\omega_1$ and $\delta\omega_2$ appear due to partial overlap of terms in the frequency domain. Note that if $\delta\omega_1$ and $\delta\omega_2$ could be determined, the obtained frequencies ω_1^* and ω_2^* could be corrected to obtain ω_1 and ω_2. For the linear correction, the FMFT is applied to the signal with frequencies ω_1^* and ω_2^* (which we have obtained by the first application of the FMFT). We obtain $\omega_1^{**} = \omega_1^* + \delta\omega_1^*$ and $\omega_2^{**} = \omega_2^* + \delta\omega_2^*$. Because ω_1^* was similar to ω_1 and ω_2^* was similar to ω_2, correction $\delta\omega_1^*$ will be similar to $\delta\omega_1$ and correction $\delta\omega_2^*$ will be similar to $\delta\omega_2$. Thus, we have obtained an estimate of $\delta\omega_1$ and $\delta\omega_2$ and can use those values to correct frequencies ω_1^* and ω_2^* and obtain better estimates of the original ω_1 and ω_1.

The latter method is called the *nonlinear correction* and uses an additional application of the FMFT on the signal with frequencies ω_1^{**} and ω_2^{**}. As in the previous case, one obtains corrections $\delta\omega_1^{**}$ and $\delta\omega_2^{**}$. These will not be exactly equal to $\delta\omega_1^*$ and $\delta\omega_2^*$ determined previously because ω_1^{**} (or ω_2^{**}) was not equal to ω_1^* (or ω_2^*). In fact, $\delta\omega_1^{**} - \delta\omega_1^*$ can be thought of as an estimate of $\delta\omega_1^* - \delta\omega_1$ and $\delta\omega_2^{**} - \delta\omega_2^*$ can be thought as an estimate of $\delta\omega_2^* - \delta\omega_2$. The idea of the nonlinear correction is then: use approximate ω_1 by $\omega_1^* - 2\delta\omega_1^* + \delta\omega_1^{**}$, and the same for second frequency. In practice, this method works if $\delta\omega_1$ and $\delta\omega_2$ are small quantities and fails if they are not.

For guidance, in most cases the FMFT method without corrections is more robust but less precise. One has to test the given application to understand whether the use of the corrections is beneficial or not.

5.3.3 Wavelets and Time-Frequency Analysis

Wavelets are useful in the study of perturbed Hamiltonian systems, mainly in connection with the numerical measure of the *instantaneous frequency*; for example, see Vela-Arevalo (2002), Vela-Arevalo & Marsden (2004), and Chandre, Wiggins & Uzer (2003). Given a complex-valued function $f(t)$, which will also be called a *signal*, it seems natural to define its instantaneous frequency $\omega(t)$ as the time derivative of the phase; in other words, by writing

$$f(t) = A(t)e^{i\varphi(t)}, \quad A(t), \varphi(t) \quad \text{real functions,}$$

then

$$\omega(t) \overset{\text{def}}{=} \frac{d\varphi(t)}{dt}. \tag{5.3.4}$$

Unfortunately, this definition is ambiguous. For example, if

$$A(t) = a + b\cos\omega_1 t, \quad \varphi(t) = \omega_2 t, \quad a, b, \omega_1, \omega_2 \quad \text{real const.,}$$

one can take both $\omega(t) = \omega_2$ and $f(t)$ as the sum of three signals, with frequencies ω_2, $\omega_2 - \omega_1$, $\omega_2 + \omega_1$, respectively. However, if the amplitude $A(t)$ is almost constant, the definition does make sense. In the sequel we will always consider this case.

In order to extract the instantaneous frequency from a signal $f(t)$, one may define the *Gabor transform*

$$Gf(t, \omega) \overset{\text{def}}{=} \int_{-\infty}^{\infty} f(t') g(t' - t) e^{-i\omega t'} dt', \tag{5.3.5}$$

where $g(t)$ is an envelope function that acts like a time window. We will take the Gaussian

$$g(t) = e^{-\frac{t^2}{2\sigma^2}},$$

where σ^2 is called the *variance*. The parameter t slides the envelope continuously, in that way localizing the signal, and the "ridge" of $|Gf|^2$ in the plane t-ω will describe the instantaneous frequency. Indeed, let us consider an integration interval small enough to have $A(t)$ and $g(t)$ almost constant but large enough to contain many oscillations of frequency ω; thus $|Gf|^2$ will be very small, except when the total phase $\varphi(t) - \omega t$ is stationary: (5.3.4) follows.

The choice of the variance σ^2 is at our disposal. Taking small values the window shrinks, improving the time localization but making the frequency determination worse; the opposite occurs, obviously, for large values. This is known as the *Heisenberg uncertainty principle*, with a clear reference to quantum mechanics. Unfortunately, in (5.3.5) the choice is *a priori*, while one would prefer an adaptive window: short for high and long for low frequencies.

This requirement is just fulfilled by the *wavelet transform*, defined as

$$Wf(t,s) = \int_{-\infty}^{\infty} f(t')\,\overline{\psi}\left(\frac{t'-t}{s}\right) dt', \quad \overline{\psi} = \text{complex conjugate}.$$

Here, $\psi(t)$ is the called the *mother wavelet*, for which we take in the sequel

$$\psi(t) = g(t)e^{i\eta t},$$

with η an arbitrary, user-defined, real parameter. Putting $s = \eta/\omega$, if $A(t) \approx$ const. $= A$, we have

$$Wf(t,\omega) = Ae^{i\omega t} \int_{-\infty}^{\infty} e^{-\frac{1}{2\sigma^2}\frac{\omega^2}{\eta^2}(t'-t)^2}\, e^{i[\varphi(t')-\omega t']}\, dt'.$$

Up to an unessential phase factor, Wf equals Gf but with the substitution $\sigma^2 \to \sigma^2\eta^2/\omega^2$: the variance, hence the window width, is rescaled as desired by the frequency value. As for $|Gf|^2$, the ridge of $|Wf|^2$ in the plane t-ω describes the time evolution of the instantaneous frequency. Note that the rescaling depends on the product $\sigma^2\eta^2$: we will take $\eta = 1$, then, as for the Gabor Transform, varying σ^2 improves the time localization to the detriment of the frequency localization or vice versa.

Let us consider a signal of the type $Z_j(t) = I_j(t)\,e^{i\varphi_j(t)}$, where I_j and φ_j (action and conjugate angle, respectively) are the numerical solution of a perturbed Hamiltonian system. We know from the KAM theorem that, for most of the initial conditions, the solution of the perturbed system evolves on tori obtained by slightly deforming the unperturbed solution

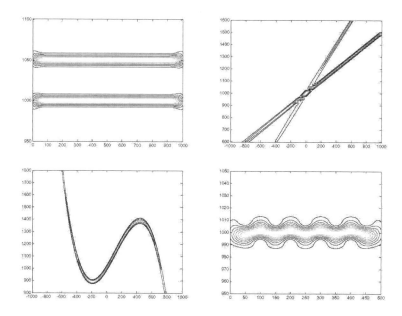

Figure 5.3: Some examples of the instantaneous frequency of a numerical signal. Here and in the subsequent figures the irregularities on the left and right borders are an artifact of the computation algorithm, and thus devoid of any meaning.

$I = \text{const.}$, $\varphi = \omega t + \text{const.}$ We are therefore in the case of almost constant amplitude, and the wavelet transform can be applied.

The numerical computation of the wavelet transformation, and thus of the instantaneous frequency, has been implemented in a MATLAB program, exploiting the "built-in" ability of MATLAB to compute the fast Fourier transform and its inverse. Indeed, the wavelet transformation may be viewed as a convolution product, whose Fourier transform is the product of the Fourier transform of the two factors. The computation is much faster with respect to the implementation of the wavelet transformation obtained by a mere implementation of the definition, i.e., by a double numerical integration on a rectangular grid in the plane $t\text{-}\omega$. The function $|Wf|^2$ is displayed by plotting its level lines on the plane $t\text{-}\omega$. Note that a judicious choice of the product value $\sigma^2\eta^2$ makes useless the "ridge extraction," implemented instead in the other quoted works, thus simplifying the program without loss of information.

In Figure 5.3 some examples are shown which illustrate the effectiveness of the program. In the first picture the signal $f(t)$ is the sum of two exponentials with constant frequencies: this is what happens in a KAM solution, when $Z_j(t) = I_j(t)e^{i\varphi_j(t)}$ can be developed in a Fourier series as a sum of signals with constant frequency. In the second picture the phases of the two signals are taken as quadratic functions of time, so that the frequencies are linear, while in the third picture the phase is a quartic polynomial, and so the frequency is cubic. Lastly, in the fourth picture we consider a time evolution of the type

$$\varphi(t) = \omega_0 t + a \sin \omega_r t, \quad \omega_0 >> \omega_r,$$

so that the instantaneous frequency is the sum of a constant term plus a sinusoid; in this case we will speak of a sinusoidal *frequency modulation*, and call ω_r the *modulating frequency*. As we will now show, this last case is particularly relevant when studying resonant orbits, or very close to a resonance.

Inside a resonance it is possible, with a linear canonical transformation, to take new action-angle variables such that one of the new angles is just the resonant angle φ_r, librating with frequency $\omega_r = 2\pi/T_r$. Coming back to the original action-angle variables, if $Z^{(j)}(t) = I_j(t)\, e^{i\varphi_j(t)}$ does not depend on the resonant variables I_r, φ_r, its dynamical evolution in the complex plane will be the composition of rotations with constant frequency. If, in contrast, the resonant motion is present, the instantaneous frequency will oscillate with frequency ω_r about some constant value ω_0; we will thus speak of *frequency modulation*. Note that the modulating frequency ω_r is in general much smaller than ω_0, since it is as small as the square root of the perturbation and exponentially small with the order of the resonance.

As in the pendulum model, the modulating frequency reaches a local maximum at the center of the resonance, vanishes on the separatrices, and tends quickly to infinity outside the resonance. At the same time, the amplitude of the modulation vanishes at the center of the resonance, reaches the maximum on the separatrices, halves when crossing them, then vanishes going away. By inspecting the instantaneous frequency it is therefore possible to discriminate between a KAM and a regular resonant torus. The frequency of the first is constant or, at most, slightly modulated if in the outer neighborhood of a resonance; that of the latter exhibits a marked modulation bearing the described characteristics. Obviously, an instantaneous frequency behaving irregularly is not only an indication of chaotic motion, but its irregularity degree is a measurement of the chaos itself.

Without entering the details of the underlying concrete physical system,[3] let us consider an example which will test the wavelet method, comparing it with the Poincaré map.

The top picture of Figure 5.4 shows the Poincaré map, G being the action and ω the angle; below the wavelet analysis for two different orbits.

---

[3]The physical system is the quadratic Zeeman problem, which will be investigated in Chapter 8. Bear in mind that here the angle ω is the argument of pericenter for the Kepler problem, not a frequency.

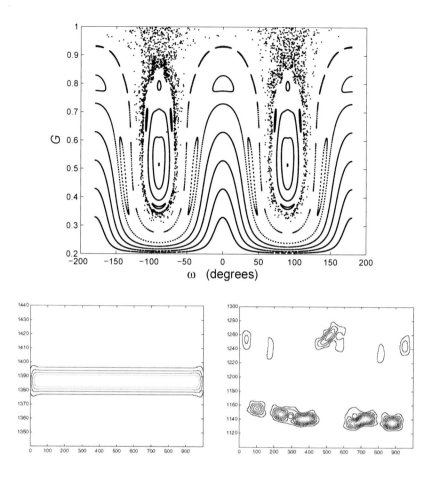

Figure 5.4: In the top picture the Poincaré map and below the instantaneous frequency for two different orbits, starting from the point $G = 0.7$, $\omega = 0$ (left) and $G = 0.95$, $\omega = \pi/2$ (right). Here ω is an angle variable (the argument of pericenter).

Poincaré map and wavelet analysis agree very well in the diagnosis of where order or chaos occur.

With Figure 5.5 we investigate the behavior of the instantaneous frequency inside a resonance and in the outer neighborhood. Remember that the irregularities on the left and right borders of every figure are an artifact of the computation algorithm, and thus devoid of any meaning. The sequence clearly shows that the instantaneous frequency behaves as expected. It is constant outside and slightly modulated approaching the resonance, then the modulation amplitude suddenly increases crossing the separatrix, and afterwards decreases until it vanishes at the center of the resonance;

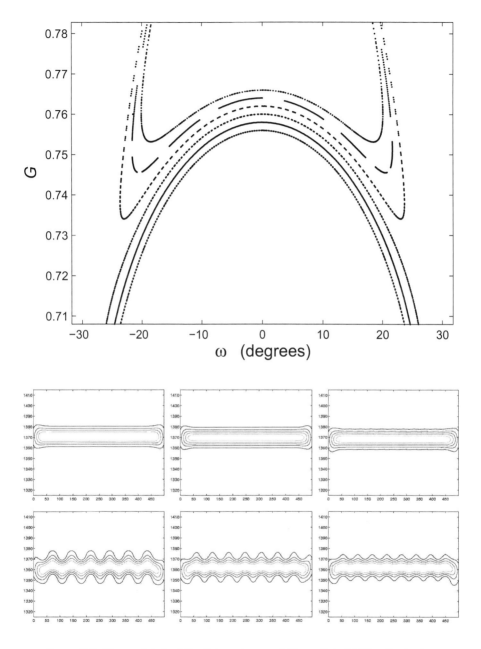

Figure 5.5: Top: a detail of the Poincaré map of Figure 5.4. Middle: the instantaneous frequency of the orbit relative to $\omega = 0$ and $G = 0.756$ (left), $G = 0.758$ (center), $G = 0.760$ (right). Bottom: $G = 0.762$ (left), $G = 0.764$ (center), $G = 0.766$ (right). The irregularities on the left and right borders of every figure are an artifact of the computation algorithm, thus devoid of any meaning.

at the same time, the frequency ω_r of the modulation grows.

5.3.4 Frequency Modulation Indicator (FMI)

The *Frequency Modulation Indicator* (FMI) is a sensitive tool that can detect and localize very thin resonances in quasi-integrable Hamiltonian systems, giving, moreover, a comprehensive picture of the whole geography of
the phase space. In some sense it is complementary to other chaos indicators, directly revealing the resonances while the chaotic zones are indirectly
marked by a blurring of the Arnold web. The FMI method has been introduced in Cordani (2008).

As we have established, inside or very near a resonance but sufficiently
far from the separatrix, the frequencies are not constant but modulated
by an approximately sinusoidal signal. We will study what happens when
performing a numerical frequency analysis inside a resonance and consequently show how to define the FMI. The following study will enable us to
explain why the FMI is such an effective tool and to understand the ap-
parence of some spurious phenomena in its graphical display.

The Fourier expansion of $\exp(ia \sin \varphi)$ is calculated, e.g., in Brouwer
& Clemence (1961, page 67) and shows that the spectrum of a frequency
modulated signal is

$$e^{i(\omega_0 t + a \sin \omega_r t)} = J_0 \, e^{i\omega_0 t} + \sum_{k=1}^{\infty} J_k \, e^{i(\omega_0 + k\omega_r)t} + \sum_{k=1}^{\infty} (-1)^k J_k \, e^{i(\omega_0 - k\omega_r)t},$$

$$(5.3.6)$$

where $J_k(a)$ is the Bessel function of order k of the first kind. Figure 5.6 confirms this analytical development: the modulating frequency ω_r becomes
very small in crossing the separatrix, so that the spectral lines become more
closely spaced approaching it. The characteristic multi-triangular shape of
the spectrum is a direct consequence of the approximately linear dependence of the function $f(k) = \log J_k(a)$ for k sufficiently large, with a slope
in inverse relation to the modulating amplitude a.

For simplicity (this does not qualitatively change our final conclusions)
we retain only the first three, most significant terms of (5.3.6):

$$f(t) = J_0 \, e^{i\omega_0 t} + J_1 \, e^{i(\omega_0 + \omega_r)t} - J_1 \, e^{i(\omega_0 - \omega_r)t}.$$

Let us take a rectangular window, of width $2T$ and centered in t_0 (again:
a different shape of the window, given for example by the Hanning filter,
does not affect our conclusions qualitatively, changing only the power of
the denominator in the fractions below). The numerical frequency analysis
algorithm FMFT consists in finding the $\overline{\omega}$ value for which the function

$$F(\omega) = \frac{1}{2T} \int_{t_0 - T}^{t_0 + T} f(t) \, e^{-i\omega t} dt$$

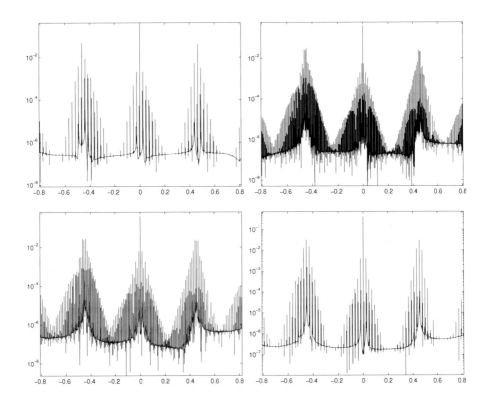

Figure 5.6: The FFT graphical display relative to four orbits as in Figure 5.5: $\omega = 0$ and $G = 0.760$ (left-top), $G = 0.761$ (right-top), $G = 0.762$ (left-bottom), $G = 0.780$ (right-bottom). As expected, the spectrum lines become more closely spaced approaching the separatrix, giving rise to a continuous spectrum inside the stochastic layer. For the characteristic multi-triangular shape of the spectrum, see the text.

takes its maximum amplitude. Taking into account that

$$\frac{1}{2T} \int_{t_0-T}^{t_0+T} e^{i\omega t}\,dt = e^{i\omega t_0}\frac{\sin(\omega T)}{\omega T},$$

and putting $(\omega_0 - \omega)T = \epsilon$ we find

$$F(\omega) = J_0\frac{\sin\epsilon}{\epsilon} + J_1 e^{i\omega_r t_0}\frac{\sin(\omega_r T + \epsilon)}{\omega_r T + \epsilon} - J_1 e^{-i\omega_r t_0}\frac{\sin(-\omega_r T + \epsilon)}{-\omega_r T + \epsilon},$$

where we have dropped the phase factor $e^{i(\omega_0-\omega)t_0}$. Take the width of the window much larger than the period of the modulating sinusoid, that is, $\omega_r T \gg \pi$, and seek an ω that satisfies $\epsilon = \mathcal{O}(\omega_r^{-1}T^{-1})$ (compare below).

Developing in a Taylor series

$$\frac{\sin(\omega_r T + \epsilon)}{\omega_r T + \epsilon} = A + B\epsilon + \mathcal{O}(\epsilon^3),$$

$$A = \frac{\sin(\omega_r T)}{\omega_r T}, \quad B = \frac{\cos(\omega_r T)}{\omega_r T} \quad A, B = \mathcal{O}(\epsilon),$$

we obtain

$$F(\omega) = J_0 \left(1 - \frac{1}{3!}\epsilon^2 \right) + J_1\, e^{i\omega_r t_0}(A + B\epsilon) - J_1\, e^{-i\omega_r t_0}(A - B\epsilon) + \mathcal{O}(\epsilon^3),$$

from which

$$|F(\omega)|^2 = \text{const.} - \frac{1}{3}J_0^2\epsilon^2 + 4J_0 J_1 B\epsilon\, \cos(\omega_r t_0) + \mathcal{O}(\epsilon^3).$$

Differentiate with respect to ϵ and equate to zero. Accordingly with the assumption on ω, finally we find

$$\frac{\omega_0 - \overline{\omega}}{\omega_r} \simeq 6\frac{J_1}{J_0}\frac{\cos(\omega_r T)}{(\omega_r T)^2}\cos(\omega_r t_0), \tag{5.3.7}$$

which is the relative "error" inherent to the algorithm FMFT. From expression (5.3.7) and applying the numerical frequency analysis to a supporting ω_0 which is ω_r frequency modulated, we can conclude that the numerical output oscillates about the "right" value ω_0, and that this oscillation is triggered by shifting the window, by changing its width $2T$ and by modifying ω_r; moreover, the amplitude of the oscillation rapidly vanishes when the width increases.

Let us fix T such that $\cos(\omega_r T) \neq 0$. Taking into account that $\frac{J_1(x)}{J_0(x)} \simeq \frac{x}{2}$ for $x \geq 0$ sufficiently small (say, $x < 0.8$), we obtain from (5.3.7)

$$\overline{\omega} \simeq \omega_0 - \eta a \cos(\omega_r t_0),$$

$$\eta = \frac{3}{T}\frac{\cos(\omega_r T)}{\omega_r T},$$

with η much smaller than ω_r and decreasing when $\omega_r T$ grows.

The recipe for the computation of the FMI is therefore the following. Perform several numerical frequency analyses, say N, each time shifting the window of some Δt_0, and take the $\overline{\omega}_{\max}$ and the $\overline{\omega}_{\min}$ among the N output values. For a suitable Δt_0 (but the choice is not very critical) and N large enough (for example, $N = 7$ is a good value) the N values will practically span the whole oscillation range, so that the difference $\overline{\omega}_{\max} - \overline{\omega}_{\min}$ will result very close to $2\eta a$. Let us define the FMT as

$$\sigma_{\text{FMI}} \stackrel{\text{def}}{=} \log\left(\frac{\overline{\omega}_{\max} - \overline{\omega}_{\min}}{\overline{\omega}_{\max} + \overline{\omega}_{\min}}\right) \simeq \log\left(a\frac{\eta}{\omega_0}\right).$$

Clearly, $\sigma_{\text{FMI}} = -\infty$ for orbits of KAM type starting sufficiently far away from a resonance of low order (the high order resonances are practically not detectable in numerical investigations). In the graphical representation we fix a cut-off, i.e., we raise all the lower values to, for example, $\sigma_{\text{FMI}} = -12$, in order to avoid the $-\infty$ singularities. Instead, inside a resonance or in a very small neighborhood, σ_{FMI} will generally be larger, growing with the amplitude a of the frequency modulation. Note, however, that the exact numerical value of σ_{FMI} is not very meaningful since it depends on ω_r, which change with the orbit; moreover, the presence of the $\cos(\omega_r T)$ term causes a sort of "ghost undulation" in the graphical display of a resonance strip and especially of its neighborhood. Examples will be given in the next section.

It should be clear that the FMI method *cannot* be considered a mere variant of the frequency analysis, its central point being the recognition that the characteristic feature of the resonances is the frequency modulation of the fundamental frequencies. In principle, another, more effective, method for computing the frequency modulation amplitude could be found, without resorting to the frequency analysis algorithm.

5.4 Some Examples

In the numerical study of a slightly perturbed Hamiltonian system one would find a method which is implementable on a computer and is able to detect the geography of the resonances in the action space, discriminating between ordered and chaotic zones. All the variational equation methods require further analytical and heavy integration work with respect to the mere integration of the equation of motion, which makes the methods practically inapplicable in some concrete and realistic case. The numerical frequency analysis, instead, requires only the integration of the equation of motion but, unfortunately, plotting the frequency values as a function of the actions does not clearly reveal the location of the resonances. (The method instead becomes effective when sectioning the energy hypersurface, displaying, at least for the 3-dimensional systems, 1-dimensional plots.) We are therefore in a disappointing situation: the effective methods are not applicable in general, while the applicable method is not effective.

Exploiting instead an intrinsic imprecision inherent to the algorithm itself of the numerical frequency analysis, hence transforming a drawback into an opportunity, we have implemented the FMI computation. We thus obtain a very sensitive tool in detecting where the fundamental frequencies are modulated, and one which localizes the resonances without resorting to the integration of the variational equations. Therefore, with the applicable method we reach the same result as that of the effective ones. Let us show some examples.

5.4.1   Symplectic Maps and the POINCARE Program

In the 6-dimensional phase space of a Hamiltonian system with three degrees of freedom, let us fix the value of the energy and consider the 5-dimensional hypersurface given by $H(q, p)$ = const. Recording the intersection points of an orbit generated by the Hamiltonian with a 4-dimensional hyperplane transverse to the flow, one performs in practice a reduction with the energy integral, thus obtainig a symplectic mapping of the hyperplane into itself. A suitably chosen 4-dimensional symplectic map can therefore "simulate" the dynamical evolution of a quasi-integrable Hamiltonian system with three degrees of freedom. The absence of integration errors ensures the objectivity of the result and that spurious artifacts do not occur.

We have tested the FMI method on three different 4-dimensional symplectic maps $x \rightarrow y$. The first two variables are action-like while the other two are angle-like variables. The results are reported in Figure 5.7, where the Arnold map is well enlightened. The 4-dimensional symplectic maps are a generalization of the well-known *standard map*, already considered in the first chapter,

$$y_1 = x_1 + \varepsilon \sin x_2,$$
$$y_2 = x_2 + y_1,$$

introduced by Chirikov (1979). The standard map describes the motion of a simple mechanical system known as the *kicked rotator*. It consists of a stick that is free of the gravitational force, which can rotate frictionlessly in a plane around an axis located in one of its tips, and which is periodically kicked on the other tip. The two variables determine the angular position of the stick and its angular momentum, respectively. The constant ε measures the intensity of the kicks.

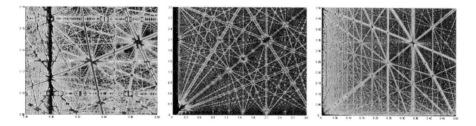

Figure 5.7: Three results of the FMI computation relative to 4-dimensional symplectic maps: see the text. The grid resolution is 500×500 and the iteration number is 10^4.

The left picture of Figure 5.7 regards the map of Guzzo *et al.* (2002, eq.

21):

$$y_1 = x_1 + \varepsilon_1 \sin(x_1 + x_3) + \varepsilon_3 \sin(x_1 + x_2 + x_3 + x_4)$$
$$y_2 = x_2 + \varepsilon_2 \sin(x_2 + x_4) + \varepsilon_3 \sin(x_1 + x_2 + x_3 + x_4)$$
$$y_3 = x_1 + x_3$$
$$y_4 = x_2 + x_4$$

with $\varepsilon_1 = 0.4$, $\varepsilon_2 = 0.3$, and $\varepsilon_3 = 0.001$. It should be compared with the two pictures of Guzzo *et al.* (2002, fig. 10), which have been obtained with the FLI method and perturbative parameter $\varepsilon_3 = 0.004$ and $\varepsilon_3 = 0.006$, respectively.

The center picture regards the map of Lukes-Gerakopoulos *et al.* (2008, eq. 39):

$$y_1 = x_1 - \varepsilon_1 \frac{\sin(x_1 + x_3)}{(\cos(x_1 + x_3) + \cos(x_2 + x_4) + 4)^2}$$
$$y_2 = x_2 - \varepsilon_1 \frac{\sin(x_2 + x_4)}{(\cos(x_1 + x_3) + \cos(x_2 + x_4) + 4)^2}$$
$$y_3 = x_1 + x_3$$
$$y_4 = x_2 + x_4$$

with $\varepsilon_1 = 0.05$ and should be compared with that of Lukes-Gerakopoulos *et al.* (2008, fig. 10(a)), which has been obtained with the APLE method.

Lastly, the right picture regards the map of Honjo & Kaneko (2008, eq. 1):

$$y_1 = x_1 + \varepsilon_1 \sin x_3 + \varepsilon_2 \sin(x_3 + x_4)$$
$$y_2 = x_2 + \varepsilon_1 \sin x_4 + \varepsilon_2 \sin(x_3 + x_4)$$
$$y_3 = x_3 + x_1 + \varepsilon_1 \sin x_3 + \varepsilon_2 \sin(x_3 + x_4)$$
$$y_4 = x_4 + x_2 + \varepsilon_1 \sin x_4 + \varepsilon_2 \sin(x_3 + x_4)$$

with $\varepsilon_1 = 0.2$ and $\varepsilon_2 = 0.0005$. It should be compared with that of Honjo & Kaneko (2008, fig. 1(b)), which has been obtained with the rotation number method and the much larger parameters $\varepsilon_1 = 0.5$ and $\varepsilon_2 = 0.1$.

The reader himself can get the pictures of Figure 5.7 by means of the program POINCARE (without accent): the three maps above are named FGL, APLE, and ROT. The value grid is chosen in the left part of the panel "Frequency Modulation Indicator", while N is fixed in "Step number for FMI".

In order to obtain good results, the computations invoked in the FMI procedure require very long times, which can be shortened if the computer is a multicore machine. Indeed, all the five programs provided in the CD are able to parallelize the computations in the following way. If you possess an n-core machine, create a subfolder Master and $n - 1$ subfolders Slave1, Slave2, . . . in an empty folder POINCARE, then copy the whole program

POINCARE identically in every folder. Start MATLAB then POINCARE from the `Master` folder, set the parameters and click on "Save setting/Save setting now". Without closing, start a *new* instance of MATLAB, then POINCARE from a folder `SlaveX`: you will notice that all the buttons of the computations are disabled while the new button "Start Slave" appears. Click on this button and POINCARE will wait for the start of the master. Redo for every slave, and lastly go back to the master and click on "Calculate . . .". The whole work will be automatically shared out among the n cores. The final result is displayed by the master.

The same procedure can be carried out for the other programs, i.e., HAMILTON, LAGRANGE, KEPLER, LAPLACE.

5.4.2   A Test Hamiltonian and the HAMILTON Program

In Guzzo *et al.* (2002) and Froeschlé *et al.* (2000) the FLI is tested on the Hamiltonian

$$H_{\text{FGL}} = \frac{1}{2}I_1^2 + \frac{1}{2}I_2^2 + I_3 + \varepsilon \left(\frac{1}{\cos \varphi_1 + \cos \varphi_2 + \cos \varphi_3 + 4} \right) \qquad (5.4.1)$$

for three values of the perturbative parameter: $\varepsilon = 0.01, 0.03, 0.04$, on a grid of 500×500 initial conditions regularly spaced on the I_1, I_2 action plane and with an integration time $t = 1000$ for every orbit. The result is reported in Guzzo *et al.* (2002, fig. 5), where global pictures with relative magnified details are shown. The Arnold web is very clearly enlightened and appears as expected: the frequency map for (5.4.1) is

$$\omega_1 = I_1, \quad \omega_2 = I_2, \quad \omega_3 = 1,$$

and the resonance lines on the plane I_1, I_2 are given by the linear equation

$$k_1 I_1 + k_2 I_2 + k_3 = 0, \quad k_1, k_2, k_3 \in \mathbb{Z},$$

i.e., by straight lines whose slopes and intersections with the axes take rational values. Obviously, this set is dense on the plane but the finite integration time allows us to visualize only those resonances whose order, defined as $\sum_i |k_i|$, is below a threshold value. The threshold grows with the integration time itself, so that in principle one is able to visualize all the resonances at various orders.

With the perturbative value $\varepsilon = 0.01$ we have analyzed the Hamiltonian (5.4.1) using the FMI method of the HAMILTON program. The first result is reported in Figure 5.8, which should be compared with the top-left picture of Guzzo *et al.* (2002, fig. 5). Our picture has been shifted in the plane in order to avoid the neighborhood of the origin, where the presence of chaos (not revealed by the FLI) and the resulting high values of the FMI would make the visualization somewhat blurred.

Figure 5.8 confirms the same resonance structure appearing in Guzzo *et al.* (2002, fig. 5), and the Arnold web is enlightened in great detail. Figure 5.9 (left) should be compared with the top-right picture of Guzzo *et al.* (2002, fig. 5); this small detail exhibits a curious structure at the crossing of the various resonances, resembling the petals of a flower. In Figure 5.9 (right) the same detail is reported but with $\varepsilon = 10^{-4}$; with a much smaller perturbation the Arnold web is still very well enlightened, but the flower structure disappears.

To complete the comparison with the FLI method one must also take into account the CPU time. With the FMI method one integrates the motion equations (5.2.1), then invokes the frequency analysis: the time it takes for the latter step is about 20–30 % of that relative to the first step. The FLI does not require the frequency analysis; however, one must also integrate the variational Equations (5.2.2). For the test Hamiltonian (5.4.1) the variational equations are very simple and the CPU times needed by the two methods are comparable. However, when the variational equations are very complicated, the FMI is surely faster.

The HAMILTON program is able to analyze a Hamiltonian also with the FFT, FMFT and wavelet tools.

To add a further Hamiltonian to the pop-up menu

i) Write a `*.m` file with the Hamiltonian equations of motion, taking one of the supplied files as a template: note that the three first variables are angles while the latters are actions.

ii) Open `Hamilton.fig` in GUIDE, double-click on the pop-up menu and add the `*.m` file to the already present list in the "String" field.

iii) Look for the string `DiffEquation` in the `Hamilton.m` file and update.

5.4.3 The Lagrange Points L_4 and L_5 and the LAGRANGE Program

The Lagrange points are two relative equilibrium points of the circular re-stricted three-body problem, a special case of the general three-body prob-lem: two bodies move in circular orbits at unitary distance about the com-mon center of mass B, according to the laws of the two-body dynamics, while a third body of negligible mass moves in their gravitational field. Let μ and $1 - \mu$ be their mass, with $\mu < 1$. In the plane of the circular orbits, rotating about the origin B of the axes xyz with unit angular speed, the two massive bodies are motionless and can be placed on the x axis at the points $\mu - 1$ and μ, respectively.

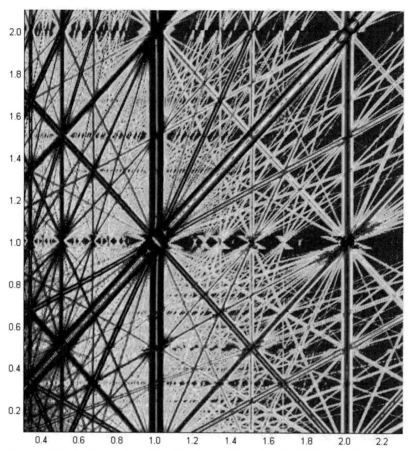

Figure 5.8: Analysis of the Hamiltonian (5.4.1) using the FMI method with $\varepsilon = 10^{-2}$.

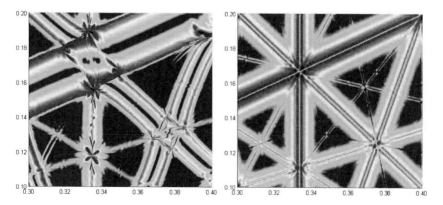

Figure 5.9: Left: detail of Figure 5.8. Right: as in the left picture but with $\varepsilon = 10^{-4}$.

The Hamiltonian describing the dynamics of the third body is

$$H = \frac{1}{2}(p_x^2 + p_y^2 + p_z^2) - xp_y + yp_x$$
$$- \frac{1 - \mu}{\sqrt{(x - \mu)^2 + y^2 + z^2}} - \frac{\mu}{\sqrt{(x - \mu + 1)^2 + y^2 + z^2}}, \qquad (5.4.2)$$

which encompasses centrifugal and Coriolis inertial forces, as well as kinetic energy and gravitational potentials. There are five equilibrium points in the rotating reference system, since the centrifugal force can balance the gravity force. Three of them are placed on the x axis and are called L_1, L_2, L_3: they are always unstable. The other two are L_4 and L_5, and along with the two fixed masses they form two equilateral triangles, the third vertex having positive and negative y coordinate, respectively.

Let us consider L_4, but similar considerations can be made for L_5. Its coordinates and momenta are

$$x^{eq} = \mu - 1/2, \quad y^{eq} = \sqrt{3}/2, \quad z^{eq} = 0,$$
$$p_x^{eq} = -\sqrt{3}/2, \quad p_y^{eq} = \mu - 1/2, \quad p_z^{eq} = 0.$$

With a translation in coordinates and momenta the equilibrium point L_4 is sent to the origin, so that the Hamiltonian becomes

$$H = \frac{1}{2}(p_1^2 + p_2^2 + p_3^2) - q_1 p_2 + q_2 p_1$$
$$- \left(\mu - \frac{1}{2}\right) q_1 - \frac{\sqrt{3}}{2} q_2 - \frac{1 - \mu}{\varrho_-} - \frac{\mu}{\varrho_+},$$
$$\text{with } \varrho_\pm = \sqrt{q_1^2 + q_2^2 + q_3^2 \pm q_1 + \sqrt{3} q_2 + 1}.$$

In order to investigate the nature of the equilibrium we must linearize the problem, developing H in a Taylor series and retaining the quadratic terms. Instead of a straightforward but lengthy calculation the following, more elegant method is preferable, which also gives easily the terms of higher order. We recall that, given two vectors $\vec{r}_0$ and $\vec{r}$, with $r < r_0$ (but, usually, $r \ll r_0$), the development

$$\frac{1}{\|\vec{r}_0 - \vec{r}\|} = \frac{1}{r_0}\left[1 + \sum_{k=1}^{\infty}\left(\frac{r}{r_0}\right)^k P_k(\cos\vartheta)\right], \qquad \cos\vartheta = \frac{\vec{r}_0 \cdot \vec{r}}{r_0 r},$$

holds, where ϑ is the angle between the two vectors, and

$$P_k(x) = \frac{1}{2^k k!}\left[\frac{d^k}{dx^k}(x^2 - 1)^k\right]$$

is the *Legendre polynomial* of degree k. To get the development of $1/\varrho_-$, take $\vec{r} \equiv (q_1\ q_2\ q_3)$ and $\vec{r}_0 \equiv (1/2 \quad -\sqrt{3}/2 \quad 0)$, which connects the origin with the mass $1 - \mu$. Taking $\vec{r}_0 \equiv (-1/2 \quad -\sqrt{3}/2 \quad 0)$, which connects the origin with the mass μ, gives the development of $1/\varrho_+$.

The Hamiltonian becomes

$$H = \frac{1}{2}(p_1^2 + p_2^2 + p_3^2) - q_1 p_2 + q_2 p_1 + \frac{1}{8}q_1^2 - \frac{5}{8}q_2^2 + \frac{1}{2}q_3^2 \qquad (5.4.3)$$

$$+ \frac{3}{4}\sqrt{3}(1 - 2\mu)q_1 q_2 + H_3 + H_4 + \cdots,$$

where H_s is a polynomial of degree s in the variables q, p. The 6×6 symmetric matrix relative to the quadratic term is

$$\mathbf{H} = \begin{pmatrix} \frac{1}{4} & \frac{3}{4}\sqrt{3}(1 - 2\mu) & 0 & 0 & -1 & 0 \\ \frac{3}{4}\sqrt{3}(1 - 2\mu) & -\frac{5}{4} & 0 & 1 & 0 & 0 \\ 0 & 0 & 1 & 0 & 0 & 0 \\ 0 & 1 & 0 & 1 & 0 & 0 \\ -1 & 0 & 0 & 0 & 1 & 0 \\ 0 & 0 & 0 & 0 & 0 & 1 \end{pmatrix}.$$

Note that the motion along the third axis is uncoupled from that on the horizontal plane.

Following Example 2.35 and Theorem 2.33, we bring the Hamiltonian (5.4.3) into the form

$$H = \sum_{h=1}^{3} \omega_h I_h + H_3 + H_4 + \ldots, \quad \text{with } \omega_3 = 1 \text{ and}$$

$$\omega_1 = \sqrt{\frac{1}{2} + \frac{1}{2}\sqrt{1 - 27\mu(1 - \mu)}}, \quad \omega_2 = -\sqrt{\frac{1}{2} - \frac{1}{2}\sqrt{1 - 27\mu(1 - \mu)}}.$$

Clearly, if $27\mu(1 - \mu) \leq 1$, i.e, $\mu \leq 0.03852\ldots$, then ω_1 and ω_2 are real and L_4 is an elliptic equilibrium point: the motion is stable in the linear approximation. But the three frequencies do not have the same sign; the truncated Hamiltonian is not a definite (positive or negative) function, so the further terms can destroy the stability. One can therefore proceed as in Section 3.5, invoking the Birkhoff Theorem 3.10 in order to reduce $H_3 + H_4$ to a polynomial quadratic in the actions, then applying the KAM and Nekhoroshev theorems; see Alfriend (1970), Alfriend (1971), Meyer & Hall (1992), Meyer & Schmidt (1986), and Benettin, Fassò & Guzzo (1998).

With the program LAGRANGE the reader can investigate numerically the motion about the point L_4. The panel "Single orbit analysis" is devoted to the numerical integration of the motion equations with consequent plotting of the dynamical evolution of position, momenta, and action-angle variables. Note that the q, p distance from the origin of the initial point plays the role

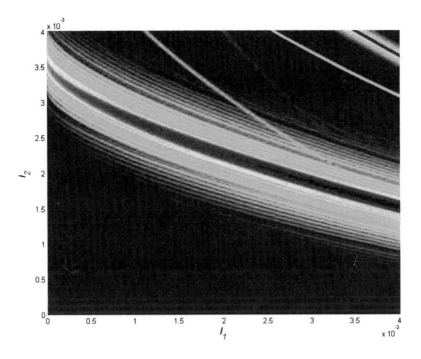

Figure 5.10: The FMI distribution about the Lagrange point L_4, where we have taken $I_3 = 0.002$. Notice the extremely low value $\approx 10^{-12}$ in the neighborhood of the origin, denoting in practice a great stability of the Lagrange point.

of perturbative parameter: starting from a point very close to the origin gives rise to actions almost constant and to an almost linear evolution of the angles. With the panel "Frequency Modulation Indicator" one can study the resonance distribution in the action space. In Figure 5.10 we report an example. In the neighborhood of the origin the FMI value is extremely low, i.e., $\approx 10^{-12}$, and below the thereshold of reliability of the computation, which denotes in practice a great stability of the Lagrange point. The Arnold web is almost absent, and the 3-dimensional system behaves in practice as having two degrees of freedom only.

CHAPTER 6

The Kepler Problem

There is a force in the earth
which causes the moon to move.

— J. KEPLER

In this chapter, we will study the group-geometrical structure of the Kepler problem and point out how this structure also turns out to be useful in the study of the perturbed case.

6.1 Basics on the Kepler Problem

The Kepler problem is the 3-dimensional system with Hamiltonian

$$H_0(\vec{q}, \vec{p}) = \frac{1}{2}p^2 - \frac{1}{q}.$$

Here and in the sequel, in the 3-dimensional case a letter without arrow denotes the norm of the relative vector.

It is a typical fact for the Hamiltonian systems that the integrals of motion play a fundamental role, so let us investigate their properties. With a direct calculation one can easily verify that the integrals of motion are the following:

(i) the Hamiltonian, due to the fact that it is time-independent;

(ii) the angular momentum vector

$$\vec{G} = \vec{q} \times \vec{p},$$

due to the rotational symmetry (Kepler's second law);

(iii) the eccentricity vector

$$\vec{E} = \vec{p} \times \vec{G} - \frac{\vec{q}}{q},$$

due to . . .? A simple and direct answer to this question does not exist, and in fact this is the starting point for the geometrical analysis of the Kepler problem.

We have seven integrals of motion, but, since the phase space is 6-dimensional, only five can be independent. In other words, two relations must exist among them. Indeed

$$\vec{G} \cdot \vec{E} = 0, \quad E^2 - 1 = 2H_0 G^2.$$

The conservation of the angular momentum ensures that the motion belongs to a plane. To find the equation of the orbit, take the scalar product of the eccentricity and position vectors and define the *true anomaly* as the angle formed by the two vectors. We obtain

$$q(f) = \frac{G^2}{E \cos f + 1}, \quad f = \text{true anomaly}, \quad E = \left\| \vec{E} \right\| = \text{eccentricity}.$$

This is the well-known generalization of the first Kepler law: the orbit is a conic, in particular an ellipse if the eccentricity is smaller than 1.

To integrate the equations of motion, we consider a far-reaching change in the independent variable, introducing the *eccentric anomaly*. Put

$$\frac{dt}{ds} = \frac{q}{\sqrt{-2H_0}}, \quad s = \text{eccentric anomaly}; \tag{6.1.1}$$

then it is easy to obtain, as a consequence of the unperturbed Hamilton equations, that

$$\frac{d^2 \vec{q}}{ds^2} + \vec{q} = \frac{\vec{E}}{2H_0}.$$

We have thus transformed the second order equation of motion into a *linear and regular* one: the introduction of the eccentric anomaly allows us to eliminate the singularity in the origin.

The last equation is that of a harmonic oscillator with a constant forcing term and can be easily integrated. The general solution is

$$\vec{q}(s) = \frac{1}{-2H_0} \frac{\vec{E}}{E} \cos s + \frac{1}{\sqrt{-2H_0}} \vec{G} \times \frac{\vec{E}}{E} \sin s - \frac{\vec{E}}{-2H_0}.$$

Due to regularization, the collision orbits, which are characterized by a null angular momentum and a unit eccentricity, are no longer forbidden, and the particle "bounces back" when it reaches the origin.

Equating the norm of both sides of this equation, we obtain

$$q(s) = a(1 - E \cos s), \quad a = \frac{1}{-2H_0}, \tag{6.1.2}$$

showing that a is the semimajor axis of the ellipse. From the definition we obtain

$$\frac{dt}{ds} = a^{\frac{3}{2}}(1 - E \cos s),$$

then, integrating,

$$t - t_p = a^{\frac{3}{2}}(s - E \sin s)$$

results, which is the celebrated *Kepler equation*. To express the position vector as a function of the time (a natural wish) one should invert this equation and substitute into the orbit equation. But the Kepler equation is a transcendental one and cannot be inverted in a closed form. An immediate consequence of this equation is Kepler's third law:

$$T = 2\pi a^{\frac{3}{2}}, \quad T = \text{period}.$$

It is usual to define the *mean anomaly*

$$l = s - E \sin s.$$

Instead of the components of the position and velocity vectors, one may use six other, more expressive, coordinates. The most popular choice is the following:

- the semimajor axis a;

- the eccentricity E;

- the longitude of ascending node Ω;

- the inclination i;

- the argument of pericenter ω;

- the true anomaly f.

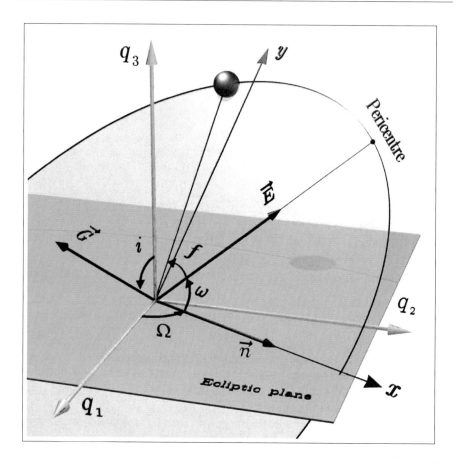

Figure 6.1: The elements of the orbit.

The first two parameters fix the size and shape of the orbit. The following three parameters are the Euler angles of the orbit and thus fix its spatial orientation. The last one fixes the position of the moving point on the orbit. These six parameters are called *Keplerian parameters*. See Figure 6.1

These Keplerian elements cannot globally parametrize the whole phase space for topological reasons, thus they are affected by some singularities:

(i) if the orbit lies in the ecliptic plane, the longitude of the ascending node and the argument of pericenter are undefined;

(ii) if the orbit is circular, the argument of pericenter and the true anomaly are undefined;

(iii) in the collision orbits, the motion is linear and the plane of the orbit is undefined.

6.1.1 Conformal Group and Geodesic Flow on the Sphere

Looking at the first integrals of the Kepler problem suggests that we face a problem which is "more symmetric" than one possessing only rotational invariance. But what about this larger symmetry? The beautiful answer was first given at quantum level (hydrogen atom) in Pauli (1926) and Fock (1935), then at the classical level in Souriau (1974), Souriau (1983), and Moser (1970); see also Guillemin & Sternberg (1990), Englefield (1972), and Cordani (2003). The point is that the Keplerian motion is equal, in some sense, to the geodesic motion on a 3-dimensional sphere, so that the phase space of the regularized Kepler problem is T^+S^3, the cotangent bundle with the null section deleted. From that an SO(4) enlarged symmetry emerges naturally. Let us review how to reach this conclusion: see Cordani (2003) for the details.

Let $\eta_{AB} = \text{diag}\,(--++++)$ be the flat metric tensor of $\mathbb{R}^{2,4}$ and $G_{AB} = -G_{BA}$ be a basis of the Lie algebra $\mathfrak{g}$ of $\mathfrak{G} = \text{SO}(2,4)$. The range of the indices is

$$
\begin{aligned}
A, B, C &= -1, 0, \ldots, 4 \\
\lambda, \mu, \nu &= 0, \ldots, 3 \\
\alpha, \beta, \gamma &= 1, \ldots, 4 \\
i, j, k &= 1, 2, 3.
\end{aligned}
$$

Then, the Lie algebra brackets are

$$[G_{AB}, G_{AC}] = \eta_{AA} G_{BC} \tag{6.1.3}$$

or zero if all indices are different. Since the action of $\mathfrak{G}$ on $\mathbb{R}^{2,4}$ is linear, it induces an action on the projective manifold of the rays through the origin; moreover, $\mathfrak{G}$ sends the null cone into itself, and thus acts transitively on the manifold M of null rays. This manifold is diffeomorphic to $S^1 \times S^3$ and is endowed with a class of pseudo-Riemannian metrics g_Γ obtained by restriction of the SO(2,4) invariant metric η on any section Γ of the null cone. The action of $\mathfrak{G}$ on M is conformal. The metrics g_Γ is conformally flat, with signature $(-+++)$, and the Lie algebra $\mathfrak{g}$ of $\mathfrak{G}$ is isomorphic to the Lie algebra of conformal vector fields on Minkowski space. We can identify the generators as follows:

$$
\begin{aligned}
G_{\mu\nu} &= \text{Lorentz group}, \qquad G_{\mu 4} + G_{-1\mu} = \text{translations}, \\
G_{\mu 4} - G_{-1\mu} &= \text{conformal translations}, \qquad G_{-14} = \text{dilations}.
\end{aligned}
$$

Let $\mathfrak{H}$ be the closed subgroup of $\mathfrak{G}$ with Lie algebra

$$\mathfrak{h} = \{G_{\mu\nu}, G_{\mu 4} - G_{-1\mu}, G_{-14}\}.$$

It is the isotropy group of the origin in $\mathbb{R}^{1,3}$. Since $M = \mathfrak{G}/\mathfrak{H}$, we can identify M with the "conformal compactification" of $\mathbb{R}^{1,3}$. In other words, one can obtain M by adding to $\mathbb{R}^{1,3}$ a null cone at infinity.

Let us now consider the symplectic action of $\mathfrak{G}$ on $T^*(\mathfrak{G}/\mathfrak{H})$. Since this action is not transitive, we may decompose $T^*(\mathfrak{G}/\mathfrak{H})$ into orbits of $\mathfrak{G}$. They are symplectic manifolds on which the group action is transitive, and so they may be identified with (covering spaces of) orbits of $\mathfrak{G}$ in $\mathfrak{g}^*$, the dual Lie algebra. The point is that to obtain the Kepler manifold T^+S^3 we must restrict ourselves to the subbundle of the null nonvanishing covectors.

Let $X_A = \eta_{AB}X^B$ be the coordinates of a point in $\mathbb{R}^{2,4}$ and Y_B the components of a covector. We can now project the "natural" moment map $T^*\mathbb{R}^{2,4} \rightarrow \mathfrak{g}^*$ given by

$$G_{AB} = Y_A X_B - Y_B X_A \tag{6.1.4}$$

to the moment map $T^+S^3 \rightarrow \mathfrak{g}^*$. Consider indeed a section Γ of the null cone of $\mathbb{R}^{2,4}$, i.e., $X^A = X^A(x^\mu)$, $\eta(X,X) = 0$, with x^μ local coordinates on the section. The metric η induces the metric

$$g_{\Gamma\mu\nu} = \frac{\partial X^A}{\partial x^\mu} \eta_{AB} \frac{\partial X^B}{\partial x^\nu}$$

on Γ. The moment map (6.1.4) is linear in the covector part, and this happens whenever the symplectic manifold is the cotangent bundle of a base manifold on which a Lie group acts. If $V_{(\xi)}$ is the vector on this manifold induced by the infinitesimal action of a 1-dimensional subgroup generated by the element ξ of the Lie algebra, the corresponding element $J_{(\xi)}$ of the moment map is $\langle y, V_{(\xi)} \rangle$, where y belongs to the cotangent space. The components of the moment map may therefore be identified with the vectors on the base manifold generated by the infinitesimal group action. Since the action of $\mathfrak{G}$ on $T^*\mathbb{R}^{2,4}$ preserves the null cone, the relative vectors are tangent to it, but not to the section Γ. Actually the group action is on the rays of the null cone, and the vectors can thus be safely projected on the tangent space to the section Γ. If V^B is a vector of $\mathbb{R}^{2,4}$ tangent to the null cone, this projection is given in coordinates by

$$V^\mu = g_\Gamma^{\mu\nu} \frac{\partial X^A}{\partial x^\nu} \eta_{AB} V^B.$$

Let us choose the section Γ as follows:

$$X^{-1} = \cos x^0$$
$$X^0 = \sin x^0$$
$$X^k = \frac{2x^k}{x^2 + 1}$$
$$X^4 = \frac{x^2 - 1}{x^2 + 1},$$

where $x^2 = \sum (x^i)^2$. Clearly, x^0 parametrizes S^1 and x^1, x^2, x^3 are *local* coordinates on $S^3 \subset \mathbb{R}^4$. The section $X^\alpha(x)$ is obtained through a stereographic projection from the North pole N of the unit sphere onto the hyperplane $x^1 x^2 x^3$: a point (the North pole) is missing from this parametrization and *restoring this point is just equivalent to the regularization of the Kepler problem.* The metric on the section Γ is

$$g_\Gamma = -y_0^2 + \frac{1}{4} y^2 (x^2 + 1)^2,$$

where the y's are the momenta canonically conjugate to the x's:

$$\{x^\mu, x^\nu\} = \{y_\mu, y_\nu\} = 0, \quad \{y_\mu, x^\nu\} = \delta_\mu^\nu.$$

After constructing the moment map $T^*\Gamma \to so^*(2,4)$, we notice that the constraint $g_\Gamma = 0$ is preserved by the conformal action, and defines a 7-dimensional presymplectic manifold. Dividing out the kernel of the presymplectic 2-form by setting for example $x^0 = 0$, then exchanging coordinates and momenta, we finally obtain the moment map

$$T^+(S^3 - \{N\}) = T^*(\mathbb{R}^3 - \{0\}) \to so^*(2,4)$$

given by

$$K_0(\vec{x}, \vec{y}) = \frac{1}{2} x (y^2 + 1),$$

$$\vec{G}(\vec{x}, \vec{y}) = \vec{x} \times \vec{y},$$

$$\vec{R}(\vec{x}, \vec{y}) = \frac{1}{2}(y^2 - 1)\vec{x} - (\vec{x} \cdot \vec{y})\vec{y}, \tag{6.1.5}$$

$$u(\vec{x}, \vec{y}) = \begin{pmatrix} \vec{u} \\ u_4 \end{pmatrix} = \begin{pmatrix} \frac{1}{2}(y^2 + 1)\vec{x} - (\vec{x} \cdot \vec{y})\vec{y} \\ -(\vec{x} \cdot \vec{y}) \end{pmatrix},$$

$$v(\vec{x}, \vec{y}) = \begin{pmatrix} \vec{v} \\ v_4 \end{pmatrix} = \begin{pmatrix} x\vec{y} \\ -\frac{1}{2} x (y^2 - 1) \end{pmatrix}.$$

We have defined

$$K_0 = -G_{-10}, \quad G_i = -\epsilon_{ihk} G_{hk}, \quad R_i = G_{i4}, \tag{6.1.6}$$
$$u_\alpha = G_{-1\alpha}, \quad v_\beta = G_{0\beta}.$$

If, at this point, the reader finds all that a bit confusing, well, no panic: the above procedure leads to the following.

First conclusion. *The above fifteen functions $K_0, G_h, R_k, u_\alpha, v_\beta$, defined by (6.1.5), satisfy the nine constraints*

$$u_\alpha u_\alpha = v_\beta v_\beta = K_0^2, \quad u_\alpha v_\alpha = 0, \tag{6.1.7}$$

$$G_i = \frac{1}{K_0} \epsilon_{ihk}(u_h v_k - u_k v_h), \quad R_k = \frac{1}{K_0}(u_4 v_k - u_k v_4), \tag{6.1.8}$$

from which

$$G^2 + R^2 = K_0^2, \quad \vec{G} \cdot \vec{R} = 0, \tag{6.1.9}$$

and have the following Poisson brackets:

$$\{G_i, G_j\} = -\epsilon_{ijk} G_k, \quad \{G_i, R_j\} = -\epsilon_{ijk} R_k, \quad \{R_i, R_j\} = -\epsilon_{ijk} G_k, \tag{6.1.10}$$

$$\{G_{\alpha\beta}, u_y\} = \eta_{\alpha y} u_\beta, \quad \{G_{\alpha\beta}, v_y\} = \eta_{\alpha y} v_\beta, \tag{6.1.11}$$

$$\{u_\alpha, u_\beta\} = \{v_\alpha, v_\beta\} = \frac{1}{K_0}(u_\alpha v_\beta - u_\beta v_\alpha), \quad \{u_\alpha, v_\beta\} = -K_0 \eta_{\alpha\beta}, \tag{6.1.12}$$

$$\{K_0, G_{\alpha\beta}\} = 0, \quad \{K_0, u_\alpha\} = v_\alpha, \quad \{K_0, v_\alpha\} = -u_\alpha. \tag{6.1.13}$$

We remark that the content of the first conclusion may also be checked with direct, but somewhat lengthy, calculations.

Let us briefly comment on this first conclusion. From (6.1.7) and (6.1.8) we infer that

(i) all fifteen functions are generated by two 4-vectors u and v which are orthogonal and of equal norm; indeed

(ii) $K_0 \vec{G}$ and $K_0 \vec{R}$ together form a 4-dimensional simple bivector, equal to the exterior product of u and v, $\vec{G}$ being the "magnetic" and $\vec{R}$ the "electric" part of the bivector;

(iii) K_0 is equal to the common norm of u and v.

Notice that relation (6.1.9) is an algebraic consequence of (6.1.7) and (6.1.8). Taking definitions (6.1.6) and (6.1.5) into account, the last four relations (6.1.10–6.1.13) are identical to (6.1.3), though more expressive. We infer

(i) from (6.1.10), that $\vec{G}$ and $\vec{R}$ are the generators of the SO(4) subgroup;

(ii) from (6.1.11), that u and v are 4-vectors under the action of the SO(4) subgroup;

(iii) from (6.1.12), that the manifold $\mathbb{R}^4 \times \mathbb{R}^4$, parametrized by u and v, has a Poisson structure of rank 6;

(iv) from (6.1.13), that K_0 may be viewed as a Hamiltonian whose phase space is the Poisson manifold of the previous item and which generates the motion

$$\begin{pmatrix} u_\alpha(x^0) \\ v_\alpha(x^0) \end{pmatrix} = \begin{pmatrix} \cos x^0 & \sin x^0 \\ -\sin x^0 & \cos x^0 \end{pmatrix} \begin{pmatrix} U_\alpha \\ V_\alpha \end{pmatrix}, \tag{6.1.14}$$

with U_α, V_α initial values;

moreover, $G_{\alpha\beta}$ is a first integral of the motion.

The motion of item (iv) is clearly the geodesic flow on the sphere S^3 (as confirmed by the expression of K_0 in the moment map), whose description we have thus placed in a group-theoretical framework. But what is the connection with the Kepler problem?

6.1.2 The Moser–Souriau Regularization

Now, a sort of miracle arises. Let us return, for a moment, to the 8-dimensional cotangent bundle $T^*\Gamma$, with local canonical coordinates x^μ, y_ν, and *before* applying the conformally invariant constraint

$$y_0 + \frac{1}{2}y(x^2 + 1) = 0, \tag{6.1.15}$$

deduced from $g_\Gamma = 0$, perform the canonical transformation

$$(\vec{q}, \vec{p}, t, p_t) = C(\vec{x}, \vec{y}, x^0, y_0)$$

given by

$$\vec{q} = -y_0 \vec{y}, \tag{6.1.16}$$

$$\vec{p} = \frac{\vec{x}}{y_0}, \tag{6.1.17}$$

$$t = -y_0^3 \left[x^0 - \frac{\vec{x} \cdot \vec{y}}{y_0} \right], \tag{6.1.18}$$

$$p_t = \frac{1}{2y_0^2}, \tag{6.1.19}$$

with inverse transformation

$$\vec{y} = \vec{q}\sqrt{2p_t},$$

$$\vec{x} = -\frac{\vec{p}}{\sqrt{2p_t}},$$

$$x^0 = (\sqrt{2p_t})^3 t - \sqrt{2p_t}(\vec{q} \cdot \vec{p}),$$

$$y_0 = -\frac{1}{\sqrt{2p_t}}.$$

C may be viewed as the composition of three canonical transformations:

(i) exchanging coordinates and momenta;

(ii) followed by (6.1.16) and (6.1.17), equivalent to an "energy rescaling";

(iii) concluding with (6.1.19).

Notice that (6.1.18) is enforced by the requirement of canonicity.
The substitution of C^{-1} into the constraint yields

$$-\sqrt{\frac{1}{2p_t}} + \frac{1}{2}q\sqrt{2p_t}\left(\frac{p^2}{2p_t}+1\right) = \sqrt{\frac{1}{2p_t}}\left[-1+\frac{1}{2}2qp_t\frac{p^2+2p_t}{2p_t}\right] = 0,$$

and hence,

$$p_t + \frac{1}{2}p^2 - \frac{1}{q} = 0;$$

i.e., $-p_t$ is equal to the Kepler unperturbed Hamiltonian $H_0(\vec{q},\vec{p})$. Moreover, it is easy to check that x^0 coincides with the eccentric anomaly defined
in (6.1.1), showing that s arises as the canonical transform of the physical
time; see Cordani (2003, page 140).

Let us transform the whole moment map $T^*\Gamma \to so^*(2,4)$ by C, restrict
to the constraint, and divide out the kernel of the presymplectic 2-form, setting $t = 0$. We get the transformed moment map $T^*(\mathbb{R}^3 - \{0\}) \to so^*(2,4)$:

$$L(\vec{q},\vec{p}) = \frac{1}{\sqrt{-2H_0}},$$

$$\vec{G}(\vec{q},\vec{p}) = \vec{q}\times\vec{p},$$

$$\vec{R}(\vec{q},\vec{p}) = \frac{1}{\sqrt{-2H_0}}\left(\vec{p}\times\vec{G}-\frac{\vec{q}}{q}\right), \qquad (6.1.20)$$

$$u(\vec{q},\vec{p}) = U\cos\Delta - V\sin\Delta,$$

$$v(\vec{q},\vec{p}) = U\sin\Delta + V\cos\Delta,$$

where

$$U = \begin{pmatrix}\frac{1}{\sqrt{-2H_0}}\left(-(\vec{q}\cdot\vec{p})\vec{p}+\frac{\vec{q}}{q}\right)\\ -(\vec{q}\cdot\vec{p})\end{pmatrix},$$

$$V = \begin{pmatrix}q\vec{p}\\ \frac{1}{\sqrt{-2H_0}}(1-p^2q)\end{pmatrix},$$

$$\Delta = \sqrt{-2H_0}(\vec{q}\cdot\vec{p}).$$

To stress that the moment map (6.1.20) is the canonical transform of
(6.1.5), we have retained the same symbols in the left members, with the sole
exception of K_0 and L for traditional reasons and notational convenience.

Second conclusion. *The fifteen functions of the moment map (6.1.20) satisfy
the same algebraic relations and, for the canonicity of the transformation,
the same Poisson brackets of the moment map (6.1.5). The symmetry properties of the Kepler problem are clearly evident: the Hamiltonian L has the*

generators of the SO(4) group as first integrals. Moreover, two 4-vectors have been constructed: they are orthogonal and of equal norm, and their dynamical evolution is an uniform rotation in a 2-plane fixed by the initial conditions.

6.1.3 The Kustaanheimo–Stiefel Regularization

In the previous subsection the Kepler problem has been regularized with the *Moser–Souriau method*. An alternative procedure is the *Kustaanheimo-Stiefel (KS) transformation*, which basically transforms the Hamiltonian $K_0(\vec{x}, \vec{y})$ of (6.1.5) into the Hamiltonian of a 4-dimensional harmonic oscillator whose motion is subject to a constraint. First, let us view how the *KS* transformation works.

DEFINITION 6.1 *The Kustaanheimo-Stiefel (KS) transformation is the map*

$$KS : \mathbb{R}^4 - \{0\} \times \mathbb{R}^4 \rightarrow \mathbb{R}^3 - \{0\} \times \mathbb{R}^3$$

given explicitly by

$$\begin{aligned}
x_1 &= 2(z_1 z_3 + z_2 z_4), \\
x_2 &= 2(z_2 z_3 - z_1 z_4), \\
x_3 &= -z_1^2 - z_2^2 + z_3^2 + z_4^2,
\end{aligned} \qquad (6.1.21)$$

from which

$$x = \|z\|^2,$$

and by

$$\begin{aligned}
y_1 &= -\frac{z_1 w_3 + z_2 w_4 + z_3 w_1 + z_4 w_2}{\|z\|^2}, \\
y_2 &= \frac{z_1 w_4 - z_2 w_3 - z_3 w_2 + z_4 w_1}{\|z\|^2}, \\
y_3 &= \frac{z_1 w_1 + z_2 w_2 - z_3 w_3 - z_4 w_4}{\|z\|^2}.
\end{aligned} \qquad (6.1.22)$$

The coordinates of the domain are subject to the constraint

$$z_1 w_2 - z_2 w_1 + z_3 w_4 - z_4 w_3 = 0. \qquad (6.1.23)$$

Because of this constraint the domain of the *KS* map is 7-dimensional, while the target space is 6-dimensional; thus the map has a 1-dimensional kernel. In fact, as one verifies, all pairs z', w' which are connected to z, w by the relations

$$
\begin{pmatrix} z_1' \\ z_2' \\ z_3' \\ z_4' \end{pmatrix} = \begin{pmatrix} \cos\beta & -\sin\beta & 0 & 0 \\ \sin\beta & \cos\beta & 0 & 0 \\ 0 & 0 & \cos\beta & -\sin\beta \\ 0 & 0 & \sin\beta & \cos\beta \end{pmatrix} \begin{pmatrix} z_1 \\ z_2 \\ z_3 \\ z_4 \end{pmatrix}, \tag{6.1.24}
$$

$$
\begin{pmatrix} w_1' \\ w_2' \\ w_3' \\ w_4' \end{pmatrix} = \begin{pmatrix} \cos\beta & -\sin\beta & 0 & 0 \\ \sin\beta & \cos\beta & 0 & 0 \\ 0 & 0 & \cos\beta & -\sin\beta \\ 0 & 0 & \sin\beta & \cos\beta \end{pmatrix} \begin{pmatrix} w_1 \\ w_2 \\ w_3 \\ w_4 \end{pmatrix}, \tag{6.1.25}
$$

are sent by the map into the same pair x, y for every β.

Let us now suppose that the z's are coordinates and the w's canonically conjugate momenta of an 8-dimensional phase space. After a canonical exchange between coordinates and momenta, let us compose the Hamiltonian $K_0(\vec{x}, \vec{y})$ with the KS map. A straightforward calculation shows that

$$
K_0\left(\vec{x}(z), \vec{y}(z, w)\right) = \frac{1}{2}\left(\|z\|^2 + \|w\|^2\right),
$$

which is the Hamiltonian of a 4-dimensional isotropic harmonic oscillator. If the constraint (6.1.23) is satisfied by the initial conditions, it will be respected by the dynamical evolution for all times.

All this machinery sounds somewhat mysterious, if we limit ourselves to the mere computations. For a better insight see Cordani (2003, Chapter VII), where it is shown that the KS transformation bases itself on the local isomorphism between SU(2,2) and SO(2,4). Let us briefly summarize the main points.

It is well known that SU(2) is the double covering of the SO(3) group. This means that the homomorphism SU(2) $\rightarrow$ SO(3) is a local isomorphism and maps SU(2) onto SO(3) in an *essentially* (2-1) fashion. The term "essentially" here refers to the fact that SU(2) is connected. There exists a higher dimensional but less known analogue, namely the homomorphism SU(2,2) $\rightarrow$ SO(2,4), which similarly maps SU(2,2) onto the identity-connected component of SO(2,4) in an essentially (2-1) fashion. For details see Guillemin & Sternberg (1990).

Let **h** be the 4×4 matrix of the quadratic form invariant under the SU(2,2) action:

$$
g^\dagger \mathbf{h} g = \mathbf{h}, \quad g \in SU(2,2).
$$

An element $\psi \in \mathbb{C}^{2,2}$ such that

$$
\psi^\dagger \mathbf{h} \psi = 0, \quad \psi \neq 0 \tag{6.1.26}
$$

is called a *null twistor* (Penrose 1974) and (Penrose 1967). Let T_0 be the space of null twistors, and assume the equivalence relation $\psi \sim \psi e^{i\beta}$, with

β a generic angle. The set $T_0/\sim$ of null twistors modulo a phase transformation is a 6-dimensional manifold, which may be equipped with the natural symplectic structure $d\Theta$, where

$$\Theta = \mathrm{Im}(\psi^\dagger \mathbf{h} d\psi). \tag{6.1.27}$$

Since the natural linear action of $SU(2,2)$ on $T_0/\sim$

$$\psi \mapsto \psi' = g\psi, \quad g \in SU(2,2)$$

is manifestly transitive and symplectic, $T_0/\sim$ is symplectomorphic to (a covering of) a coadjoint orbit $\mathcal{O}$ of $SU(2,2)$. The moment map

$$j : T_0/\sim \to \mathcal{O} \subset su^*(2,2)$$

which connects the two manifolds is given by

$$j(\psi) = -i\psi\psi^\dagger \mathbf{h} \quad \text{with } \psi^\dagger \mathbf{h}\, \psi = 0. \tag{6.1.28}$$

We verify that (6.1.28) is just an injective Ad^*-equivariant moment map $T_0/\sim \to su^*(2,2)$. Indeed

(i) $j(\psi) \in su^*(2,2)$ since

$$j^\dagger(\psi)\mathbf{h} + \mathbf{h}\, j(\psi) = 0, \quad \mathrm{Tr}\, j(\psi) = -i\psi^\dagger \mathbf{h}\, \psi = 0;$$

(ii) taking $T_0/\sim$ as the domain, we make j injective since ψ and ψ' have the same image if and only if $\psi' = \psi e^{i\beta}$;

(iii) the diagram

$$
\begin{array}{ccc}
 & SU(2,2) & \\
T_0/\sim & \xrightarrow{\hspace{2cm}} & T_0/\sim \\
\Big\downarrow{\scriptstyle j} & & \Big\downarrow{\scriptstyle j} \\
 & \mathbf{Ad}^* & \\
su^*(2,2) & \xrightarrow{\hspace{2cm}} & su^*(2,2)
\end{array}
$$

is commutative: if $g \in SU(2,2)$, by definition $g^\dagger \mathbf{h}\, g = \mathbf{h}$, so that $g^{-1} = \mathbf{h}^{-1} g^\dagger \mathbf{h}$, and the coadjoint action is given by

$$j \mapsto j' = g\, j\, g^{-1} = -ig\psi\psi^\dagger \mathbf{h}\mathbf{h}^{-1}g^\dagger \mathbf{h} = -i\, (g\psi)\, (g\psi)^\dagger\, \mathbf{h}.$$

Take

$$\psi = \begin{pmatrix} z \\ iw \end{pmatrix}, \quad z \in \mathbb{C}^2 - \{0\}, \quad w \in \mathbb{C}^2, \quad h = \begin{pmatrix} 0 & 1 \\ 1 & 0 \end{pmatrix};$$

then

$$j = \begin{pmatrix} zw^\dagger & izz^\dagger \\ iww^\dagger & -wz^\dagger \end{pmatrix}. \tag{6.1.29}$$

Let s_k be the three Pauli matrices

$$s_1 = \begin{pmatrix} 0 & 1 \\ 1 & 0 \end{pmatrix}, \quad s_2 = \begin{pmatrix} 0 & i \\ -i & 0 \end{pmatrix}, \quad s_3 = \begin{pmatrix} -1 & 0 \\ 0 & 1 \end{pmatrix},$$

which satisfy

$$s_h s_k + s_k s_h = 2\delta_{hk} 1, \quad s_h^\dagger = s_h, \quad \mathrm{Tr}\, s_h = 0.$$

The basis in $su(2,2)$ is given by

$$g_{hk} = \frac{1}{2} \begin{pmatrix} -s_h s_k & 0 \\ 0 & -s_h s_k \end{pmatrix}, \quad g_{0k} = \frac{1}{2} \begin{pmatrix} -s_k & 0 \\ 0 & s_k \end{pmatrix},$$

$$g_{-1,4} = \frac{1}{2} \begin{pmatrix} 1 & 0 \\ 0 & -1 \end{pmatrix},$$

$$g_{0,4} = \frac{1}{2} \begin{pmatrix} 0 & i1 \\ -i1 & 0 \end{pmatrix}, \quad g_{k4} = \frac{1}{2} \begin{pmatrix} 0 & is_k \\ is_k & 0 \end{pmatrix},$$

$$g_{-1.0} = \frac{1}{2} \begin{pmatrix} 0 & i1 \\ i1 & 0 \end{pmatrix}, \quad g_{-1,k} = \frac{1}{2} \begin{pmatrix} 0 & is_k \\ -is_k & 0 \end{pmatrix};$$

then it is simply a matter of calculation to check that

$$z = \begin{pmatrix} z_1 + iz_2 \\ z_3 + iz_4 \end{pmatrix}, \quad w = \begin{pmatrix} w_1 + iw_2 \\ w_3 + iw_4 \end{pmatrix}, \quad z_1, \dots, w_4 \in \mathbb{R},$$

and (6.1.21), (6.1.22) exhibit the equivalence between the two moment maps (6.1.29) and (6.1.5). Notice that the constraint (6.1.23) is a rephrasing of (6.1.26) and (6.1.25) of the equivalence relation $\sim$. The canonicity of the coordinates $z_1, \dots, w_4$ is induced by the 1-form (6.1.27).

6.1.4 Action-Angle Variables

Let us consider an n-dimensional integrable system, and suppose that the Hamiltonian is a function of $n - d$ action variables only. The dynamics takes place on $(n - d)$-dimensional tori, and the system is said to be *d-fold totally degenerate*. If $d = n - 1$, then all the orbits are periodic, and the system is *completely totally degenerate*: it is the case of the Kepler problem with negative energy.

The geometry of the totally degenerate systems is richer than that of the nondegenerate ones. As pointed out in Fassò (1996), the general situation is the following. The phase space of a d-fold totally degenerate Hamiltonian system (minus, in case, a subset of zero measure, comprising equilibrium points and separatrices) is fibered by $(n - d)$-dimensional invariant tori, filled by quasi-periodic motions, while the base space is an $(n + d)$-dimensional Poisson manifold of rank $2d$. The point is that the $2d$-dimensional symplectic leaves of this Poisson manifold are not in general homeomorphic to $\mathbb{R}^d \times \mathbb{T}^d$.

For example, in the 3-dimensional Kepler problem the fiber is the circle S^1 and the base, i.e., the space of the orbits, the 5-dimensional Poisson manifold spanned by the two orthogonal vectors $\vec{G}$ and $\vec{R}$. The symplectic leaves, i.e., the spaces of the orbits for a fixed energy, have topology $S^2 \times S^2$, as explained below after Equation (6.1.32).

It is clear that in the totally nondegenerate case the Lagrangian fibration in n-dimensional tori is intrinsic, so that the action-angle variables are essentially unique, up to the equivalence relations mentioned at the end of Subsection 2.3.9. In contrast, in the totally degenerate case only the fibration in $(n - d)$-dimensional tori is intrinsic; to complete this last to a Lagrangian fibration, one selects a piece of the symplectic leaf, homeomorphic to a cylinder $\mathbb{R}^d \times \mathbb{T}^d$, and attaches the factor $\mathbb{T}^d$ to the $(n - d)$-dimensional invariant tori. This operation is in general, i.e., when the symplectic leaves are not homeomorphic to $\mathbb{R}^d \times \mathbb{T}^d$, *neither global nor unique*, and every such choice of action-angle variables inevitably develops singularities.

In the case of the Kepler problem, one must add to L and l two other pairs of action-angle variables parametrizing the symplectic leaf $S^2 \times S^2$, but this parametrization is clearly only local. There are many possible choices, which are in general difficult to work out explicitly. Here we will consider the well-known Delaunay variables, and then the much less popular, but equally significant, Pauli variables.

Incidentally, there is a strict relation between total degeneration and over-integrability. We say that a Hamiltonian is *over-integrable* when it admits several systems of n inequivalent first integrals in involution. The level surfaces of these inequivalent systems are, by definition, different families of n-dimensional tori, whose intersection is just the family of $(n - d)$-dimensional tori on which the dynamics takes place. Whatever the n-tuple of the first integrals in involution, they will, in general, lose their independence in a subset of the phase space of zero measure; this is obviously due to the fact that one is looking for an object, a Lagrangian fibration in tori, that does not exist globally.

Delaunay Variables

The Delaunay action-angle variables are

$$L \text{ and } l, \quad G \text{ and } \omega, \quad G_3 \text{ and } \Omega$$
$$\text{with } \Theta = L \, dl + G \, d\omega + G_3 \, d\Omega.$$

They are so popular that sometimes they are improperly called "the" action-angle variables of the Kepler problem, suggesting erroneously that they are unique.

Their common derivation is reported in many books, for example, Goldstein (1980) and Cordani (2003), and follows basically the route traced when defining the action-angle coordinates in Subsection 2.3.9. However, this derivation is not trivial, and the involved integrations require some skill.

To get a geometric insight into this derivation, let us consider the geodesic flow on the 3-dimensional sphere S^3, which is equivalent to the Kepler problem. Notice that the following argument can be extended to a generic dimension. The moment map $T^*S^3 \rightarrow so^*(4)$, from the cotangent space to the dual of the Lie algebra of the rotation group in 4 dimensions, is given by the angular momentum $x \wedge y$, with $x, y \in \mathbb{R}^4$, $\|x\| = 1$ and $\langle x, y \rangle = 0$. The norm K_0 of this angular momentum has vanishing Poisson bracket with all the generators of the rotation group, in particular with the norm of the projection of this angular momentum into linear subspaces of $\mathbb{R}^4$: this explains why K_0, G, G_3 are in involution. Moreover, K_0, G, G_3 generate three distinct and commuting rotations in the three planes containing the total 4-dimensional angular momentum and its 3-dimensional and 2-dimensional projections, respectively: the angles parametrizing the three rotations are just s, ω, Ω. Lastly C, the canonical transformation (6.1.16–6.1.19), sends K_0, s in L, l.

The Delaunay variables are not well suited when the eccentricity E and the inclination i become very small, since they are plagued with the same singularities of the Keplerian elements. This implies that the variables L, G, G_3 differ in quantities that are of the order of the square of E and i. These three variables become identical if E and i vanish, and the angles l, ω, Ω become undetermined. In order to escape from this difficulty, Poincaré introduced the canonical variables

$$\Lambda = L, \qquad\qquad\qquad \lambda = l + \omega + \Omega,$$
$$\chi_{ecc} = \sqrt{2(L-G)} \cos(\omega + \Omega), \quad \eta_{ecc} = -\sqrt{2(L-G)} \sin(\omega + \Omega), \quad (6.1.30)$$
$$\chi_{inc} = \sqrt{2(G-G_3)} \cos\Omega, \qquad \eta_{inc} = -\sqrt{2(G-G_3)} \sin\Omega.$$

To pass from Delaunay to Poincaré variables, one composes two canonical transformations. The first is

$$
\begin{pmatrix} \Lambda \\ \rho_1 \\ \rho_2 \end{pmatrix} = \begin{pmatrix} 1 & 0 & 0 \\ 1 & -1 & 0 \\ 0 & 1 & -1 \end{pmatrix} \begin{pmatrix} L \\ G \\ G_3 \end{pmatrix},
$$

$$
\begin{pmatrix} \lambda \\ \varphi_1 \\ \varphi_2 \end{pmatrix} = \begin{pmatrix} 1 & 1 & 1 \\ 0 & -1 & -1 \\ 0 & 0 & -1 \end{pmatrix} \begin{pmatrix} l \\ \omega \\ \Omega \end{pmatrix},
$$

which is canonical since each of the two unimodular matrices is the transpose of the inverse of the other. The latter canonical transformation is of the type

$$
\chi = \sqrt{2\rho} \cos \varphi,
$$
$$
\eta = \sqrt{2\rho} \sin \varphi,
$$

analogous to the change in the plane from polar to Cartesian coordinates, which explains why the singularity in the origin is removed.

We remark that the three new actions Λ, ρ_1, ρ_2 are functions of the old ones, so that the Poincaré variables are different coordinates on the *same* Delaunay tori: the foliation in tori is identical for the two coordinate systems.

Pauli Variables

In contrast, Pauli action-angle variables parametrize different sets of tori. They were introduced in Cordani (2003), following the general method; here we adopt a different approach which is surprisingly simple and suggests that Pauli action-angle variables are, in some sense, the "natural" ones for the 3-dimensional Kepler problem.

We will work in the SU(2,2) framework of the previous subsection but with a more convenient basis, obtained with the linear transformation

$$
\mathbf{T} = \frac{1}{\sqrt{2}} \begin{pmatrix} \mathbf{1} & \mathbf{1} \\ \mathbf{1} & -\mathbf{1} \end{pmatrix},
$$

acting on $\mathbb{C}^{2,2}$, where $\mathbf{1}$ is the 2×2 unit matrix. It induces the transformations

$$
\mathbf{h} \mapsto \mathbf{H} = (\mathbf{T}^\dagger)^{-1} \mathbf{h} \, \mathbf{T}^{-1} = \begin{pmatrix} \mathbf{1} & 0 \\ 0 & -\mathbf{1} \end{pmatrix},
$$

$$
\mathbf{g}_{AB} \mapsto \mathbf{G}_{AB} = \mathbf{T} \, \mathbf{g}_{AB} \mathbf{T}^{-1}.
$$

The explicit expression of the elements of this basis is

$$\mathbf{G}_{hk} = \frac{1}{2}\begin{pmatrix} -s_h s_k & 0 \\ 0 & -s_h s_k \end{pmatrix}, \quad \mathbf{G}_{k,4} = \frac{1}{2}\begin{pmatrix} is_k & 0 \\ 0 & -is_k \end{pmatrix},$$

$$\mathbf{G}_{-1,0} = \frac{1}{2}\begin{pmatrix} i1 & 0 \\ 0 & -i1 \end{pmatrix},$$

$$\mathbf{G}_{0h} = \frac{1}{2}\begin{pmatrix} 0 & -s_h \\ -s_h & 0 \end{pmatrix}, \quad \mathbf{G}_{0,4} = \frac{1}{2}\begin{pmatrix} 0 & -i1 \\ i1 & 0 \end{pmatrix},$$

$$\mathbf{G}_{-1,h} = \frac{1}{2}\begin{pmatrix} 0 & -is_h \\ is_h & 0 \end{pmatrix}, \quad \mathbf{G}_{-1,4} = \frac{1}{2}\begin{pmatrix} 0 & 1 \\ 1 & 0 \end{pmatrix}.$$

$$(6.1.31)$$

Write every element of $\psi \in \mathbb{C}^{2,2}$ in exponential form: $\psi_k = \varrho_k e^{i\varphi_k}$ with $k = 1,\ldots,4$ not summed, so that $\mathrm{Im}(\overline{\psi}_k \, d\psi_k) = \varrho_k^2 d\varphi_k$. The 1-form (6.1.27) becomes

$$\Theta = \mathrm{Im}(\psi^\dagger \mathbf{H} d\psi) = \varrho_1^2 d\varphi_1 + \varrho_2^2 d\varphi_2 - \varrho_3^2 d\varphi_3 - \varrho_4^2 d\varphi_4.$$

Comparing the diagonal elements of the moment map $J(\psi) = -i\psi\psi^\dagger \mathbf{H}$ with the corresponding ones of the basis (6.1.31), we get

$$\Theta = \frac{1}{2}(K_0 + G_3 + R_3)d\varphi_1 + \frac{1}{2}(K_0 - G_3 - R_3)d\varphi_2$$

$$+ \frac{1}{2}(-K_0 + G_3 - R_3)d\varphi_3 + \frac{1}{2}(-K_0 - G_3 + R_3)d\varphi_4$$

$$= \frac{1}{2}K_0 d(\varphi_1 + \varphi_2 - \varphi_3 - \varphi_4) + S_3 d(\varphi_1 - \varphi_2) + D_3 d(\varphi_3 - \varphi_4).$$

We have defined

$$\vec{S} = \frac{1}{2}\left(\vec{G} + \vec{R}\right), \quad \vec{D} = \frac{1}{2}\left(\vec{G} - \vec{R}\right), \tag{6.1.32}$$

which we call *Pauli vectors* since they were used by the homonymous physicist to quantize the hydrogen atom. The Pauli vectors span two 2-dimensional spheres of radius $\frac{1}{2}\frac{1}{\sqrt{-2H_0}}$ and, as a direct consequence of the so(4) Lie algebra generated by $\vec{G}$ and $\vec{R}$, they satisfy the Lie algebra of SO(3):

$$\{S_1, S_2\} = -S_3, \quad \{S_2, S_3\} = -S_1, \quad \{S_3, S_1\} = -S_2,$$
$$\{D_1, D_2\} = -D_3, \quad \{D_2, D_3\} = -D_1, \quad \{D_3, D_1\} = -D_2, \tag{6.1.33}$$
$$\text{and} \quad \{\vec{S}, \vec{D}\} = 0.$$

These relations give to $\mathbb{R}^3 \times \mathbb{R}^3$ the structure of a Poisson manifold, with $S^2 \times S^2$ playing the role of symplectic leaves, and they are a consequence of the fact that SO(4) is the direct product of two copies of SO(3).

By a direct comparison with the moment map $J(\psi) = -i\psi\psi^\dagger H$ we get

$$s_P \overset{\text{def}}{=} \frac{1}{2}(\varphi_1 + \varphi_2 - \varphi_3 - \varphi_4) = \frac{1}{2}\arctan\left(\frac{2(u_3 v_3 + u_4 v_4)}{v_3^2 + v_4^2 - u_3^2 - u_4^2}\right),$$

$$\phi_S \overset{\text{def}}{=} \varphi_1 - \varphi_2 = \arctan\frac{S_2}{S_1},$$

$$\phi_D \overset{\text{def}}{=} \varphi_3 - \varphi_4 = \arctan\frac{D_2}{D_1},$$

which allows us to write the symplectic 1-form as

$$\Theta = K_0\, ds_P + S_3\, d\phi_S + D_3\, d\phi_D,$$

while the canonical transformation C gives

$$\Theta = L\, dl_P + S_3\, d\phi_S + D_3\, d\phi_D.$$

We recall that one uses the two moment maps (6.1.5) and (6.1.20), respectively, in order to make these last two expressions explicit.

We can therefore claim that L, S_3, D_3 are actions of the Kepler problem, with l_P, ϕ_S, ϕ_D the corresponding conjugate angles.

6.2 The Perturbed Kepler Problem

Our basic idea in order to normalize the perturbed Hamiltonian of the Kepler problem is to use the two 4-vectors u and v as global coordinates, because of their simple unperturbed dynamical evolution. But they are complicated functions of the position $\vec{q}$ and momentum $\vec{p}$, and it is generally impossible to explicitly obtain the expression of a generic perturbation. Instead, it is immediate to find from (6.1.5):

$$\vec{x} = \vec{u} - \vec{R}, \quad \vec{y} = \frac{\vec{v}}{x}, \quad x = K_0 + v_4. \tag{6.2.1}$$

This suggests using (6.1.5), instead of (6.1.20). Let us view how.

Consider the two Hamiltonians

$$H(\vec{x}, \vec{y}) = \frac{1}{2}y^2 - \frac{1}{x} + \varepsilon H_p(\vec{x}, \vec{y}), \tag{6.2.2}$$

$$K(\vec{x}, \vec{y}) = \frac{1}{2}x(y^2 + 1) + \varepsilon x H_p(\vec{x}, \vec{y}). \tag{6.2.3}$$

The reader will recognize the Hamiltonian of the Kepler problem in the unperturbed part of the first expression and the component K_0 of the moment map (6.1.5) in that of the latter.

Then consider the Hamilton equations generated by (6.2.2):

$$\frac{d\vec{x}}{dt} = \vec{y} + \varepsilon\frac{\partial H_p}{\partial\vec{y}}, \quad \frac{d\vec{y}}{dt} = -\frac{\vec{x}}{x^3} - \varepsilon\frac{\partial H_p}{\partial\vec{x}},$$

and those generated by (6.2.3), but with a new, for the moment unspecified, "false time"

$$\frac{d\vec{x}}{d\tau} = x\left(\vec{y} + \varepsilon\frac{\partial H_p}{\partial\vec{y}}\right), \quad \frac{d\vec{y}}{d\tau} = -x\left(K(\vec{x},\vec{y})\frac{\vec{x}}{x^3} + \varepsilon\frac{\partial H_p}{\partial\vec{x}}\right).$$

The Hamiltonian (6.2.3), along with the Hamiltonian (6.2.2), is a first integral of the motion, so that, suitably rescaling coordinates and momenta (as specified below), we can get

$$K(\vec{x},\vec{y}) = 1. \tag{6.2.4}$$

Now, defining

$$dt = x\,d\tau, \tag{6.2.5}$$

the two systems of Hamilton equations turn out to be identical. The normalization condition (6.2.4) implies

$$H(\vec{x},\vec{y}) = -\frac{1}{2}, \tag{6.2.6}$$

and the false time coincides with the eccentric anomaly to the vanishing of the perturbation. This is not surprising: for the rescaled Hamiltonian (6.2.4), the transformations (6.1.16), (6.1.17) reduce to an exchange between coordinates and momenta, while the infinitesimal version of (6.1.18) is just (6.2.5); lastly, (6.1.19) yields (6.2.6).

The abovely mentioned rescaling is easily found. Before exchanging coordinates and momenta, apply the covector dilation

$$\mathcal{R}_{xy} : x^\mu \to x^\mu, \; y_\nu \to \frac{y_\nu}{\lambda}, \quad \lambda \in \mathbb{R},$$

which leaves the constraint (6.1.15) invariant. Imposing

$$C \circ \mathcal{R}_{xy} = \mathcal{R}_{qp} \circ C,$$

this covector dilation induces the rescaling

$$\mathcal{R}_{qp} : \vec{q}, \vec{p}, t, p_t \to \frac{\vec{q}}{\lambda^2}, \lambda\vec{p}, \frac{t}{\lambda^3}, \lambda^2 p_t. \tag{6.2.7}$$

Let us suppose (this hypothesis is not very restrictive in practice) that the perturbation Hamiltonian is a homogeneous function of position and velocity:

$$H_p(\mu\vec{q}, \nu\vec{p}) = \mu^{d_q}\nu^{d_p} H_p(\vec{q},\vec{p}), \quad \mu, \nu, d_q, d_p \in \mathbb{R},$$

or, at most, a sum of homogeneous terms. The total Hamiltonian is a first integral of the motion, whose numerical value h is fixed by the initial condition:

$$\frac{1}{2}p^2 - \frac{1}{q} + \varepsilon H_p(\vec{q}, \vec{p}) = h.$$

Remembering the rescaling described above, put

$$\vec{q} = \lambda^2 \vec{x}, \quad \vec{p} = \frac{\vec{y}}{\lambda}, \quad \lambda = \frac{1}{\sqrt{-2h}}, \tag{6.2.8}$$

to obtain

$$\frac{1}{2}y^2 - \frac{1}{x} + \varepsilon_r H_p(\vec{x}, \vec{y}) = -\frac{1}{2}, \quad \varepsilon_r = \varepsilon \lambda^{2d_q + 2 - d_p}. \tag{6.2.9}$$

The total Hamiltonian has been rescaled to the required value, provided that the perturbative parameter is also suitably rescaled, as we suppose from now on. Therefore, integrating the flow of the Hamiltonian (6.2.3) is equivalent to the integration of the original problem.

Let us consider the solutions of the perturbed problem. As finding an *exact* analytical solution is generally out of the question, we can proceed with

(i) a normalization (or averaging) method, truncated at some order, or

(ii) numerical integration.

6.3 Normal Form of the Perturbed Kepler Problem

Let us apply the machinery of the reduction to normal form. Multiply the total Hamiltonian (6.2.3) from the left with the operator $\exp(\varepsilon \mathcal{L}_{\chi_1})$ where χ_1 is an unknown function, and retain the terms up to second order:

$$\exp(\varepsilon \mathcal{L}_{\chi_1})(K_0 + \varepsilon K_p)$$
$$= K_0 + \varepsilon(\mathcal{L}_{\chi_1} K_0 + K_p) + \varepsilon^2 \left(\mathcal{L}_{\chi_1} K_p + \frac{1}{2}\mathcal{L}_{\chi_1}^2 K_0 \right) + \mathcal{O}(\varepsilon^3).$$

Since $\mathcal{L}_{\chi_1} K_0 = -\mathcal{L}_{K_0} \chi_1$, we are led to solve the equation

$$\mathcal{L}_{K_0} \chi_1 = K_p - \overline{K}_1, \tag{6.3.1}$$

where $\overline{K}_1$ is the mean value of K_p:

$$\overline{K}_1 = \frac{1}{2\pi} \int_0^{2\pi} \phi_s^* K_p \, ds.$$

With ϕ_s we denote the dynamical flow generated by K_0 while $\phi_s^* K_p$ is the pull-back. The solution of (6.3.1) is given by

$$\chi_1 = \frac{1}{2\pi} \int_0^{2\pi} s\, \phi_s^*(K_p - \overline{K}_1)\, ds;$$

in fact,

$$\mathcal{L}_{K_0} \chi_1 = \frac{1}{2\pi} \int_0^{2\pi} s\, \frac{d}{ds}\, \phi_s^*(K_p - \overline{K}_1)\, ds$$

$$= \frac{1}{2\pi} \left[s\, \phi_s^*(K_p - \overline{K}_1) \Big|_0^{2\pi} - \int_0^{2\pi} \phi_s^*(K_p - \overline{K}_1)\, ds \right]$$

$$= K_p - \overline{K}_1,$$

using the periodicity of $K_p\, (u(s), v(s))$.

We have normalized the perturbed Hamiltonian to first order. Iterating the procedure and dropping the terms $\mathcal{O}(\varepsilon^{r+1})$, we obtain the truncated normalized Hamiltonian $K^{(r)}$ to order r

$$K^{(r)} = K_0 + \sum_{i=1}^r \varepsilon^i \overline{K}_i, \quad \{K_0, \overline{K}_i\} = 0\; \forall i.$$

The normalized Hamiltonian $K^{(r)}$ is invariant under the dynamical flow generated by K_0, and hence must be a function of only the first integrals $K_0, \vec{G}, \vec{R}$.

The normalization method does not lead, in general, to an integrable truncated Hamiltonian because of the complete total degeneration of the Kepler problem, but it allows us to lower the dimensions of the phase space from 6 to 4. For the complete integrability one needs another first integral: this case, frequently met in practice, will be investigated later.

In the generic case, we will exploit the existence of critical points[1] of the averaged Hamiltonian. Let us consider the averaged truncated Hamiltonian $K^{(r)}$ which induces a dynamical flow on the space of the orbits with topology $S^2 \times S^2$. Every point of the space of orbits is, by definition, an elliptic orbit in the original physical space, while a trajectory in the orbit space describes a continuous variation and deformation of the physical ellipse. It is obvious that a critical point of $K^{(r)}$ corresponds to a fixed ellipse, and thus to a periodic orbit. Although all the orbits of the unperturbed system are periodic, this need not be the case for the perturbed truncated case and a periodic solution will branch off only from a, in general finite, number of points, namely from the critical points. In Moser (1970) it is proved that, if

---

[1] Let us remember that a critical point of a vector field f is a point for which $f = 0$. Analogously, a critical point for a Hamiltonian H is a point for which $dH = 0$, i.e., the Hamiltonian vector field vanishes.

the critical points are nondegenerate[2] and if the neglected terms in the full perturbed Hamiltonian are small, this fact continues to be true for the full, not truncated motion. More exactly:

THEOREM 6.2 *Consider a perturbed Hamiltonian $h = h_0 + \varepsilon h_1 + \mathcal{O}(\varepsilon^2)$. Let M be the hypersurface $h_0 = constant$ and ϕ_t the flow on it generated by the vector field $\Omega^\sharp (dh_0)$, where $\Omega^\sharp$ is the inverse matrix of the symplectic form. Suppose that the orbits of ϕ_t all have period 2π, and let $\overline{M}$ be the quotient manifold with respect to the flow. Then, to every nondegenerate critical point $p^* \in \overline{M}$ of the averaged Hamiltonian*

$$\overline{h}_1 = \frac{1}{2\pi} \int_0^{2\pi} \phi_t^* h_1 \, dt,$$

there corresponds a periodic solution of the vector field $\Omega^\sharp (dh)$ that branches off from the orbit representing p^* and has period close to 2π.

The theorem can be applied to the normalized nontruncated Hamiltonian

$$K = K_0 + \sum_{i=1}^{r} \varepsilon^i \overline{K}_i + \mathcal{O}(\varepsilon^{r+1})$$

noticing that $K_0 = \frac{1}{2}x(y^2 + 1)$ is the square root of the Hamiltonian of the geodesic motion on the sphere S^3: it is clear that all the orbits have period 2π, with the "time" s denoting the angle of the point on the great circle.

Hence, we are led to find the critical points of the Hamiltonian $\sum \varepsilon^i \overline{K}_i$, defined on $\overline{M} = S^2 \times S^2$, or $\overline{M} = S^2$ in the plane case. This gives a lower bound for the number of periodic orbits of the perturbed Kepler problem. Indeed, the sum of the indices of the nondegenerate critical points of a vector field on a manifold $\overline{M}$ is equal to the Euler–Poincaré characteristic[3] of $\overline{M}$; see for example Milnor (1965) or Perko (1991). If the vector field is Hamiltonian, the nondegenerate critical points can only be elliptic or hyperbolic. Since the Euler-Poincaré characteristic of $\overline{M} = S^2 \times S^2$ is four, on $\overline{M}$ there exist at least four elliptic points. Therefore, the above topological argument ensures that, for a generic perturbation of the Kepler problem, at least four periodic orbits of the averaged truncated Hamiltonian exist, while the Moser theorem states that these orbits continue to persist for the full Hamiltonian.

---

[2] A critical point x^* of a vector field with components $f_k(x_1, \ldots, x_n)$, $k = 1, \ldots, n$ is said to be nondegenerate if $\det \left(\dfrac{\partial f_k}{\partial x_h} \right)_{x=x^*} \neq 0$.

[3] The Euler–Poincaré characteristic of a 2-dimensional compact manifold is given by $F - L + V$, where, for a given triangulation, F is the number of triangles, L of edges and V of vertices. It can be shown that it is a topological invariant. It is immediate to check that for the sphere it is equal to 2.

The existence of an elliptic equilibrium point for the averaged Hamiltonian allows one to respond to a basic question: What kind of action-angle variables do we utilize? As one verifies, the tori related to the Delaunay variables (the action-angle par excellence of the Kepler problem) are *not* slightly deformed by a generic, even if very small, perturbation but may be completely deformed. More exactly, while the trajectory in the complex plane of the variable Le^{il} (or Le^{is}) always resembles a circle, those in the planes of $Ge^{i\omega}$ and $G_3 e^{i\Omega}$ are in general radically different. This is not surprising, since only the first circle is intrinsically defined by the dynamics. The reason is the degeneration of the unperturbed Hamiltonian, and to escape the problem we must remove the cause.

We proceed as follows. Let us apply the first step of the general procedure leading to the progressive elimination of the angles in the perturbation Hamiltonian, but restrict ourselves to the elimination of the mean anomaly l only. Dropping the prime, we have

$$H = H_0 + \varepsilon \overline{H_p} + \mathcal{O}(\varepsilon^2), \quad \text{with } \overline{H_p} = \frac{1}{2\pi} \int_0^{2\pi} H_p \, dl,$$

the integration being performed along the unperturbed motion. The averaged Hamiltonian $\overline{H_p}$ is a function defined on the base of the fibration by circles of the phase space, and thus on the 5-dimensional Poisson manifold having $S^2 \times S^2$ as symplectic leaves and L as Casimir invariant. A family of Hamiltonian systems turns out to be so defined, all isomorphic and parametrized by L. They have the symplectic manifold $S^2 \times S^2$ as phase space and $\overline{H_p}(L, \vec{S}, \vec{D})$ as Hamiltonian. We know that, for any H_p, these Hamiltonian systems have at least four elliptic nondegenerate equilibrium points.

Consider one of these stationary points, which we may safely rotate with a canonical transformation at the two North poles. We can take local canonical coordinates $\chi_1, \eta_1, \chi_2, \eta_2$ of the Poincaré type, centered on this point, by putting

$$
\begin{aligned}
S_1 &= \frac{\chi_1}{2}\sqrt{2L - (\chi_1^2 + \eta_1^2)}, & D_1 &= \frac{\chi_2}{2}\sqrt{2L - (\chi_2^2 + \eta_2^2)}, \\
S_2 &= \frac{\eta_1}{2}\sqrt{2L - (\chi_1^2 + \eta_1^2)}, & D_2 &= \frac{\eta_2}{2}\sqrt{2L - (\chi_2^2 + \eta_2^2)}, \\
S_3 &= \frac{1}{2}L - \frac{1}{2}(\chi_1^2 + \eta_1^2), & D_3 &= \frac{1}{2}L - \frac{1}{2}(\chi_2^2 + \eta_2^2);
\end{aligned}
\tag{6.3.2}
$$

indeed, the Poisson brackets (6.1.33) result, assuming that

$$\{\chi_i, \chi_j\} = \{\eta_i, \eta_j\} = 0, \quad \{\eta_i, \chi_j\} = \delta_{ij}.$$

In order to deduce (6.3.2), recall that $\phi_S = \arctan(S_2/S_1)$, hence $\{\phi_S, S_3\} =$

1, and that $\left\| \vec{S} \right\| = \left\| \vec{D} \right\| = L/2$; then define

$$\chi_1 = \sqrt{L - 2S_3} \cos \phi_S, \quad \eta_1 = \sqrt{L - 2S_3} \sin \phi_S, \quad \Rightarrow \quad \{\eta_1, \chi_1\} = 1,$$

so that the left column of (6.3.2) follows. The right column follows analogously.

Substitute (6.3.2) into $\overline{H_p}(L, \vec{S}, \vec{D})$; this Hamiltonian has at least four elliptic stationary points and can be expanded into a Taylor series in the neighborhood of one of these points. Consider the constant (which does not generate any dynamics) and the quadratic terms only; the linear equations of motion $\underline{\dot{x}} = \Omega H \underline{x}$ (Ω is the inverse of the canonical symplectic matrix) generated by the quadratic Hamiltonian

$$H = \frac{1}{2}\underline{x}^t H \underline{x}, \quad \underline{x}^t = (\chi_1, \eta_1, \chi_2, \eta_2), \quad H = 4 \times 4 \text{ symmetric matrix},$$

can be reduced to those of two uncoupled harmonic oscillators by a linear canonical transformation $\chi, \eta \mapsto q, p$: see Theorem 2.33. Define the action-angle variables

$$J_k = \frac{1}{2}(q_k^2 + q_k^2), \quad \varphi_h = \arctan \frac{p_h}{q_h}, \quad h, k = 1, 2$$

and put $L = J_0$. Going back to the Taylor expansion of the averaged Hamiltonian, in the new variables we get

$$\overline{H_p} = a(J_0) + v_1(J_0)J_1 + v_2(J_0)J_2 + \cdots ,$$

where the dots stand for a polynomial in q and p variables, whose first terms are cubic. Provided v_1 and v_2 are rationally independent, the Birkhoff Theorem 3.10 on page 124 guarantees the existence of a canonical transformation that takes $\overline{H_p}$ into a power series in the variables J_1, J_2, with coefficients depending on J_0.

We are finally left with a total Hamiltonian of the type

$$H = -\frac{1}{2J_0^2} + \varepsilon \left[a(J_0) + \sum_h v_h(J_0)J_h + \frac{1}{2}\sum_{hk} v_{hk}(J_0)J_hJ_k + \cdots \right] + \mathcal{O}(\varepsilon^2),$$

which is the sum of an integrable "extended" Hamiltonian

$$H_0^{\text{ext}} = -\frac{1}{2J_0^2} + \varepsilon \left[a(J_0) + \sum_h v_h(J_0)J_h + \frac{1}{2}\sum_{hk} v_{hk}(J_0)J_hJ_k \right]$$

plus perturbative terms. These last are the terms $\mathcal{O}(\varepsilon^2)$ and those, linear in ε, coming from the Taylor expansion of $\overline{H_p}$, for which the role of perturbation parameter is played by the size of the neighborhood of the origin.

Moreover, if

$$\det(v_{hk}) \neq 0 \quad \text{and/or} \quad \det \begin{pmatrix} v_{hk} & v_h \\ v_k & 0 \end{pmatrix} \neq 0$$

then the integrable Hamiltonian H_0^{ext} satisfies the condition(s) required by the KAM and/or isoenergetic KAM theorem in a small neighborhood of the origin: cf Corollary 3.7 and (3.2.25), respectively. Besides, if the quasi-convexity (quasi-concavity) property (3.3.2) holds,[4] the Nekhoroshev theorem can also be applied.

The fundamental frequencies are

$$\omega_0 = \frac{\partial H_0^{\text{ext}}}{\partial J_0} = \frac{1}{J_0^3} + \mathcal{O}(\varepsilon),$$

$$\omega_1 = \frac{\partial H_0^{\text{ext}}}{\partial J_1} = \varepsilon \left[v_1(J_0) + v_{11}(J_0)J_1 + v_{12}(J_0)J_2 \right],$$

$$\omega_2 = \frac{\partial H_0^{\text{ext}}}{\partial J_2} = \varepsilon \left[v_2(J_0) + v_{21}(J_0)J_1 + v_{22}(J_0)J_2 \right].$$

The first frequency ω_0 is very close to that of the unperturbed case, while the ratios ω_1/ω_0 and ω_2/ω_0 are small of order ε.

It is important to stress however that, in practice, i.e., implementing the KEPLER program, we do *not* perform the Birkhoff normalization which is surely fully legitimate but, as it will be made clear by considering concrete examples in Chapter 8, gives poor results. The basic reason is that a complete normalization by means of the Birkhoff theorem "kills" excessively the features of the problem, giving rise to a dynamical evolution which is unable to capture the essence of the perturbed motion. The alternative method we will adopt consists in performing a double rotation in the $S^2 \times S^2$ space, so that one of the stationary points is moved to the two North poles; the system thus acquires an axial symmetry and the method of Section 6.5 can be applied. The method entails the elimination of only one angle, instead of two: the normalized Hamiltonian will still depend on one remaining angle, and thus will be integrable, and a global geometrical method can be invoked.

6.4 Numerical Integration

The KEPLER program (which will be described in the next chapter) provides two methods for the numerical integrations: perturbative, based on Moser–Souriau regularization, and non-perturbative, based on the Kustaanheimo–Stiefel regularization.

---

[4]In the present case: if the matrix (v_{hk}) has nonnegative or nonpositive eigenvalues.

6.4.1 Perturbative Method

Let us consider the dynamical system

$$\frac{dx}{dt} = f_0(x) + \varepsilon f_p(x), \quad x \in \mathbb{R}^N, \quad f_0, f_p : \mathbb{R}^N \to \mathbb{R}^N,$$

and assume that we know the general solution of the unperturbed part

$$x = x(t, X), \tag{6.4.1}$$

where $X \in \mathbb{R}^N$ are the integration constants. Suppose now that these integration constants are functions of time in such a way that (6.4.1) is a solution of the perturbed equation. Thus

$$\frac{dX}{dt} = \varepsilon \frac{\partial X}{\partial x} f_p(x(t, X)),$$

so that the numerical errors do not affect the integration of the known and predominant part.

We put (6.2.1) into (6.2.3), obtaining

$$K(u, v) = K_0(u, v) + \varepsilon K_p(u, v), \quad K_p = x H_p.$$

The Hamilton equations are

$$\frac{du_\alpha}{d\tau} = \{K, u_\alpha\}, \quad \frac{dv_\alpha}{d\tau} = \{K, v_\alpha\}. \tag{6.4.2}$$

We stress that the constraints (6.1.7) are preserved by the solutions of these Hamilton equations, whatever the explicit form of $K(u, v)$ will be. Equations (6.4.2) are eight first order equations in normal form, but the right-hand side member is the sum of a "large" vector field, whose integral flow is exactly known, and a "small" one. Therefore,

$$\frac{dU_\alpha}{d\tau} = \varepsilon\{K_p(u(\tau), v(\tau)), U_\alpha\}, \quad \frac{dV_\alpha}{d\tau} = \varepsilon\{K_p(u(\tau), v(\tau)), V_\alpha\}, \tag{6.4.3}$$

where U, V satisfy the same Poisson brackets as do u, v, and $u(\tau), v(\tau)$ is the general unperturbed solution (6.1.14).

Let us summarize how to proceed in order to implement the perturbative method.

- The starting point is the Hamiltonian $H(\vec{q}, \vec{p})$ and the initial conditions. Compute its constant value h and the scale factor $\lambda = 1/\sqrt{-2h}$.

- Rescale the initial conditions with (6.2.8).

- Define two 4-vectors u, v according to (6.1.5). They satisfy the constraints (6.1.7) and have the Poisson brackets described in the First conclusion. Compute their initial values using the rescaled conditions of the previous item.

- Integrate Equations (6.4.3) with the initial values computed in the previous item and the rescaled value (6.2.9) of the perturbative parameter.

- Insert the values of $U(\tau)$ and $V(\tau)$ computed in the previous item into Equations (6.1.14), with x^0 replaced by τ.

- Compute with (6.2.1) the physical rescaled coordinates as functions of τ; in case, integrate the rescaled (6.2.5)

$$t = \lambda^3 \int x(\tau)\, d\tau$$

to obtain the relation between τ and t.

- Finally, invert (6.2.8) and return to the physical, not rescaled coordinates.

6.4.2 Non-Perturbative Method

Alternatively, we directly integrate the equations regularized with the KS transformation, i.e., substituting (6.1.21) and (6.1.22) into (6.2.3). We get eight equations in the variables z_i, w_k, $i, k = 1, \ldots, 4$ satisfying the constraint (6.1.23). The unperturbed part is the Hamiltonian of a 4-dimensional harmonic oscillator. The point is that there exist some ODE solvers which integrate almost *exactly* the motion equations of the harmonic oscillator, such as IRK–Gauss and Tom which come with the KEPLER program. The non-perturbative method is well suited particularly when the perturbation is not too small.

6.5 Reduction under Axial Symmetry

Unlike the general case of the previous section, the axisymmetric perturbations lead, after normalization, to a Liouville integrable problem: besides K and K_0, the projection of the angular momentum on the symmetry axis is also a first integral, clearly in involution with the other two. This allows us to study the perturbed motion in a graphical and global way. The basic idea is to exploit the axial symmetry in order to further reduce the manifold of the orbits, which has $S^2 \times S^2$ topology, thus obtaining a symplectic manifold which turns out to be homeomorphic (but in general not diffeomorphic) to S^2. Then, the level surfaces of the twice reduced Hamiltonian intersect this symplectic manifold and give rise to a family of curves describing the perturbed motion.

Our first task is therefore to obtain the symplectic reduction, $\overline{\overline{M}}$ say, of the manifold of the orbits $\overline{M} \simeq S^2 \times S^2$ with respect to the circle or S^1-action of the axial symmetry. We may always assume the symmetry to be about the third axis, thus generated by the action of G_3. Following the Marsden-Weinstein theorem, we should consider the moment map $\Gamma : \overline{M} \to$ so$^*(2)$ given by G_3, choose some value,[5] with $-K_0 \le G_3 \le K_0$, and, on the 3-dimensional manifold $\Gamma^{-1}(G_3)$, divide out the flow generated by the action of G_3. Unfortunately, for $G_3 = 0$ this flow has fixed points, as is clear from Figure 6.2, so that, in this case, the reduced space will no longer be a *differentiable* manifold.

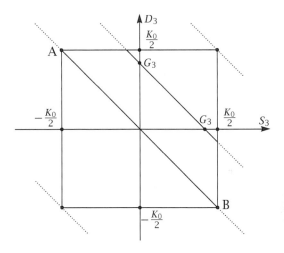

Figure 6.2: Once the value of G_3 is chosen, the admissible values of S_3, D_3 are those lying on the straight line $S_3 + D_3 = G_3$ and falling on and internally to the square. The manifold $\Gamma^{-1}(G_3)$ is empty for $|G_3| > K_0$ and trivial (one point) for $|G_3| = K_0$. For $-K_0 < G_3 < K_0$ and $G_3 \ne 0$, the rotational action has no fixed points, while for $G_3 = 0$ the two points $A \equiv \left(0, 0, -\frac{K_0}{2}, 0, 0, \frac{K_0}{2}\right)$ and $B \equiv \left(0, 0, \frac{K_0}{2}, 0, 0, -\frac{K_0}{2}\right)$ are fixed points.

This fact has been pointed out by Cushman (1991), who uses the general technique of "singular reduction" to construct $\overline{\overline{M}}$ for every value of G_3. To achieve the same goal, we prefer an *ad hoc* but very simple procedure. We do not directly reduce the symplectic manifold $S^2 \times S^2$, but, rather, the 6-dimensional Poisson manifold so$^*(4)$, of which $S^2 \times S^2$ is a symplectic leaf.

---

[5] To lighten the notation and at the cost of a slight abuse of language, we denote with G_3 (as for K_0) both the function and its value, the context making clear the sense.

In this way, the $G_3 = 0$ value is no longer a "bad" one. The Poisson reduction yields a 4-dimensional Poisson manifold of rank 2. The two Casimir invariants descend to the reduced manifold, and the twofold reduced Kepler manifold $\overline{\overline{M}}$ appears as a symplectic 2-dimensional leaf.

Let us view the details. Under the S^1-action on so*(4) of G_3, this last and R_3 stay unchanged, the action being effective only on G_1, G_2, R_1, R_2 not all null, and thus on $\mathbb{R}^4 - \{0\}$. To reduce the S^1-action, one implements the Hopf fibration:

$$\frac{\mathbb{R}^4 - \{0\}}{S^1} \simeq \frac{\mathbb{R}^+ \times S^3}{S^1} \simeq \mathbb{R}^+ \times S^2 \simeq \mathbb{R}^3 - \{0\}.$$

Explicitly, define the spinor, i.e., an element of $\mathbb{C}^2 - \{0\}$,

$$\psi = \begin{pmatrix} \psi_1 \\ \psi_2 \end{pmatrix} = \begin{pmatrix} G_1 + iG_2 \\ R_1 + iR_2 \end{pmatrix}, \quad \psi \neq 0, \tag{6.5.1}$$

on which the S^1-action of G_3 is given by $\psi \mapsto \psi e^{i\varphi}$. To reduce this action we must first find the 3-dimensional manifold that parametrizes the orbits of the action itself.

PROPOSITION 6.3 *The Hermitian matrices with null trace*

$$\psi\psi^\dagger - \frac{1}{2}(\psi^\dagger\psi)1_2 \stackrel{def}{=} \frac{1}{2}\begin{pmatrix} \xi_3 & \xi_4 + i\xi_2 \\ \xi_4 - i\xi_2 & -\xi_3 \end{pmatrix}, \tag{6.5.2}$$

$$(\xi_2, \xi_3, \xi_4) \in \mathbb{R}^3 - \{0\},$$

are in 1-1 correspondence with the orbits of the S^1-action on $\mathbb{C}^2 - \{0\}$.

Proof. Put $\psi = \begin{pmatrix} \sqrt{\varrho_1}e^{i\theta_1} \\ \sqrt{\varrho_2}e^{i\theta_2} \end{pmatrix}$. The S^1-action leaves invariant ϱ_1, ϱ_2 and $\theta_1 - \theta_2$, hence this triplet identifies the orbit univocally. Given this triplet, i.e., the orbit, the matrix is fixed unambiguously. Vice versa, given the matrix, i.e., ξ_2, ξ_3, ξ_4, by the definition of ψ we write

$$\xi_3 = \varrho_1 - \varrho_2,$$
$$\xi_4 + i\xi_2 = 2\sqrt{\varrho_1\varrho_2}e^{i(\theta_1 - \theta_2)}.$$

From the latter equation we get $\theta_1 - \theta_2$. Moreover,

$$\varrho_2 = \varrho_1 - \xi_3,$$
$$\varrho_1\varrho_2 = \frac{1}{4}(\xi_4^2 + \xi_2^2),$$

so that a direct calculation allows us to find unambiguously the real non negative values of ϱ_1 and ϱ_2.   **QED**

Insert (6.5.1) into (6.5.2); then, putting $G_3 = $ constant, we obtain the Poisson reduction with respect to the S^1-action:

$$so^*(4) \simeq \mathbb{R}^2 \times (\mathbb{C}^2 - \{0\}) \to \mathbb{R} \times (\mathbb{R}^3 - \{0\}) : (\vec{G}, \vec{R}) \mapsto (\xi_1, \xi_2, \xi_3, \xi_4),$$

where

$$\begin{aligned}
\xi_1 &= R_3, & \xi_2 &= 2(G_2 R_1 - G_1 R_2), \\
\xi_3 &= G_1^2 + G_2^2 - R_1^2 - R_2^2, & \xi_4 &= 2(G_1 R_1 + G_2 R_2).
\end{aligned} \tag{6.5.3}$$

The two Casimir invariants of $so^*(4)$ are

$$G^2 + R^2 = G_3^2 + \xi_1^2 + \sqrt{\xi_2^2 + \xi_3^2 + \xi_4^2},$$

$$\vec{G} \cdot \vec{R} = G_3 \xi_1 + \frac{1}{2} \xi_4,$$

while the manifold $\overline{M} \simeq S^2 \times S^2$ is characterized by

$$G^2 + R^2 = K_0^2, \quad \vec{G} \cdot \vec{R} = 0.$$

The second relation allows us to eliminate $\xi_4 = -2G_3 \xi_1$ in the first relation, so that the twofold reduced Kepler manifold turns out to be a 2-dimensional algebraic manifold imbedded in $\mathbb{R}^3$:

$$\left(K_0^2 + G_3^2 - \xi_1^2\right)^2 - \xi_2^2 - \xi_3^2 = 4K_0^2 G_3^2. \tag{6.5.4}$$

We say that Equation (6.5.4) is of the "R-type," since it is symmetric about the $\xi_1 = R_3$ axis. See Figure 6.3.

From the definitions (6.5.3), and bearing the Poisson brackets (6.1.10) in mind, one calculates

$$\{\xi_1, \xi_2\} = -2\xi_3, \quad \{\xi_1, \xi_3\} = 2\xi_2, \quad \{\xi_2, \xi_3\} = 4\xi_1(\xi_1^2 - K_0^2 - G_3^2). \tag{6.5.5}$$

Putting

$$C(\xi) = \left(K_0^2 + G_3^2 - \xi_1^2\right)^2 - \xi_2^2 - \xi_3^2, \tag{6.5.6}$$

we find that the Poisson brackets (6.5.5) can be written in the compact form

$$\{\xi_i, \xi_j\} = \epsilon_{ijh} \frac{\partial C(\xi)}{\partial \xi_h}.$$

Plainly, $C(\xi)$ is a Casimir function:

$$\{\xi_i, C(\xi)\} = \{\xi_i, \xi_h\} \frac{\partial C(\xi)}{\partial \xi_h} = \epsilon_{ijh} \frac{\partial C(\xi)}{\partial \xi_h} \frac{\partial C(\xi)}{\partial \xi_j} = 0.$$

Summing up, we have proved the following proposition.

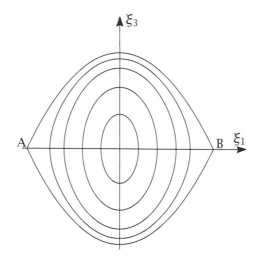

Figure 6.3: Section on the plane $\xi_1\xi_3$ of the algebraic manifold $\overline{\overline{M}}$ of the R-type, for $K_0 = 1$ and various values of $G_3 = 0.8, 0.6, 0.4, 0.25, 0$. The manifold has rotational symmetry around the axis ξ_1. Notice, for $G_3 = 0$, the two singular points A and B on axis ξ_1, which make the manifold homeomorphic, but nondiffeomorphic, to the sphere.

PROPOSITION 6.4 *The symplectic reduction of the manifold $S^2 \times S^2$ with respect to the S^1-action generated by G_3 is the 2-dimensional algebraic manifold (6.5.4), which is a symplectic leaf of $\mathbb{R}^3$ equipped with the Poisson structure (6.5.5). Varying G_3, these symplectic leaves are diffeomorphic (for $G_3 \neq 0$) but only homeomorphic (for $G_3 = 0$) to the 2-dimensional sphere S^2.*

We will introduce now another parametrization of $\overline{\overline{M}}$ called of G-type and characterized by the symmetry about the third axis. Put

$$\eta_1 = \xi_1 \sqrt{\frac{1}{2}(\xi_3 - \xi_1^2 + K_0^2 + G_3^2)},$$

$$\eta_2 = \frac{1}{2}\xi_2,$$

$$\eta_3 = \sqrt{\frac{1}{2}(\xi_3 - \xi_1^2 + K_0^2 + G_3^2)},$$

whose (formal) inverse relations are

$$\xi_1 = \frac{\eta_1}{\eta_3}, \quad \xi_2 = 2\eta_2, \quad \xi_3 = 2\eta_3^2 + \frac{\eta_1^2}{\eta_3^2} - (K_0^2 + G_3^2).$$

Clearly, this change of coordinates is a diffeomorphism for $\eta_3 \neq 0$. Recalling the definitions (6.5.3), one easily verifies that $\eta_3 = G$, the norm of the total angular momentum: this explains the choice of the name "G-type." In particular, the condition $\eta_3 \neq 0$ holds if $G_3 \neq 0$, so that manifolds diffeomorphic to S^2 are transformed into manifolds again diffeomorphic to S^2.

Let us examine how the algebraic manifold and the Poisson structure change when passing from the R-type to the G-type. Substituting the above transformation into the definition (6.5.6) we find

$$C(\eta) = -4 \left[\eta_1^2 + \eta_2^2 + \eta_3^4 - (K_0^2 + G_3^2)\eta_3^2 \right],$$

and the equation of $\overline{\overline{M}}$ is written as

$$\eta_1^2 + \eta_2^2 + \left[\eta_3^2 - \frac{1}{2}(K_0^2 + G_3^2) \right]^2 = \frac{1}{4}(K_0^2 - G_3^2)^2. \qquad (6.5.7)$$

See Figure 6.4.

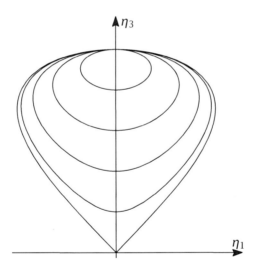

Figure 6.4: Section on the plane $\eta_1\eta_3$ of the algebraic manifold $\overline{\overline{M}}$ of the G-type for $K_0 = 1$ and various values of $G_3 = 0.8, 0.6, 0.4, 0.2, 0$. The manifold has rotational symmetry about the axis η_3.

The Poisson structure is

$$\{\eta_r, \eta_s\} = \frac{\partial \eta_r}{\partial \xi_i} \frac{\partial \eta_s}{\partial \xi_h} \epsilon_{ihk} \frac{\partial C(\eta)}{\partial \eta_t} \frac{\partial \eta_t}{\xi_k} = \det\left(\frac{\partial \eta}{\partial \xi}\right) \epsilon_{rst} \frac{\partial C(\eta)}{\partial \eta_t},$$

and since

$$\det\left(\frac{\partial \xi}{\partial \eta}\right) = \det\begin{pmatrix} \dfrac{1}{\eta_3} & 0 & -\dfrac{\eta_1}{\eta_3^2} \\ 0 & 2 & 0 \\ 2\dfrac{\eta_1}{\eta_3^2} & 0 & 4\eta_3 - 2\dfrac{\eta_1^2}{\eta_3^3} \end{pmatrix} = 8,$$

we find

$$\{\eta_r, \eta_s\} = \frac{1}{8}\epsilon_{rst}\frac{\partial C(\eta)}{\partial \eta_t}.$$

We write the transformed Poisson structure explicitly as

$$\{\eta_1, \eta_2\} = -2\eta_3\left[\eta_3^2 - \frac{1}{2}(K_0^2 + G_3^2)\right],$$
$$\{\eta_3, \eta_2\} = \eta_1, \qquad\qquad\qquad\qquad (6.5.8)$$
$$\{\eta_3, \eta_1\} = -\eta_2.$$

Owing to the rotational symmetry about the ξ_1 and the η_3 axis, respectively, we can immediately find action-angle coordinates in the two cases. In fact, from (6.5.5) and (6.5.8) we deduce

$$\left\{\xi_1, -\frac{1}{2}\arctan\frac{\xi_2}{\xi_3}\right\} = 1, \quad \left\{\eta_3, -\arctan\frac{\eta_1}{\eta_2}\right\} = 1. \qquad (6.5.9)$$

In a slightly different form this result was already known. With regard to the first bracket, we write

$$\arctan\frac{S_2}{S_1} - \arctan\frac{D_2}{D_1} = \arctan\frac{S_2 D_1 - S_1 D_2}{S_1 D_1 + S_2 D_2}$$
$$= \arctan\frac{2(G_1 R_2 - G_2 R_1)}{G_1^2 + G_2^2 - R_1^2 - R_2^2} = -\arctan\frac{\xi_2}{\xi_3};$$

we recall that $\xi_1 = R_3 = S_3 - D_3$. With regard to the second bracket, we find

$$\tan\omega = \frac{R_3 G}{-G_2 R_1 + G_1 R_2} = -\frac{\eta_1}{\eta_2}; \qquad (6.5.10)$$

we recall that $\eta_3 = G$. We have thus found again the third pair of action-angle variables for the Pauli and Delaunay parameters respectively.

Let us now consider the G-type parametrization, but what we will say can be, with minor changes, repeated for the R-type. If the full Hamiltonian is axisymmetric, the perturbing term turns out to be defined on the twofold reduced Kepler manifold $\overline{\overline{M}}$, once averaged with respect to the unperturbed evolution. Therefore, the Hamiltonian can be written as

$$K = K_0 + \sum_{i=1}^{r} \epsilon^i \overline{\overline{K}}_i(K_0, G_3, \eta_1, \eta_2, \eta_3) + \mathcal{O}(\epsilon^{r+1}) = K^{(r)} + \mathcal{O}(\epsilon^{r+1})$$

with $\{\overline{K}_i, K_0\} = 0 = \{\overline{K}_i, G_3\}$. Disregarding the term $\mathcal{O}(\varepsilon^{r+1})$, we are left with the truncated Hamiltonian $K^{(r)}$, which forms a system of three first integrals in involution with K_0 and G_3: the system is Liouville integrable.

Considering K_0 and G_3 as parameters, the equation $K^{(r)} = $ constant describes a family of 2-dimensional surfaces which intersect the reduced Kepler manifold $\overline{\overline{M}}$ in a family of curves. A point on one of these intersection curves represents an elliptic orbit rotating around the x_3 axis with frequency $\frac{\partial K^{(r)}}{\partial G_3}$, while the motion on the intersection curve describes the change in the eccentricity and the inclination of the elliptic orbit. The interesting fact is that this description is *global*, and may reveal some features hidden in a *local* description, as is usual in standard perturbation theory.

Anyway, the local description can be easily recovered. To this end, remembering (6.5.7) and (6.5.10), we put

$$\eta_1 = -\sin\omega\sqrt{\frac{1}{4}(K_0^2 - G_3^2)^2 - \left[G^2 - \frac{1}{2}(K_0^2 + G_3^2)\right]^2},$$

$$\eta_2 = \cos\omega\sqrt{\frac{1}{4}(K_0^2 - G_3^2)^2 - \left[G^2 - \frac{1}{2}(K_0^2 + G_3^2)\right]^2},$$

$$\eta_3 = G,$$

and substitute in $K^{(r)}$. We recall that $\{G, \omega\} = 1$.

Analogously,

$$\xi_1 = R_3,$$

$$\xi_2 = -\sin 2\varphi\sqrt{(K_0^2 + G_3^2 - R_3^2)^2 - 4K_0^2 G_3^2},$$

$$\xi_3 = \cos 2\varphi\sqrt{(K_0^2 + G_3^2 - R_3^2)^2 - 4K_0^2 G_3^2},$$

with $\{R_3, \varphi\} = 1$.

CHAPTER 7

The KEPLER Program

What is to be done?

— LENIN

In this chapter we will describe how to use the KEPLER program, trying to be self-contained and without referring excessively to the underlying mathematical structure. The command statements are gathered in four windows, which one selects in the pop-up menu in the left-top corner. See the Appendix A for snapshots of the MATLAB programs in the CD.

7.1 First Window: Single Orbit Analysis

The window of the program is formed by seven panels, to be read like a printed page: from left to right and from top to bottom.

7.1.1 Perturbation Hamiltonian

In the left column of this panel, the user can choose the perturbation Hamiltonian H_p as a linear combination of five single Hamiltonians, i.e.,

$$H_p = \varepsilon_1 H_1 + \varepsilon_2 H_2 + \varepsilon_3 H_3 + \varepsilon_4 H_4 + \varepsilon_5 H_5,$$

the coefficients ε being the perturbative parameters to be fixed in the right column. Every Hamiltonian must be homogeneous in the components of the position and momentum vectors, with each homogeneity degree completely arbitrary. When the integration process starts, the perturbative parameters are suitably rescaled by the program, taking into account initial conditions and homogeneity degrees: see (6.2.8) and (6.2.9). If these rescaled, i.e., effective, values exceed a threshold, the user is warned that the hypothesis of small perturbation is no longer satisfied. In this case, proceed with caution, looking at the integration errors calculated in the fifth panel and comparing with the result of the non-perturbative method.

Below we will describe the perturbation Hamiltonians that come with KEPLER. To add some other perturbation Hamiltonians, proceed as follows.

(i) Write a file containing the function which describes the perturbation; it is convenient to take one of the files in the directory H_perturbation as a template.

(ii) Open Kepler.fig in Guide, right-click on the first pop-up menu of the panel, and left-click on Property inspector.

(iii) Click on "String", then add your file name.

(iv) Click on "UserData", then add the *total degree* of the Hamiltonian, which is defined as: position degree minus half of the momentum degree.

(v) Repeat for the other four pop-up menus in the panel.

Zeeman Effect

The perturbing Hamiltonian

$$H_p = q_1 p_2 - q_2 p_1$$

is that of the Zeeman effect: a weak constant magnetic field directed along the third axis and acting on the hydrogen atom. Alternatively, it describes the inertial forces, i.e., centrifugal and Coriolis, in a rotating reference system. This perturbation is Liouville integrable: the elliptic orbit rotates uniformly around the vertical axis and all the Kepler elements stay constant except the longitude of the ascending node, which evolves uniformly with time. This Hamiltonian, along with the other integrable ones, can be used to test the precision of the program.

The file Zeeman3.m contains this Hamiltonian, while the similar files Zeeman1.m and Zeeman2.m are relative to magnetic fields directed along the first and the second axis, respectively.

Stark Effect

The perturbing Hamiltonian

$$H_p = q_3$$

is that of the Stark effect, a constant electric field directed along the third axis and acting on the hydrogen atom. The perturbation is Liouville integrable: besides the Hamiltonian and the third component of the angular momentum, also $E_3 - \frac{1}{2}\varepsilon(q_1^2 + q_2^2)$ is a first integral, as one may check by clicking on the "User Functions" button and choosing the file `1st_Integral_Stark3.m`. We recall that E_3 is the third component of the eccentricity vector.

   With initial conditions belonging to a vertical plane, the axial symmetry ensures that the dynamical evolution belongs to this plane. As the reader can verify, the orbit fills the area enclosed by the arcs of two confocal parabolas.

   The perturbing Hamiltonian is in the file `Stark3.m`. The similar files `Stark1.m` and `Stark2.m` are relative to electric fields directed along the first and the second axis, respectively.

Quadratic Zeeman Effect

The perturbing Hamiltonian

$$H_p = q_1^2 + q_2^2$$

describes the second order effect of a constant magnetic field directed along the third axis, when the observer is rotating about the same axis. See Cushman (1991) or Cordani (2003, page 311). The perturbation is not integrable.

   This Hamiltonian is in the file `QZ3.m`, while the similar files `QZ1.m` and `QZ2.m` are relative to magnetic fields directed along the first and the second axis, respectively.

Euler Problem

The perturbing Hamiltonian

$$H_p = -\frac{1}{\sqrt{q_1^2 + q_2^2 + (q_3 - d)^2}}$$

(see the file `Euler3.m`) is that of the Euler problem: a secondary small mass ε is placed at the point $P = (0, 0, d)$. At the prompt, the user can choose the numerical value of the distance d. The perturbation is Liouville integrable and, besides the Hamiltonian and the third component of angular momentum, also

$$E_3 - \frac{1}{d}G^2 + \varepsilon \frac{q_3 - d}{\sqrt{q_1^2 + q_2^2 + (q_3 - d)^2}}$$

is a first integral: check with the file 1st_Integral_Euler3.m. The numerical output of KEPLER may be compared with that of EULER, a program written in the MAPLE language, which integrates the problem *analytically* with $0 < \varepsilon \leq 1$. See the appendix at the end of Chapter 2 and Cordani (2003, page 179). Remark that the fixed point P is a nonregularized singularity and that the regularization method fails when the moving point approaches P too closely. In this case, try with the standard method.

With initial conditions belonging to a vertical plane, the axial symmetry ensures that the dynamical evolution lies in this plane. As the reader can verify, the orbit fills the area enclosed by the arcs of an ellipse and a hyperbola which are confocal.

The file Euler1.m contains analogously the Hamiltonian of a point of mass ε placed in P = $(d, 0, 0)$. Along with the Hamiltonians in Zeeman3.m and p_2.m, one can form the Hamiltonian of the restricted circular three-body problem

$$H = \frac{1}{2}p^2 - \frac{1}{q} - \varepsilon \frac{1}{\sqrt{(q_1 - d)^2 + q_2^2 + q_3^2}}$$
$$+ \sqrt{\frac{1 + \varepsilon}{d^3}}(q_1 p_2 - q_2 p_1) - \frac{\varepsilon}{\sqrt{d(1 + \varepsilon)}} p_2.$$

Remark that only when the moving point stays much closer to the primary in the origin than to the fixed point P are we in a perturbative situation. As an example, put $d = 5.2$ with Euler1.m, then 3body_SJM.mat allows us to load the numerical values of the Sun-Jupiter-Mercury system.

The Lagrange points L_4 and L_5 are placed in

$$q_1 = \frac{d}{2}, \quad q_2 = \pm \frac{\sqrt{3}}{2}d,$$
$$p_1 = \pm \frac{\sqrt{3}}{2}\sqrt{\frac{1 + \varepsilon}{d}}, \quad p_2 = -\frac{1}{2}\sqrt{\frac{1 + \varepsilon}{d}} + \frac{\varepsilon}{\sqrt{d(1 + \varepsilon)}}.$$

Oscillator

The perturbing Hamiltonian

$$H_p = \frac{1}{2}(k_1 q_1^2 + k_2 q_2^2 + k_3 q_3^2)$$

is that of an anisotropic harmonic oscillator. In the file Oscillator.m the user may change the anisotropy coefficients. The problem is Liouville integrable.

p_2

This Hamiltonian, when combined with Euler1.m and Zeeman3.m, provides the perturbation Hamiltonian of the restricted circular three-body problem.

Satellite

The perturbing Hamiltonian

$$H_p = \frac{q_3^2 - \frac{1}{2}(q_1^2 + q_2^2)}{(q_1^2 + q_2^2 + q_3^2)^{5/2}}$$

is that of an artificial satellite about a nonspherical primary; only the second harmonic is considered, but obviously other terms may be added. Only the Hamiltonian and the third component of angular momentum are first integrals, and the problem is not Liouville integrable. See Cushman (1983), Deprit (1981), and Coffey, Deprit & Deprit (1994). The origin is a non-regularized singularity.

Hill Potential

The Hill potential
$$H_p = -2q_1^2 + q_2^2 + q_3^2$$

takes into account the perturbation of the Sun on the Moon, and it must be combined with the Zeeman effect (rotating reference system) to describe the motion of the Moon around the Earth. Only Hamiltonian and third component of angular momentum are first integrals, and the problem is not Liouville integrable. See Kummer (1983) and Cordani (2003, page 302).

Inverse Quadratic and Cubic Potentials

The potentials
$$H_p = \frac{1}{q^2} \quad \text{and} \quad H_p = \frac{1}{q^3}$$

are clearly Liouville integrable. The origin is a non-regularized singularity.

Anisotropic

The perturbing Hamiltonian
$$H_p = p_3^2$$

gives what is called the *anisotropic Kepler problem*.

7.1.2   Initial Conditions

In the second panel, "Initial conditions", the user can fix the initial conditions, choosing between position and velocity or Keplerian parameters. Once one of the two subpanels "Position and velocity" or "Kepler parameters" has been changed, the other must be updated before continuing, by clicking on the corresponding button "Update". Notice that Longitude of the ascending node, Argument of pericenter, Inclination, and True anomaly are expressed in degrees.

If the "Update" button in the seventh panel is enabled, the third button, "Update Pauli v. and Energy", allows one to synchronize the values of the two normalized Pauli vectors $\vec{S}, \vec{D}$ and of the energy with the initial conditions. We recall that giving normalized Pauli vectors and energy is equivalent to fixing $\vec{G}$ and $\vec{R}$, and hence to fixing an elliptic orbit.

7.1.3   ODE Solver

In the third panel, "ODE solver", the user can choose the method for the numerical integration with relative options. Several methods are supplied in the pop-up menu; see the last section of Chapter 4. Some methods integrate with an adaptive, i.e., variable step, but the output can be obtained at regularly spaced points by fixing the "Output Step" value; carrying out the frequency analysis of the seventh panel of the window does require indeed a fixed stepsize. The Output Step and the Stepsize for the fixed-step integrators are defined as the ratio: total numbers of cycles over total number of steps. The adaptive step, though useful, is not strictly necessary if a regularization method of the fifth panel is used, as is strongly recommended.

In the "Options" subpanel the user can choose the relative and absolute tolerance in the integration process, and the initial, maximal, and output step for ode113, ode45, Dop853, and Odex. These methods are adaptive; thus the integration step is chosen automatically by the algorithm and is not affected by the user-selected value of the output step. All the other methods instead have a fixed integration step, which can be chosen by fixing the corresponding "StepSize" value.

The field "Project on the constraints every n revolutions" allows one to restore the constraints (6.1.7) for the perturbative method and the constraint (6.1.23), or equivalently (6.1.26), for the non-perturbative method. If the field is left blank, the projection on the constraints is never performed.

7.1.4   Revolution Number

The fourth panel allows one to choose the total number of revolutions in the first subpanel, and the revolution interval to display in the second subpanel: this is useful in the case of long integration, when plotting the whole solution would generate an overloaded figure. The third subpanel is useful in the Long Frequency Analysis (LFA), or in general for *very* long orbits, to avoid MATLAB going out of memory. *Be aware that clearing the memory entails a total loss of information.* The LFA procedure is explained below, in Subsection 7.5.2.

7.1.5 Regularization Method

The fifth panel, "Regularization method", is the core of the program. The button "Integrate" performs the integration of the chosen Hamiltonian with the parameter values fixed in the previous panels. The user can choose between two methods: Perturbative and Non-perturbative, as explained in Section 6.4. Once the procedure is complete, several buttons are enabled, allowing the display of the dynamical evolution of some variables.

Space and Momentum Trajectories

By clicking on this button, two figures are displayed relative to the trajectories of the momentum and position vectors. If "Output Step" in the "ODE solver" panel is large, the trajectories may appear broken. Choose a smaller output step to fix the problem.

Kepler Parameters vs False Time

The dynamical evolution of Kepler parameters is displayed. We recall that the false time is basically (i.e., to the vanishing of the perturbation) the eccentric anomaly.

Kepler Parameters vs Time

As above, but with the true time as the independent variable.

Time vs False Time

The relation between true time and false time is shown, which gives an idea of the regularity of the problem: the more linear the relation, the more regular the problem. For example, compare the same perturbation but with low and high eccentricity.

Integration Errors

Using the Perturbative method four figures are displayed. The first two show how much the constraints (6.1.7) are violated, the third shows the error affecting the numerical value of the rescaled Hamiltonian (6.2.3), and the fourth shows the error affecting the numerical value of the true Hamiltonian. We remark that possible peaks in the fourth figure (always in correspondence of close encounters with the singularity in the origin) are not worrisome, since they are not caused by the integration procedure (as they would be, instead, for the third figure) and do not propagate.

When using the Non-perturbative method three figures are displayed. The first shows how much the constraint 6.1.23 is violated; the other two are similar to the third and fourth figures of the Perturbative method.

False Time vs Step Number

If the field "Output Step" of the ODE Solver panel is left blank and an adaptive integration method is used, the figure allows one to see how the regularization process is effective. Comparing with the corresponding "Time vs Step Number" in the next panel "Standard method", the user can appreciate the advantage of the regularization methods, in particular when dealing with high eccentricity orbits: besides the almost linearity of the relation, the total step number, and consequently the total integration time, are decisively smaller. The almost linearity shows that the adaptive step mechanism does not need to be invoked in practice during the numerical integration, since the regularization process alone is able to dilate the independent variable in the neighborhood of the central singularity.

User Functions

This button allows one to display the evolution of a dynamical variable chosen by the user. Selecting one of the files in the folder `UserFunctions` `\UserActionAngle`, the result is stored in memory and added to the further computations. To clear the memory, click on "Clear Append" in the third or fourth window.

7.1.6   Standard Method

The sixth panel, "Standard method", directly integrates the six Hamilton equations, without regularization and taking the time as an independent variable. This is useful, in particular, for

(i) highly non-perturbative cases, when the perturbation parameter is very large;

(ii) collision orbits, arising with nonregularizable singularities;

(iii) unbounded cases, when the actual motion is no longer a slightly deformed ellipse and the regularization method fails.

If the perturbation does not depend on the momenta, the user can also choose an ODE solver of the second order, clicking on the corresponding radiobutton.

7.1.7 Action-Angle Variables and Time-Frequency Analysis on KAM Tori

Lastly, the seventh panel allows us to analyze how the fundamental frequencies of a quasi-integrable motion change, thus detecting the transition from order to chaos. The two 3-dimensional Pauli vectors, defined in (6.1.32) on page 192, are the basic tool. They span two spheres and, once energy value is fixed, they are in a 1-1 correspondence with elliptic orbits, so that a stationary point for these two vectors (say, a relative equilibrium point) locates a periodic elliptic orbit in the physical space. Remark that, since a periodic orbit of the perturbed Kepler problem is in general a slightly deformed ellipse, it is described by a periodic motion of the Pauli vectors, covering a closed curve with the same frequency of the physical motion. The "mean diameter" of the closed curve is generally very small and gives an idea of how much the physical trajectory differs from an ellipse. We will still call this small deformed curve the "relative equilibrium point".

Five types of action-angle variables can be calculated and plotted.

(i) The classical Delaunay variables $L, G, G_3, l, \omega, \Omega$ are traditionally considered the action-angle variables par excellence of the Kepler problem. However, because of the complete total degeneration of the problem, in general they are not slightly deformed by a generic small perturbation but completely distorted, and the description by tori is lost. As shown in Section 6.5, the Delaunay and Pauli action-angle variables are well suited for perturbations with axial symmetry about the third axis, and thus with the relative equilibrium point placed at the two North poles. In the case of axial symmetry, when G_3 is a first integral, the system has two degrees of freedom, and the tori can be visualized by clicking on the button "Delaunay Action-Angle".

(ii) The rotated Delaunay action-angle variables are defined as in the previous item, but after a rotation moving the relative equilibrium point to the North poles of the two spheres spanned by the Pauli vectors. When the perturbation does not possess any symmetry, we can reduce it anyway to the symmetric case by means of this suitable rotation, i.e., a symplectomorphism, in the $S^2 \times S^2$ space. The *rotated* Delaunay action-angle variables are therefore used for any perturbation, when the relative equilibrium point is placed in generic position.

(iii) The Pauli action-angle variables are defined as

$$L, S_3, D_3, l_P, \varphi_S = \arctan \frac{S_2}{S_1}, \varphi_D = \arctan \frac{D_2}{D_1}.$$

The first three variables are the actions, and the other three are the corresponding angles. The angle l_P, canonically conjugate to L, differs from the mean anomaly l by a quantity commuting with L. The precise definition requires an excursion into the relation between the Lie group SO(2,4) and its double covering SU(2,2), as explained in Subsection 6.1.4.

(iv) The rotated Pauli action-angle variables are defined as in the previous item, but after a rotation moving the relative equilibrium point to the North poles of the two spheres spanned by the Pauli vectors. The *rotated* Pauli action-angle variables are therefore used for any perturbation, when the relative equilibrium point is placed in generic position.

(v) Action-angle variables defined by the user. Selecting one of the files in the folder UserFunctions\UserActionAngle, the result is stored in memory and added to the further computations. To clear the memory, click on "Clear Append" in the third or fourth window.

Let us examine how to use the seventh panel. When the integration is completed, some new buttons in the panel are enabled. By clicking on "Pauli vectors", not only is the dynamical evolution of these vectors plotted but, if not already present, another subpanel, entitled "Normalized Pauli vectors of equilibrium point", appears. It shows the mean value $s_1, s_2, s_3, d_1, d_2, d_3$ of the 3 + 3 components of the vectors, along with the energy value: the idea is that the Pauli vectors rotate approximately about their equilibrium point. By clicking on "Update", the initial conditions are synchronized with those corresponding to the mean values of the Pauli vectors and to the energy just now computed. If "Update" is disabled, select "no" in the subpanel "Disable update?", then redo "Pauli vectors". Redoing the integration, the Pauli vectors should now rotate closer to their equilibrium point so that, by repeating the procedure three or four times, one should reach the very neighborhood of a physical periodic orbit.

To check how close one is to the physical periodic orbit, inspect the plot of space and momentum trajectories or, more effectively, look at the motion of the rotated Pauli vectors. We recall that for periodic orbits the, practically unattainable, optimum would be to get Pauli vectors describing two very small closed curves. To this end, it is convenient to tune the initial conditions delicately, then update the Pauli vectors and successively proceed to a new integration. Note, however, that it is not mandatory to find exactly the periodic orbit; but, for what follows, only the two centers of the motion of the Pauli vectors are relevant.

Once these two centers are found, it is convenient to freeze the values of the Pauli vectors by selecting "yes" in the "Disable update?" subpanel. Then, you may explore the neighborhood of the physical periodic orbit by slightly changing the initial conditions, integrating again and clicking on "Rotated Pauli vectors": the projections of the rotated Pauli vectors on the plane 1-2 will cover the neighborhood of the origin.

At this point, the user can analyze the spectrum of the three action variables with three methods. As the frequency unit we take the frequency of a circular unperturbed Keplerian orbit with unitary radius.

(i) The Fast Fourier Transform (FFT) gives an overall image of the whole spectrum but with the shortcoming that for a better resolution one should excessively lengthen the integration interval, thus calculating in this manner only an *averaged* frequency.

(ii) The Frequency Modified Fourier Transform (FMFT) allows a much more precise calculation of the frequencies in a narrow interval chosen by the user, who has at his disposal three methods of increasing precision: see Subsection 5.3.2. The output frequencies are ordered starting from the largest amplitude and are displayed in decreasing order. The user can choose the number of frequencies to be displayed and their range. *Note that if one fixes a large number of frequencies and a narrow range, the time needed for the computation may grow excessively, which sometimes requires you to kill the FMFT process directly from the operating system.*

(iii) The Wavelet Transform (WT) allows the display of the time evolution of the instantaneous frequency; see Subsection 5.3.3. The user can choose the range of the frequency and the variance: in order to optimize the look of the graphical output, it is advisable to take the value of the variance about 30 times that of the frequency.

If all the calculated frequencies turn out to be a linear combination with integer coefficients of three fundamental frequencies, the spectrum is said to be *regular* and the motion is *ordered*, taking place on an invariant KAM torus.

Selecting the corresponding radiobutton, the FFT, FMFT, or WT can be applied to the four systems of action-angle variables of Delaunay and Pauli. Moreover, the user may write and analyze custom functions.

7.2   Second Window: "Global Analysis: Poincaré Section"

To open this window select "Global analysis: Poincaré section" in the popup menu in the left-top corner. Some nonrelevant panels are closed and a new specific panel appears.

The second window is devoted to the study of the global structure of the whole phase space, or of a portion of it, for systems with two degrees of freedom: planar systems or with axial symmetry. In the latter case the third component G_3 of the angular momentum is a first integral, whose value is fixed by the user. Once the user has chosen "2D system" or "3D with axial symmetry", two tools are offered.

(i) The classical Poincaré map in the plane G-ω with fixed energy chosen by the user, which can fix, moreover, the initial value of the argument of the pericenter ω, the min and max values of the angular momentum G, and the number of sections (= step number + 1). If the check box "Append" is selected, the sections under computation are kept in memory and automatically added to the subsequent output. The button "Clear" empties the memory.

(ii) The frequency analysis computed along a section G_3 = constant with fixed energy. The first output figure displays the values of ω_L and ω_G as a function of G. The second output figure displays the ratio $\left| \frac{\omega_L}{\omega_G} \right|$: this is useful in detecting the resonances, which reveal themselves as plateaus in correspondence of rational values. The range of displayed values of ω_L and ω_G can be chosen in the two subpanels "L-frequency" and "G-frequency". Choosing the null value for the step number, the max G field is disabled; checking "Frequency map (or single orbit)" and unchecking "Poincarè map", then clicking on "Calculate and plot" forces a jump to the first window (with initial values corresponding to the selected values of min G, G_3, Total energy, and ω, while Ω and f are null by default) to compute the relative orbit. Afterwards one can perform the numerical frequency or wavelet analysis and compare it with the results of the Poincaré map.

7.3 Third Window: "Global Analysis: Rotated Delaunay"

To open this window select "Global analysis: Rotated Delaunay" in the popup menu in the left-top corner. Some nonrelevant panels are closed, and a new specific panel appears.

The third window is devoted to the study of the global structure of the whole phase space, or of a portion of it, for generic systems with three degrees of freedom. The surface of constant total energy in the action space is parametrized by G and G_3 and covered with a rectangular grid. The button "Calculate . . ." performs the numerical analysis of the frequencies on the points of the grid, then the result is displayed by clicking on the four buttons in the subpanel ". . . and plot". In particular, for the button "FMI" (the meaning of the other three is evident) we recall the definition: once an orbit is computed, perform several numerical frequency analyses, say N, each time shifting the window of some Δt_0, and take the largest $\overline{\omega}_{max}$ and smallest $\overline{\omega}_{min}$ of the N output values. For a suitable Δt_0 (but the choice is not very critical) and N large enough, the difference $\overline{\omega}_{max} - \overline{\omega}_{min}$ will result proportional to the amplitude of the frequency modulation, which in turn indicates the presence of a resonance. N may be chosen in the subpanel "Step number for FMI". Let us define the Frequency Modulation Indicator (FMI) as

$$\sigma_{FMI} \stackrel{def}{=} \log \left(\frac{\overline{\omega}_{max} - \overline{\omega}_{min}}{\overline{\omega}_{max} + \overline{\omega}_{min}} \right).$$

Clearly, $\sigma_{FMI} = -\infty$ for orbits of KAM type starting sufficiently far away from a resonance. In the graphical representation we fix a cut-off, i.e., we raise all the lower values to, for example, $\sigma_{FMI} = -12$. Instead, inside a resonance or in its very neighborhood, σ_{FMI} will generally be larger, growing with the amplitude of the frequency modulation. However, although the FMI is very effective in showing *where* the resonances are, it requires much longer integration times to give reliable informations of what happens *inside* the resonances. To this end, it may be more advantageous to compute the frequency values along suited sections: see below.

The frequencies are progressively numbered as 1,2,3. If the 3×3 matrix on the subpanel "Unimodular transformation" is the identity, they coincide with $\omega_L, \omega_G, \omega_{G_3}$, respectively. But the action-angle variables are defined up to a unimodular transformation, represented by a matrix with integer entries and unit determinant. Inside a resonance, characterized by $\vec{k} \cdot \vec{\omega} = 0$, one can thus define three new angles such that one of them coincides with the resonant angle $\vec{k} \cdot \vec{\varphi}$. We recall that a resonant angle performs librations, i.e., oscillations, instead of rotations. The dynamical evolution of the three new action-angle variables can be viewed by clicking on "User Function" of the panel "Regularization method" of the first window, then selecting a file of the type AADelaunayprime_x.m or AAPauliprime_x.m in the UserActionAngle directory. Notice that in the thin stochastic layer surrounding the separatrices between libration and circulation zones, the point "hesitates," choosing randomly between the two behaviors: this is the source of the chaos.

For a more precise local analysis, the user can compute the frequencies along a section G = constant or G_3 = constant: to this end, select the null value for the total step number of G or G_3, respectively, then click on "Calculate . . .". The final output can be viewed with the buttons of the subpanel ". . . and plot". Selecting, instead, the null value for the total step number of G and G_3, one jumps to the first window, to start the computation of the relative orbit.

In order to obtain good results, the computations invoked in the third and fourth windows require very long times, which can be shortened if the computer is a multicore machine. KEPLER, along with the other four supplied programs, is able to parallelize the computations in the following way. If you possess an n-core machine, in an empty folder Kepler create a subfolder Master and $n - 1$ subfolders Slave1, Slave2, . . . , then copy the whole program KEPLER identically in every folder. Start MATLAB then KEPLER from the Master folder, set the parameters, and click on "File/Save setting now". *Without closing*, start a new instance of MATLAB, then KEPLER from a folder SlaveX: you will notice that all the buttons of the computations are disabled while the new button "Start Slave" appears. Click on this button and KEPLER will wait for the start of the master. Redo for every slave, and lastly go back to the master and click on "Calculate . . .". The whole work will be automatically shared among the n cores. The final result is displayed by the master.

We remark that once the single figures of the final result have been displayed (having checked "Show also single figures" in the menu "Window"), by clicking inside a picture visualizes the G and G_3 values of the selected point; then "Calculate ..." allows one to jump to the first window and calculate the relative orbit. This ability is also preserved for the pictures saved in the MATLAB format *.fig.

7.4   Fourth Window: "Global Analysis: Rotated Pauli"

To open this window select "Global analysis: Rotated Pauli" in the popup menu in the left-top corner. Some nonrelevant panels will be closed, and a new specific panel will appear.

The fourth window is devoted to the study of the global structure of the whole phase space, or of a portion of it, for generic systems with three degrees of freedom but with the actions S and D replacing G and G_3. The use of the fourth window is very similar to that of the third one.

7.5   Menu

7.5.1   File

Clicking on "Load new initial data" the user may choose a file *.mat in the folder Init and load some predefined perturbations and initial conditions. With "Save current initial data" the current data can be saved in the same format and folder.

If "Save on exit" is checked (default), the current perturbation, initial conditions, and other settings are saved into the `Setting.mat` file when KEPLER is closed and automatically reloaded when it is opened.

"Save setting now" allows one to save in `Setting.mat` without closing KEPLER: it forces all the further instances of KEPLER to share the same setting in the parallel processes with a multicore machine.

7.5.2 Window

Clicking on "Close all figures" causes all the open pictures to be simultaneously closed.

If "Show also single figures" is checked (*not* default) and a displayed figure is a collective one, the single figures are also displayed separately.

If "Show waitbars" is checked (default), a bar shows what percentage of the integration procedure is complete, as the calculation proceeds. However, showing the bar slows down the calculation significantly ($10 \div 20\%$). In some cases, the program must do other work after the completion of the integration process and the box is not closed immediately: be patient.

If "Show slow waitbars" is checked (*not* default), the waitbar is not updated continuously but only after the number of revolutions selected in the subpanel "Clear memory and in case update LFA" of the third panel is completed. The slow waitbar avoids the slowdown of the usual waitbar and is suitable for very long calculations.

If "Show warning boxes" is checked (default), in some circumstances the user is warned, for example, that the perturbation is too large. Unchecking the item also entails that the question box on the distance from the origin of the secondary mass for the Euler problem will no longer pop up.

If "Watch falling into singularities" is checked (default), the program monitors continuously if the two constraints (6.1.7) are respected; if they are not, it aborts the procedure. Surely this happens when the moving point is falling into nonregularized singularities, avoiding a MATLAB crash, but, sometimes, it also happens inopportunely (sorry!). If you are sure that the problem is regular, uncheck the item.

If "Watch output reliability" is checked (*not* default), the program monitors continuously if the Hamiltonian is conserved, if it is not, it aborts the procedure. Typically, this last happens for the unbounded motion, a situation that the Regularization method is unable to treat. Unfortunately, when the item is checked the integration time increases dramatically, so it is preferable to leave it unchecked and to control the integration errors; try also with the Standard method.

If "Performance registry" is checked (default), some information regarding the integration procedure will be recorded in the file `Perform.txt`.

If "Longtime Frequency Analysis (LFA) for Delaunay" is checked (*not* default), KEPLER analyzes how the frequencies of a *very* long orbit, typically of millions of revolutions, change in time in order to eventually detect the Arnold diffusion. The frequencies are ω_L, ω_G, ω_{G_3}, but the user can choose a linear combination of them in the "Unimodular transformation" subpanel. The total number of revolutions is divided into successive blocks, each one containing the number of revolutions indicated in the subpanel "Clear memory and in case update LFA". For every block, a numerical analysis of the frequencies is performed on a number of revolutions indicated in the second subpanel "Longtime Frequency Analysis (LFA)" of the panel "Revolution number". Then the relative figure (displayed automatically at the beginning of the process) is updated by adding the new value. Clearly, the revolution number in the second subpanel cannot exceed the number in the third subpanel of the "Revolution number" panel.

For "Longtime Frequency Analysis (LFA) for Pauli", things run similarly.

7.5.3   Figure

This menu contains some graphical utilities. See also the respective *.m files in the folder "Util" for more information.

No degree symbol

If this item is *not* checked (default), the small circle of the degree symbol is displayed in the figures. Sometimes this gives errors in the axis labels after using the zoom. Check the item if you intend to use the zoom or, alternatively, enter the code and comment the rows Degree_x, Degree_y, Degree_z.

Large marker size

If this item is checked (*not* default) the markers (in Poincaré section and in the visualization of the tori for systems with two degrees of freedom) appear larger than the default. This is useful, for example, when the user wants to see only a few points.

Save figure as...

This is used to save an open figure in the formats EPS, PDF, PNG, JPEG.

Mouse track

Once it is active, the mouse position is constantly tracked and printed on the figure title. Right-clicking deactivates the function.

Scroll plot

A scroll subwindow is added to an open picture.

Magnifying glass

A magnification box is created under the mouse position when clicking. Press $+/-$ while the mouse button is pressed to increase/decrease magnification and $>/<$ to increase/decrease box size.

7.5.4 Normal Form

In the next chapter some examples of perturbations of the Kepler problem will be considered. In particular, the normal form of the Stark–Quadratic–Zeeman problem, the circular restricted three-body and the motion about an oblate satellite will be calculated. Here the reader can generate the relative graphic output; see the next chapter for more information.

7.5.5 Help

The menu contains the brief introduction to KEPLER "Help for the impatient", along with commands to enter and quit context-help mode, some information on KEPLER, and various acknowledgments.

CHAPTER 8

Some Perturbed Keplerian Systems

First hang-gliding rule:
to take off, run against the wind.

We now possess all the necessary tools to study some interesting Keplerian perturbed systems: the Stark-Quadratic-Zeeman problem, the circular restricted three-body problem, and the motion of a satellite around an oblate primary. In all three cases we will first find the normal integrable form, comparing the relative motion with the "true" one obtained by numerical integration. Several concrete examples will be given, showing in general a very good agreement between the analytical and numerical results. What the normal integrable form is not able to show is the presence of resonances, which are just the indicators of nonintegrability. Then, with the Frequency Modulation Indicator (FMI) we will analyze how order, chaos, and resonances are localized in action space, thus completing the study of the three quasi-integrable systems.

8.1 The Stark-Quadratic-Zeeman (SQZ) Problem

Let us consider the classical (not quantum) model of the hydrogen atom in a constant electric and magnetic field. We assume that the two fields are sufficiently weak so that we can apply the perturbative methods. The spherical symmetry of the unperturbed problem allows us to take a magnetic field directed along the x_3 axis and an electric field lying in the plane $x_1 x_3$. The Hamiltonian is

$$
\begin{aligned}
H_{SQZ} &= \frac{1}{2}\left[(p_1 - \mathcal{B}q_2)^2 + (p_2 + \mathcal{B}q_1)^2 + p_3^2\right] - \frac{1}{q} + \vec{\mathcal{E}} \cdot \vec{q} \\
&= \frac{1}{2}p^2 - \frac{1}{q} + \mathcal{B}(q_1 p_2 - q_2 p_1) + \frac{1}{2}\mathcal{B}^2(q_1^2 + q_2^2) + \mathcal{E}_1 q_1 + \mathcal{E}_3 q_3, \quad (8.1.1)
\end{aligned}
$$

where $\mathcal{B}$ is the half of the value of the magnetic field and $\vec{\mathcal{E}} = (\mathcal{E}_1 \ \mathcal{E}_2 \ \mathcal{E}_3)$ is the electric field. We suppose $\mathcal{B}$ and $\|\vec{\mathcal{E}}\|$ small, but not so small that $\mathcal{B}^2$ is negligible; the adjective "quadratic" refers to this fact.

8.1.1 First Order Normal Form

It is convenient to put

$$
\alpha_1 = -\frac{3}{2}\mathcal{E}_1, \quad \alpha_3 = -\frac{3}{2}\mathcal{E}_3, \quad \beta = \mathcal{B} \quad \text{with } \alpha_1, \alpha_3, \beta = \mathcal{O}(\varepsilon).
$$

Following the method of Section 6.2, let us switch to the equivalent Hamiltonian:

$$
\begin{aligned}
K &= K_0 + K_1 + K_2, \\
K_0 &= \frac{1}{2}x(y^2 + 1), \\
K_1 &= -\frac{2}{3}\alpha_1 x x_1 - \frac{2}{3}\alpha_3 x x_3 + \beta x(x_1 y_2 - x_2 y_1), \\
K_2 &= \frac{1}{2}\beta^2 x(x_1^2 + x_2^2).
\end{aligned}
$$

We first average the K_1 Hamiltonian, for a moment neglecting K_2, which is of the second order. Recalling the moment map (6.1.5) we get

$$
\begin{aligned}
x x_1 &= (K_0 + v_4)(u_1 - R_1) = -K_0 R_1 + u_1 v_4 + \text{n.a.t.}, \\
x x_3 &= (K_0 + v_4)(u_3 - R_3) = -K_0 R_3 + u_3 v_4 + \text{n.a.t.},
\end{aligned}
$$

where "n.a.t." (null averaged terms) denote terms whose dynamical mean value is null: indeed, because of (6.1.14), the odd monomials in u and v are annihilated by the averaging process. The terms containing only K_0, R_k, G_h are already constant. Taking into account the unperturbed solution (6.1.14), let us consider the quadratic monomials:

$$
u_\alpha u_\beta = U_\alpha U_\beta \cos^2 s + V_\alpha V_\beta \sin^2 s + (U_\alpha V_\beta + U_\beta V_\alpha) \sin s \cos s,
$$

whose mean value is clearly

$$\overline{u_\alpha u_\beta} = \frac{1}{2}(U_\alpha U_\beta + V_\alpha V_\beta);$$

analogously,

$$\overline{v_\alpha v_\beta} = \frac{1}{2}(U_\alpha U_\beta + V_\alpha V_\beta),$$

so that, from the expression of the first integral $G_{\alpha\beta} = \frac{1}{K_0}(v_\alpha u_\beta - v_\beta u_\alpha)$, it results that

$$\overline{u_\alpha u_\beta} = \overline{v_\alpha v_\beta} = \frac{1}{2}\sum_\gamma G_{\alpha\gamma} G_{\beta\gamma}.$$

A similar procedure yields

$$\overline{u_\alpha v_\beta} = -\frac{1}{2}K_0 G_{\alpha\beta}.$$

The averaged value of the first order perturbed Hamiltonian is

$$\overline{K_1} = K_0(\alpha_1 R_1 + \alpha_3 R_3 + \beta G_3)$$
$$= K_0[\alpha_1 S_1 + (\alpha_3 + \beta)S_3 - \alpha_1 D_1 - (\alpha_3 - \beta)D_3],$$

which is a linear expression in $\vec{S}$ and $\vec{D}$

$$\overline{K_1} = \vec{W}_S \cdot \vec{S} + \vec{W}_D \cdot \vec{D},$$
$$\vec{W}_S = K_0(\alpha_1 \ 0 \ \alpha_3 + \beta),$$
$$\vec{W}_D = K_0(-\alpha_1 \ 0 \ -\alpha_3 + \beta).$$

The motion, induced by the first order perturbed Hamiltonian on the reduced phase space $\overline{M} \simeq S^2 \times S^2$ spanned by $\vec{S}$ and $\vec{D}$, is a uniform rotation with angular velocity $\vec{W}_S$ and $\vec{W}_D$:

$$\frac{d\vec{S}}{dt} = \{\overline{K_1}, \vec{S}\} = \vec{W}_S \times \vec{S},$$
$$\frac{d\vec{D}}{dt} = \{\overline{K_1}, \vec{D}\} = \vec{W}_D \times \vec{D}.$$

Four equilibrium positions are present, in correspondence with the four intersection couples of the two rotation axes with the respective sphere. They are all elliptic, but two angular velocities relative to two equilibrium positions have the same sign, while those relative to the other two have opposite sign. This implies that, when the further perturbative terms are considered, the first two equilibrium positions will be surely stable, while the other two may become unstable.

REMARK 8.1 The linear dependency on $\vec{S}$ and $\vec{D}$ of $\overline{K_1}$ is a peculiar characteristic of the SQZ problem, while in general $\overline{K_1}$ and, consequently, the two angular velocities $\vec{W}_S$ and $\vec{W}_D$ are nontrivial functions of the Pauli vectors.

8.1.2 Second Order Nonintegrable Normal Form

Let $\chi = \mathcal{O}(\varepsilon)$ be an unknown function generating a canonical transformation through exponentiation and consider the perturbative development

$$\exp \mathcal{L}_\chi K = \left(1 + \mathcal{L}_\chi + \frac{1}{2}\mathcal{L}_\chi^2 + \mathcal{O}(\varepsilon^3)\right)(K_0 + K_1 + K_2)$$

$$= K_0 + (K_1 - \mathcal{L}_{K_0}\chi) + \left(K_2 + \mathcal{L}_\chi K_1 + \frac{1}{2}\mathcal{L}_\chi^2 K_0\right) + \mathcal{O}(\varepsilon^3).$$

Choosing a generating function χ which satisfies the homological equation

$$\mathcal{L}_{K_0}\chi = K_1 - \overline{K_1},$$

we get

$$\exp \mathcal{L}_\chi K = K_0 + \overline{K_1} + \left[K_2 + \frac{1}{2}\{\chi, K_1 + \overline{K_1}\}\right] + \mathcal{O}(\varepsilon^3).$$

The homological equation is easily solved; we obtain

$$\chi = \frac{2}{3}K_0(\alpha_1 v_1 + \alpha_3 v_3) + \left(\frac{2}{3}\alpha_1 R_1 + \frac{2}{3}\alpha_3 R_3 + \beta G_3\right)u_4$$

$$- \frac{1}{3}(\alpha_1 u_1 + \alpha_3 u_3)u_4$$

$$K_1 + \overline{K_1} = -\frac{2}{3}\alpha_1 K_0 u_1 + \frac{5}{3}\alpha_1 K_0 R_1 - \frac{2}{3}\alpha_1 u_1 v_4 + \frac{2}{3}\alpha_1 R_1 v_4 - \frac{2}{3}\alpha_3 K_0 u_3$$

$$+ \frac{5}{3}\alpha_3 K_0 R_3 - \frac{2}{3}\alpha_3 u_3 v_4 + \frac{2}{3}\alpha_3 R_3 v_4 + 2\beta K_0 G_3 + \beta G_3 v_4.$$

A lengthy but straightforward calculation gives

$$\{\chi, K_1 + \overline{K_1}\} = \frac{4}{9}\alpha_1^2\left[-K_0\left(2R_1^2 + \frac{1}{2}R_2^2 + \frac{1}{2}R_3^2\right)\right] + \beta^2\left[-K_0 G_3^2\right]$$

$$- \frac{4}{9}\alpha_1^2\left[K_0\left(2R_1^2 + \frac{1}{2}R_2^2 + \frac{1}{2}R_3^2\right)\right] - \frac{4}{9}\alpha_3^2\left[K_0^3 + K_0(G_1^2 + G_2^2 + R_3^2)\right]$$

$$+ \frac{4}{9}\alpha_3^2\left[-K_0\left(\frac{1}{2}R_1^2 + \frac{1}{2}R_2^2 + 2R_3^2\right)\right] - \frac{4}{9}\alpha_1^2\left[K_0^3 + K_0(G_2^2 + G_3^2 + R_1^2)\right]$$

$$+ \frac{2}{9}\alpha_3^2\left[-K_0\left(\frac{1}{2}G_1^2 + \frac{1}{2}G_2^2 + R_3^2\right)\right] - \frac{5}{9}\alpha_3^2\left[\frac{1}{2}K_0(G_1^2 + G_2^2 - R_1^2 - R_2^2)\right]$$

$$- \frac{5}{9}\alpha_1^2\left[\frac{1}{2}K_0(G_2^2 + G_3^2 - R_2^2 - R_3^2)\right] + \frac{2}{9}\alpha_1^2\left[-K_0\left(\frac{1}{2}G_2^2 + \frac{1}{2}G_3^2 + R_1^2\right)\right]$$

$$+ \frac{4}{9}\alpha_3^2\left[-2K_0 R_3^2\right] + \frac{4}{9}\alpha_1^2\left[-2K_0 R_1^2\right] - \frac{4}{9}\alpha_3^2\left[K_0\left(\frac{1}{2}R_1^2 + \frac{1}{2}R_2^2 + 2R_3^2\right)\right]$$

$$-\frac{4}{9}\alpha_1\alpha_3\left[K_0(-G_1G_3+R_1R_3)\right]+\frac{4}{9}\alpha_1\alpha_3\left[-2K_0R_1R_3\right]$$

$$+\frac{2}{3}\alpha_1\beta\left[K_0\left(\frac{1}{2}G_1R_3-2R_1G_3\right)\right]-\frac{4}{9}\alpha_1\alpha_3\left[K_0(-G_1G_3+R_1R_3)\right]$$

$$+\frac{4}{9}\alpha_1\alpha_3\left[-\frac{3}{2}K_0R_1R_3\right]+\frac{2}{3}\alpha_3\beta\left[-\frac{3}{2}K_0G_3R_3\right]-\frac{4}{9}\alpha_1\alpha_3\left[\frac{3}{2}K_0R_1R_3\right]$$

$$+\frac{4}{9}\alpha_1\alpha_3\left[-2K_0R_1R_3\right]+\frac{2}{3}\alpha_1\beta\left[-\frac{3}{2}K_0R_1G_3\right]-\frac{4}{9}\alpha_1\alpha_3\left[\frac{3}{2}K_0R_1R_3\right]$$

$$+\frac{4}{9}\alpha_1\alpha_3\left[2K_0R_1R_3\right]+\frac{2}{3}\alpha_3\beta\left[-\frac{3}{2}K_0G_3R_3\right]$$

$$-\frac{2}{3}\alpha_1\beta\left[2K_0R_1G_3-\frac{1}{2}K_0G_1R_3\right]+\frac{2}{3}\alpha_1\beta\left[-\frac{3}{2}K_0R_1G_3\right]$$

$$-\frac{2}{3}\alpha_3\beta\left[\frac{3}{2}K_0R_3G_3\right]+\frac{2}{3}\alpha_3\beta\left[-\frac{3}{2}K_0R_3G_3\right]$$

$$-\frac{5}{9}\alpha_1\alpha_3\left[\frac{1}{2}K_0(-G_1G_3+R_1R_3)\right]+\frac{2}{9}\alpha_1\alpha_3\left[K_0\left(\frac{1}{2}G_1G_3-R_1R_3\right)\right]$$

$$-\frac{2}{3}\alpha_1\beta\left[\frac{1}{2}K_0(G_1R_3-R_1G_3)\right]-\frac{5}{9}\alpha_1\alpha_3\left[\frac{1}{2}K_0(-G_1G_3+R_1R_3)\right]$$

$$+\frac{2}{9}\alpha_1\alpha_3\left[K_0\left(\frac{1}{2}G_1G_3-R_1R_3\right)\right]+\text{n.a.t.}$$

Averaging over the unperturbed dynamical evolution we get

$$\frac{2}{K_0}\overline{\left[K_2+\frac{1}{2}\{\chi,K_1+\overline{K_1}\}\right]}=$$

$$\left(\frac{1}{2}\beta^2-\frac{1}{6}\alpha_3^2\right)G_1^2+\left(\frac{1}{2}\beta^2-\frac{5}{6}\alpha_1^2-\frac{5}{6}\alpha_3^2\right)G_2^2-\frac{5}{6}\alpha_1^2G_3^2$$

$$+\left(\frac{5}{2}\beta^2-\frac{10}{3}\alpha_1^2-\frac{1}{6}\alpha_3^2\right)R_1^2+\left(\frac{5}{2}\beta^2-\frac{1}{6}\alpha_1^2-\frac{1}{6}\alpha_3^2\right)R_2^2$$

$$-\left(\frac{1}{6}\alpha_1^2+\frac{10}{3}\alpha_3^2\right)R_3+\frac{5}{3}\alpha_1\alpha_3G_1G_3-\frac{43}{9}\alpha_1\alpha_3R_1R_3$$

$$+\frac{1}{3}\alpha_1\beta G_1R_3-\frac{13}{3}\alpha_1\beta G_3R_1-4\alpha_3\beta G_3R_3.$$

Instead of the old Hamiltonian K, we are thus led to consider the new averaged Hamiltonian,

$$\overline{K}(\vec{G},\vec{R})=K_0+\overline{K_1}+\overline{\left[K_2+\frac{1}{2}\{\chi,K_1+\overline{K_1}\}\right]} \qquad (8.1.2)$$

(terms of third order have been neglected), whose 4-dimensional phase space $\overline{M}\simeq S^2\times S^2$ is spanned by $\vec{G},\vec{R}$ or, equivalently, by $\vec{S},\vec{D}$. Notice that the first term K_0 is an additive first integral which does not generate any dynamics and may therefore be ignored. This new Hamiltonian system admits only one first integral, the Hamiltonian $\overline{K}$ itself, and is *not* integrable:

recall that the nonintegrability of the averaged Hamiltonian results from the complete total degeneration of the unperturbed Kepler problem.

The Hamiltonian $\overline{K}$ itself has the structure of a quasi-integrable system, the first order $\overline{K_1}$ playing the role of integrable part, perturbed by the term $[\ldots]$ which is of second order. Moreover, the unperturbed motion, due to $\overline{K_1}$ alone, is known: it is a uniform rotation on the two 2-spheres of $\overline{M}$ with angular velocity $\vec{W}_S$ and $\vec{W}_D$, respectively. Hence, the natural next step would consist of two further distinct averaging processes over the perturbing term $[\ldots]$, as already described in Section 6.3.

In effect, this is done in Von Milczewski & Uzer (1997a), but the outcome is disappointing. Clearly, the procedure is fully legitimate and coherent with the dictates of the perturbation theory, but the averaged system turns out to be very poor and unable to reproduce the key features of the true dynamics. Indeed, after the double averaging process, the perturbing second order part $[\ldots]$ depends on the same terms appearing in $\overline{K_1}$, and the resulting motion is not qualitatively different from that generated by the first order Hamiltonian. In some sense, the double averaging smooths out the perturbation excessively and "kills" the distinctive features of the problem. This statement will be more clear below, after the introduction of an alternative method which, in contrast, captures very well how the true motion evolves, obviously up to the fast oscillations smoothed out by the averaging processes.

This alternative method is a straightforward generalization of the technique in Cushman & Sadovskií (2000), regarding the crossed SQZ, i.e., with orthogonal electric and magnetic fields. It basically consists in performing only *one* averaging, so that the Hamiltonian will still depend on one angle, yet will nevertheless be integrable. Invoking geometrical considerations, the averaged evolution of the relevant pair of action-angle variables is therefore found.

8.1.3 Second Order Integrable Normal Form

The starting point of the alternative method is a double rotation, one for each of the two 2-spheres of $\overline{M}$, in order to align the two angular velocities $\vec{W}_S$ and $\vec{W}_D$ with the third axis. Clearly, rotations preserve the symplectic structure of a sphere.

REMARK 8.2 After the rotations, the Hamiltonian $\overline{K_1}$ is linear in S_3 and D_3, thus diagonal quadratic in the Poincaré variables $\chi_1, \eta_1, \chi_2, \eta_2$. The two rotations are sufficient to get the result claimed in Theorem 2.33, but this fact is peculiar to the SQZ problem, as noticed in Remark 8.1. Therefore the linearity implies that the projections of the two Pauli vectors on the plane 1-2 performs two (slightly deformed) circles. For a generic perturbation, the first order term also contains products of the type $\chi\eta$, and the projections of the Pauli vectors will cover the neighborhood of the origin. Try adding for example a potential $1/q^2$ to the SQZ Hamiltonian.

The unperturbed Hamiltonian $\overline{K_1}$ becomes axially symmetric so that, after a further averaging of the perturbing term $\overline{[\ldots]}$ with respect to the rotations about the third axis, the whole Hamiltonian also acquires the rotational symmetry. The third component G_3' of the rotated angular momentum becomes a first integral which, with the total Hamiltonian, makes the system completely integrable. This averaged system is defined on $\overline{M}$ symplectically reduced under the action of the rotation group generated by G_3'. The point is that we are able to perform this reduction explicitly, as already explained in Section 6.5. Let us view the details.

Let ϑ_S, ϑ_D be the angles, respectively, between $\vec{W}_S, \vec{W}_D$ and the third axis. Then

$$\sin\vartheta_S = \frac{\alpha_1}{\sqrt{\alpha_1^2 + (\alpha_3 + \beta)^2}}, \qquad \cos\vartheta_S = \frac{\alpha_3 + \beta}{\sqrt{\alpha_1^2 + (\alpha_3 + \beta)^2}},$$

$$\sin\vartheta_D = -\frac{\alpha_1}{\sqrt{\alpha_1^2 + (\alpha_3 - \beta)^2}}, \qquad \cos\vartheta_D = -\frac{\alpha_3 - \beta}{\sqrt{\alpha_1^2 + (\alpha_3 - \beta)^2}}.$$

Define

$$c_+ = \frac{1}{2}(\cos\vartheta_S + \cos\vartheta_D), \qquad c_- = \frac{1}{2}(\cos\vartheta_S - \cos\vartheta_D),$$

$$s_+ = \frac{1}{2}(\sin\vartheta_S + \sin\vartheta_D), \qquad s_- = \frac{1}{2}(\sin\vartheta_S - \sin\vartheta_D).$$

Denote for a moment with a prime the rotated variables; we get

$$\begin{aligned} G_1 &= c_+ G_1' + c_- R_1' + s_+ G_3' + s_- R_3', \\ G_2 &= G_2', \\ G_3 &= -s_+ G_1' - s_- R_1' + c_+ G_3' + c_- R_3', \\ R_1 &= c_- G_1' + c_+ R_1' + s_- G_3' + s_+ R_3', \\ R_2 &= R_2', \\ R_3 &= -s_- G_1' - s_+ R_1' + c_- G_3' + c_+ R_3'. \end{aligned}$$

Put these expressions into the perturbative term. Reordering and *dropping the primes,* we find that the perturbative term is transformed as

$$\frac{2}{K_0}\overline{\left[K_2 + \frac{1}{2}\{\chi, K_1 + \overline{K_1}\} \right]} \Longrightarrow$$

$$P_1 G_1^2 + P_2 G_2^2 + P_3 R_1^2 + P_4 R_2^2 + P_5 R_3^2 + P_6 G_1 R_1 + P_7 G_3 R_3$$

$$+ \text{ n.a.t.} + \text{const.},$$

where "n.a.t." denotes terms which will be annihilated in the subsequent averaging.

We have defined

$$
\begin{aligned}
P_1 = {} & \left(\frac{1}{2}\beta^2 - \frac{1}{6}\alpha_3^2\right)c_+^2 - \frac{5}{6}\alpha_1^2 s_+^2 + \left(\frac{5}{2}\beta^2 - \frac{10}{3}\alpha_1^2 - \frac{1}{6}\alpha_3^2\right)c_-^2 \\
& - \left(\frac{1}{6}\alpha_1^2 + \frac{10}{3}\alpha_3^2\right)s_-^2 - \frac{5}{3}\alpha_1\alpha_3 c_+ s_+ + \frac{43}{9}\alpha_1\alpha_3 c_- s_- \\
& - \frac{1}{3}\alpha_1\beta c_+ s_- + \frac{13}{3}\alpha_1\beta c_- s_+ - 4\alpha_3\beta s_+ s_-,
\end{aligned}
$$

$$
P_2 = \frac{1}{2}\beta^2 - \frac{5}{6}\alpha_1^2 - \frac{5}{6}\alpha_3^2,
$$

$$
\begin{aligned}
P_3 = {} & \left(\frac{1}{2}\beta^2 - \frac{1}{6}\alpha_3^2\right)c_-^2 - \frac{5}{6}\alpha_1^2 s_-^2 + \left(\frac{5}{2}\beta^2 - \frac{10}{3}\alpha_1^2 - \frac{1}{6}\alpha_3^2\right)c_+^2 \\
& - \left(\frac{1}{6}\alpha_1^2 + \frac{10}{3}\alpha_3^2\right)s_+^2 - \frac{5}{3}\alpha_1\alpha_3 c_- s_- + \frac{43}{9}\alpha_1\alpha_3 c_+ s_+ \\
& - \frac{1}{3}\alpha_1\beta c_- s_+ + \frac{13}{3}\alpha_1\beta c_+ s_- - 4\alpha_3\beta s_+ s_-,
\end{aligned}
$$

$$
P_4 = \frac{5}{2}\beta^2 - \frac{1}{6}\alpha_1^2 - \frac{1}{6}\alpha_3^2,
$$

$$
\begin{aligned}
P_5 = {} & \left(\frac{1}{2}\beta^2 - \frac{1}{6}\alpha_3^2\right)s_-^2 - \frac{5}{6}\alpha_1^2 c_-^2 + \left(\frac{5}{2}\beta^2 - \frac{10}{3}\alpha_1^2 - \frac{1}{6}\alpha_3^2\right)s_+^2 \\
& - \left(\frac{1}{6}\alpha_1^2 + \frac{10}{3}\alpha_3^2\right)c_+^2 + \frac{5}{3}\alpha_1\alpha_3 c_- s_- - \frac{43}{9}\alpha_1\alpha_3 c_+ s_+ \\
& + \frac{1}{3}\alpha_1\beta c_+ s_- - \frac{13}{3}\alpha_1\beta c_- s_+ - 4\alpha_3\beta c_+ s_+,
\end{aligned}
$$

$$
\begin{aligned}
P_6 = {} & \left(\beta^2 - \frac{1}{3}\alpha_3^2\right)c_+ c_- - \frac{5}{3}\alpha_1^2 s_+ s_- + \left(5\beta^2 - \frac{20}{3}\alpha_1^2 - \frac{1}{3}\alpha_3^2\right)c_+ c_- \\
& - \left(\frac{1}{3}\alpha_1^2 + \frac{20}{3}\alpha_3^2\right)s_+ s_- - \frac{5}{3}\alpha_1\alpha_3(c_+ s_- + c_- s_+) \\
& + \frac{43}{9}\alpha_1\alpha_3(c_- s_+ + c_+ s_-) - \frac{14}{3}\alpha_1\beta(c_+ s_+ + c_- s_-) - 4\alpha_3\beta(s_+^2 + s_-^2),
\end{aligned}
$$

$$
\begin{aligned}
P_7 = {} & \left(\beta^2 - \frac{1}{3}\alpha_3^2\right)s_+ s_- - \frac{5}{3}\alpha_1^2 c_+ c_- + \left(5\beta^2 - \frac{20}{3}\alpha_1^2 - \frac{1}{3}\alpha_3^2\right)s_+ s_- \\
& - \left(\frac{1}{3}\alpha_1^2 + \frac{20}{3}\alpha_3^2\right)c_+ c_- + \frac{5}{3}\alpha_1\alpha_3(c_+ s_- + c_- s_+) \\
& - \frac{43}{9}\alpha_1\alpha_3(c_- s_+ + c_+ s_-) - 4\alpha_1\beta(c_+ s_+ + c_- s_-) - 4\alpha_3\beta(c_+^2 + c_-^2).
\end{aligned}
$$

Averaging under the action of G_3 we get

$$
\frac{2}{K_0}\overline{\left[K_2 + \frac{1}{2}\{\chi, K_1 + \overline{K_1}\}\right]} \Longrightarrow
$$

$$
\frac{1}{2}(P_1 + P_2)(G_1^2 + G_2^2) + \frac{1}{2}(P_3 + P_4)(R_1^2 + R_2^2) + P_5 R_3^2 - \frac{1}{2}(P_6 - 2P_7)G_3 R_3.
$$

The averaged Hamiltonian (8.1.2) is defined on $\overline{M} \simeq S^2 \times S^2$, i.e., on the space of the orbits of the Kepler problem with a fixed energy. After the averaging under the G_3-action, the Hamiltonian becomes a function on the symplectic reduction $\overline{\overline{M}}$ of $\overline{M}$ with respect to the axial symmetry. The double reduced Hamiltonian will be named $\overline{\overline{K}}$ and, bearing Section 6.5 in mind, we can express it as a function of the variables ξ_1, ξ_2, ξ_3 or, alternatively, of η_1, η_2, η_3. We recall that

$$R_3 = \xi_1, \quad G_1^2 + G_2^2 = \frac{1}{2}(K_0^2 - G_3^2 - \xi_1^2 + \xi_3), \quad R_1^2 + R_2^2 = \frac{1}{2}(K_0^2 - G_3^2 - \xi_1^2 - \xi_3),$$

from which we obtain the final result:

$$\overline{\overline{K}}(\xi) = a\xi_3 + b\xi_1^2 + c\xi_1, \quad \text{where} \tag{8.1.3}$$

$$a = \frac{K_0}{8}(P_1 + P_2 - P_3 - P_4),$$

$$b = -\frac{K_0}{8}(P_1 + P_2 + P_3 + P_4 - 4P_5),$$

$$c = \frac{K_0}{2}\left[\sqrt{\alpha_1^2 + (\alpha_3 + \beta)^2} - \sqrt{\alpha_1^2 + (\alpha_3 - \beta)^2} - \frac{1}{2}(P_6 - 2P_7)G_3\right].$$

Considering K_0 and G_3 as parameters, the equation $\overline{\overline{K}}(\xi) = \text{constant}$ (R-type parametrization) describes a family of 2-dimensional surfaces which intersect the reduced Kepler manifold $\overline{\overline{M}}$ in a family of curves. We recall that $\overline{\overline{M}}$ is represented by Equation (6.5.4) on page 205. A point on one of these intersection curves represents an elliptic orbit rotating around the physical x_3 axis, while the motion on the intersection curve describes variations of the parameters of the elliptic orbit. The interesting fact is that this description is *global*, and may reveal some features which, as is usual in the standard perturbation theory, are hidden in a *local* description.

Arguing in the same manner, but with the η-variables and the G-type parametrization, we get

$$\overline{\overline{K}}(\eta) = 2a\eta_3^2 + (a + b)\frac{\eta_1^2}{\eta_3^2} + c\frac{\eta_1}{\eta_3}.$$

Lastly, recall that, from (6.5.9), we can deduce the evolution of the two pairs of action-angle variables, R_3, φ_R (R-type) and G, ω (G-type), respectively, from these intersection curves. Notice that the other two action variables L and G_3 stay unchanged in the approximate dynamics generated by the double averaged Hamiltonian $\overline{\overline{K}}$.

Detecting the qualitative nature of the surfaces $\overline{\overline{K}}(\xi) = \text{constant}$ from (8.1.3) is immediate: the ξ_2 coordinate is absent, the surfaces are cylindrical, and their intersections with the plane $\xi_1 \xi_3$ are parabolas. What is much more annoying is finding the numerical value of the parameters a, b, c "by hand," because of their complicated expressions. The program KEPLER therefore provides a tool which displays pictures of the surfaces and their intersections with the reduced Kepler manifold, along with the dynamics of the action-angle variables.

To this end, open KEPLER, then click on "Stark Quadratic Zeeman" in the menu "Normal form". A window opens where the user can fix the value of G_3, Zeeman3 = $\mathcal{B}$, Stark1 = $\mathcal{E}_1$ and Stark3 = $\mathcal{E}_3$. Clicking on "<< Update", the numerical values of the two normalized Pauli vectors $\vec{s} = \vec{W}_S/W_S$ and $\vec{d} = \vec{W}_D/W_D$ appear: they provide a good approximation of the relative equilibrium point. Clicking repeatedly on "Alternate signs" allows one to cycle over the four different equilibrium positions. Clicking on the just enabled six buttons displays respectively: two 3-dimensional pictures of $\overline{\overline{M}}$ intersected by the surfaces $\overline{\overline{K}}(\xi)$ = constant or $\overline{\overline{K}}(\eta)$ = constant, then a vertical section of these two pictures along with a family of intersecting surfaces, and lastly the dynamics of the corresponding action-angle variables. Moreover, the window offers some other fields to fill in order to make the graphical appearance better: try it and experiment.

Let us now study some particular cases and compare the approximate dynamics generated by the normal form with the true motion given by numerical integration. We will also show how to use the tools offered by KEPLER: the Poincaré section, the numerical frequency analysis and the Frequency Modulation Indicator (FMI).

8.1.4 The Quadratic Zeeman (QZ) Problem

Taking a null electric field in (8.1.1), we get the quadratic Zeeman (QZ) problem. Often the linear term in $\mathcal{B}$ is neglected because it does not affect the significant part of the dynamical evolution. The total Hamiltonian is reduced to

$$H_{QZ} = \frac{1}{2}p^2 - \frac{1}{q} + \frac{1}{2}\mathcal{B}^2(q_1^2 + q_2^2).$$

Notice that, from the physical point of view, the linear term can be eliminated by also passing to a reference system uniformly rotating around the third axis. The two angles ϑ_S and ϑ_D are null, and the system has axial symmetry.

The QZ problem was well studied in the 1980s, becoming very popular for its several attractive features. It is not an abstract model but a real physical system that can and has been investigated in the laboratory; moreover, the axial symmetry implies the conservation of G_3, thus the reduction in practice to two degrees of freedom and the possibility to use the Poincaré section. Lastly, it exhibits a typical mixing of order and chaos, making it a paradigmatic example of a quasi-integrable Hamiltonian system. The quantum features are also very interesting, but we will not touch the non-classical aspects. See Friedrich & Wintgen (1989), for a dated review, and Cushman (1991).

Up to inessential additive constants, the double averaged integrable normal form (8.1.3) becomes simply

$$\overline{\overline{K}}(\xi) = -\frac{1}{4}K_0\beta^2(2\xi_3 + 3\xi_1^2),$$

showing that the intersection parabolas with the $\xi_1\xi_3$ plane have a fixed shape, independent of the strength of the magnetic field. Despite the remarkable simplification in the Hamiltonian expression, we see that the QZ problem exhibits the essential features of the full problem.

In Figure 8.1 we consider[1] the case $G_3 = 0.2$, where G_3 is normalized: recall that $0 \le G_3 \le K_0$. On the top, the intersection between the reduced Kepler manifold $\overline{\overline{M}}$ in the R-case and various levels of $\overline{\overline{K}}(\xi)$ is shown, while on the bottom the analytical result (left) is compared with the numerical one (right): clearly, the averaged normal form reproduces the real behavior of the system very well. We recall that $\overline{\overline{M}}$ is rotationally symmetric around the ξ_1 axis, the action value R_3 is given by ξ_1, and the canonically conjugate angle is given by the projection on the plane $\xi_2\xi_3$. Figure 8.2 differs from the previous only in the choice $G_3 = 0.8$: again, no relevant difference between the analytical and numerical calculations is noticeable.

In Figures 8.3 and 8.4 the same analysis is pursued in the G-case. Now, $\overline{\overline{M}}$ is rotationally symmetric around the η_3 axis, the action value is given by $\eta_3 = G$ and the canonically conjugate angle, i.e., the argument of pericenter ω, is given by the projection on the plane $\eta_1\eta_2$. The result of the averaged normal form (bottom-left) is compared with the Poincaré section of the true system (bottom-right), where we have chosen $\mathcal{B}^2 = 0.04$: in spite of the non smallness of the perturbation, the two results agree very well.

Figures 8.1–8.4 are representative of the two typical topologies of the QZ problem. We will consider the G-case, but a similar analysis can be carried out in the R-case. When G_3 lies below a critical value, the argument of pericenter ω can librate or circulate depending on the initial conditions, and the transition between the two regimes is marked by a hyperbolic equilibrium point on $\overline{\overline{M}}$. Besides a hyperbolic point at the North pole and an elliptic one at the South pole, two other elliptic equilibrium points exist in symmetric position, which coalesce with the hyperbolic point when G_3 tends to the critical value. When G_3 grows and crosses the critical value, $\overline{\overline{M}}$ shrinks and is reduced to a point for $G_3 = 1$, so that ω can only circulate. Equating the curvature at the North pole of $\overline{\overline{M}}$ with that of the level surface of the

---

[1] In all the following examples we assume Total Energy = -0.5.

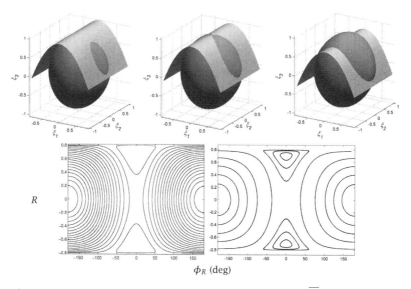

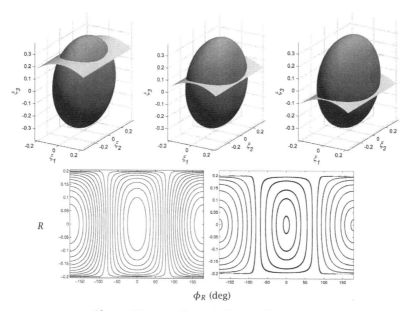

Figure 8.1: On the top, the twofold reduced Kepler manifold $\overline{\overline{M}}$ (R-case) for $G_3 = 0.2$ and its intersection with various levels of the averaged Hamiltonian of the QZ problem; notice the hyperbolic point in the central picture. On the bottom, the averaged result (left) is compared with the numerical output of the true system (right).

Figure 8.2: As in Figure 8.1 but with $G_3 = 0.8$.

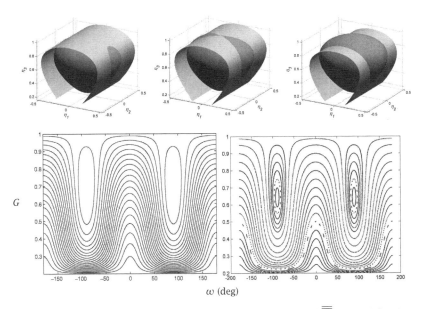

Figure 8.3: On the top, the twofold reduced Kepler manifold $\overline{\overline{M}}$ (G-case) for $G_3 =$ 0.2 and its intersection with various levels of the averaged Hamiltonian of the QZ problem. Notice the hyperbolic point in the central picture. On the bottom, the averaged result (left) is compared with the Poincaré section of the true system (right) with $\mathcal{B}^2 = 0.04$.

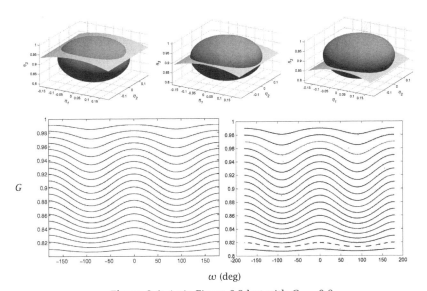

Figure 8.4: As in Figure 8.3 but with $G_3 = 0.8$.

Hamiltonian passing through the same point, one finds the critical value $G_3^{\mathrm{crit}} = 1/\sqrt{5} = 0.4472\ldots$.

The averaged Hamiltonian gives reliable results only when the perturbation is sufficiently small. Figure 5.4 on page 160 shows the Poincaré section for the QZ problem with $G_3 = 0.2$, thus as in Figure 8.3, but with $\mathcal{B}^2 = 0.4$. A comparison between the two pictures shows, as expected, three main differences: a shift of the elliptic points, the arising of chaos around the hyperbolic point and the separatrices, and the presence of resonance islands. Clearly, by thickening the computed orbits one could also enlighten the chaos which develops around hyperbolic points and separatrices created by the resonances.

However, chaos and resonance islands are present for *every* non-null value of the perturbation. What happens is that the width of a resonance increases with the square root of the perturbation and is exponentially small with the order $|k|$ of the resonance $k \cdot \omega = 0$ itself (we recall that $|k| = \sum_i |k_i|$). It follows that the Poincaré section is able to display only the low order resonances, provided the perturbation is not too small. Moreover, even if a resonance and its surrounding chaos could in principle be visualized, one has no guiding rule to sort out the right initial conditions, while thickening the computed orbits quickly yields an overloaded and unreadable drawing.

Much better results are obtained with the FMI method. In Figure 8.5 (top) the values σ_{FMI} of the FMI for the QZ problem are reported. Here the value $\mathcal{B}^2 = 0.4$ has been chosen, as in Figure 5.4. The computation has been performed on a fixed energy surface and on a grid of 400×400 points. Dark blue corresponds to lower values of σ_{FMI}, thus to KAM tori, while light blue, yellow, and red correspond to higher values, thus to resonances or chaos. We have chosen $\omega = \pi/2$ in order to also display the libration part, as suggested by Figure 5.4. For the symmetry of the picture only the part with $G_3 > 0$ is shown.

Notice how the resonance lines are locally almost parallel and without reciprocal crossing. It follows that there is no Arnold web. This is due to the fact that the problem has in effect two degrees of freedom only, so that the resonances in the 2-dimensional frequency space are straight lines through the origin with rational slope.

Figure 8.5 (top) appears clearly divided into two zones by a line surrounded by chaos, which starts from the origin and ends at the point $G_3 = G_3^{\mathrm{crit}}$; the libration zone lies on the right and the circulation zone on the left. The two zones are characterized by the fact that, moving on a vertical straight line, one encounters the same resonance twice in the libration and only once in the circulation zone, as one can ascertain by comparing Figures 8.3 and 8.4 and bearing in mind that we have chosen $\omega = \pi/2$. Using the Frequency Modified Fourier Transform, one determines the ratio between the frequencies of the mean anomaly and the argument of pericenter; in

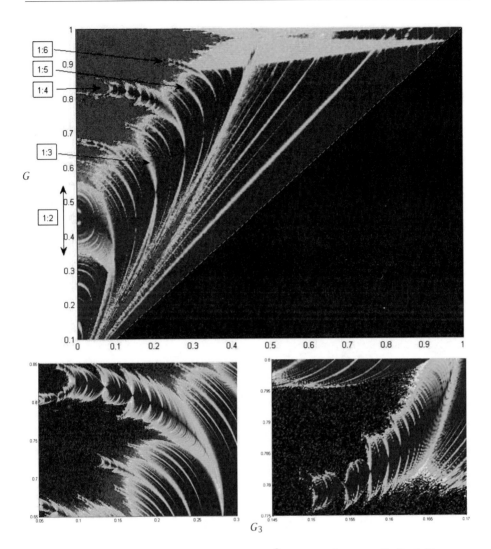

Figure 8.5: FMI for the QZ problem with $\mathcal{B}^2 = 0.4$ and $\omega = \pi/2$. Dark blue correspond to lower values of σ_{FMI}, thus to KAM tori, while light blue, yellow, and red correspond to higher values, thus to resonances or chaos. There is no Arnold web, because the problem has in effect two degrees of freedom only, so that the resonances in the 2-dimensional frequency space are straight lines through the origin with rational slope.

Figure 8.5 (top) we have labeled the resonances in the libration zone from 1:2 to 1:6. Notice that the resonances 1:2, 1:4, 1:6, have an enlarged upper branch, because, referring to the pendulum model, they are crossed in the elliptic equilibrium point, where their width is maximum. Instead, the resonances 1:3, 1:5, are crossed in the hyperbolic equilibrium point where their width is minimum. In Figure 8.5 (bottom-left and bottom-right) further details of the large upper branch relative to the resonance 1:4 are magnified, showing the auto-similarity of the structure, with resonances inside resonances inside resonances, and so on. Here, the display is up to resonances of the third/fourth level but, by increasing the resolution and lengthening the integration time, in principle one can reach all the subsequent levels.

Producing these pictures is somewhat time expensive, but they allow us, for example, to pick out the initial conditions to get the Poincarè section of very inner resonances, or to check with a wavelet analysis that every subsequent level of resonance adds a new modulating frequency. The reader is therefore invited to also explore the phase space with Poincaré section, tori visualization ("Delaunay Action-Angle" in the first KEPLER window), frequency analysis, and wavelets, comparing the results. To this end, start KEPLER, then Figure 8.5 in the CD; clicking on a picture will automatically select the coordinates G and G_3 of the point, copying them automatically on the relative field of the second and third windows.

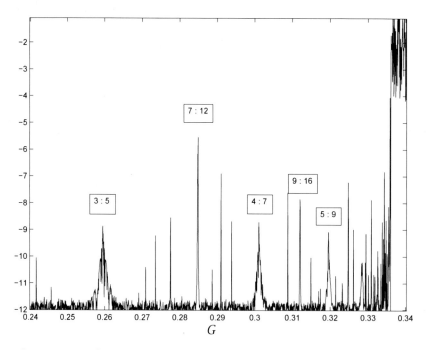

Figure 8.6: FMI for a vertical section along $G_3 = 0.01$ of Figure 8.5. Notice how resonances scarcely visible in Figure 8.5 are now well englightened.

A less expensive procedure, which applies however only after one has obtained a comprehensive view, is the computation of the FMI along a horizontal or vertical straight line, as shown in Figure 8.6.

8.1.5 The Parallel (SQZp) Problem

Taking the electric field directed as the magnetic field in (8.1.1), we get the Stark–Quadratic–Zeeman-parallel (SQZp) problem:

$$H_{SQZp} = \frac{1}{2}p^2 - \frac{1}{q} + \mathcal{B}(q_1 p_2 - q_2 p_1) + \frac{1}{2}\mathcal{B}^2(q_1^2 + q_2^2) + \mathcal{E}_3 q_3.$$

Clearly, the two angles ϑ_S and ϑ_D are still null and the system has axial symmetry.

The problem has been investigated in Deprit, Lanchares, Iñarrea, Salas & Sierra (1996), Salas, Deprit, Ferrer, Lanchares & Palacián (1998), and Salas & Lanchares (1998) but with an analytical method. The main difference with respect to the previous QZ case is that the c coefficient in (8.1.3) is now present, so the intersection parabolas with the plane $\xi_1 \xi_3$ are shifted along the ξ_1 axis. In Figure 8.7 the G-case is shown.

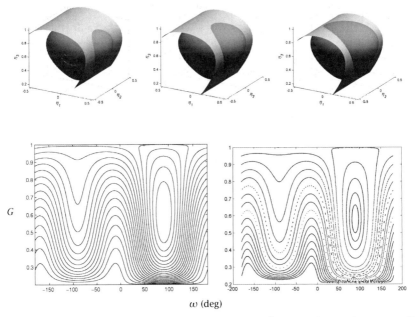

Figure 8.7: As in Figure 8.3, with $G_3 = 0.2$ and $\mathcal{B}^2 = 0.04$, but with a non null electric field $\mathcal{E}_3 = 0.01$.

In Figure 8.8 (left) we show the distribution of the FMI in the presence of an electric field, with $\mathcal{E} = 0.06$ and $\mathcal{B}^2 = 0.04$. The computation has been performed for the full range, i.e., $-1 \le G_3 \le 1$, for the asymmetry of the problem. The rather small values of the perturbation show the power of the FMI method. The resonances are clearly recognized, and magnifying further details, as in Figure 8.8 (right), allows us to penetrate the structure very deeply. We stress that all the resonances enlightened in these pictures would appear extremely thin in a Poincarè section, and in practice they can be found only if one knows very well where to seek and at the cost of great magnification.

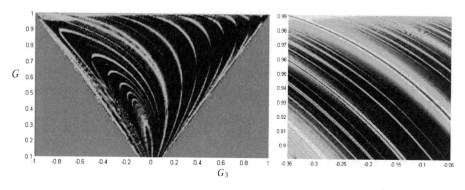

Figure 8.8: FMI in presence of parallel electric field, with $\mathcal{E} = 0.06$ and $\mathcal{B}^2 = 0.04$. Left: the whole system. Right: a detail. Compare with the Poincaré section in Figure 8.7, where no resonance is visible.

Also in this case the system is in practice 2-dimensional and the resonance lines tend to be parallel, without reciprocal crossing. It follows that there is no Arnold web.

8.1.6 The Crossed (SQZc) Problem

When electric and magnetic fields are orthogonal to one another, the Stark-Quadratic-Zeeman system is said to be *crossed*. The two vectors $\vec{W}_S$ and $\vec{W}_D$ belong, as in the general case, to the plane of the two physical fields, but in this case they also have the same norm and form the same angle, with opposite sign, with respect to the magnetic field: $\vartheta_S = -\vartheta_D$. The first order averaged Hamiltonian is reduced to a term proportional to the rotated G_3, so that the rotated Delaunay action-angle variables are recommended. The corresponding frequencies will be denoted $\omega_L, \omega_G, \omega_{G_3}$, respectively. Because α_3, s_+, c_- are null, the coefficient c also vanishes, and the Hamiltonian (8.1.3) is similar to that of the QZ problem: the intersection parabolas on the plane $\xi_1 \xi_3$ are symmetric with respect to the ξ_3 axis, whereas the concavity value $-b/a$ is no longer constant.

The SQZc problem was well studied in the 1990s, and even confining ourselves to the classical (i.e., non-quantum) problem the number of articles is huge. We quote Von Milczewski & Uzer (1997a), Von Milczewski & Uzer (1997b), and Cushman & Sadovskií (2000) as a small sample and refer to the bibliography in these articles. Here we follow closely Cordani (2008).

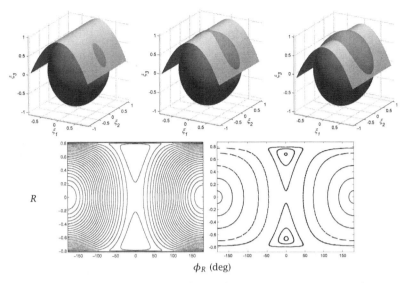

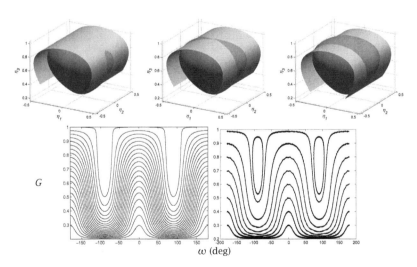

Figure 8.9: As in Figure 8.1 (R-case) but with $\mathcal{B} = 0.02$, $\mathcal{E}_1 = 0.004$, and $\mathcal{E}_3 = 0$. Notice that now the dynamical variables are *rotated*. To get the numerical output of the bottom-right picture when the system has three degrees of freedom and the Poincaré section cannot be used, see "User Functions" on page 218.

Figure 8.10: As in Figure 8.3 (G-case) but with $\mathcal{B} = 0.02$, $\mathcal{E}_1 = 0.004$, and $\mathcal{E}_3 = 0$. Notice that now the dynamical variables are *rotated*.

In Figures 8.9 and 8.10 we consider the case $\mathcal{B} = 0.02$, $\mathcal{E}_1 = 0.004$, and $\mathcal{E}_3 = 0$, with $G_3 = 0.2$, where it is understood that $\vec{R}$ and $\vec{G}$ have been *rotated*, as explained previously. The close resemblance with the corresponding Figures 8.1 and 8.3 of the QZ case shows how the rotation of the two 2-spheres is able to "absorb" the presence of an orthogonal electric field. The value of the electric and magnetic field has been taken somewhat small, in order to restrict the amplitude of the fast oscillations in the numerical output.

In the following numerical example, regarding the FMI distribution, we choose $\mathcal{B} = 0.2$ and $\mathcal{E}_1 = 0.06$. In Figure 8.11 (top) the whole phase space is displayed, with the value of the total energy fixed to -0.5. Taking into account that $\sigma_{\mathrm{FMI}} \approx -5$ is approximately the threshold below which the corresponding motion is in practice regular, one clearly recognizes three zones. The right-top corner always corresponds to the relative equilibrium point (since we are considering rotated variables) and is filled by KAM tori; the left-top corner displays a very sharply delimited Chirikov zone containing some small islands of stability; lastly, the central area is occupied by the Arnold web, and some distorted images of Figure 3.2 appear. Figures 8.11 (middle and bottom) show some details with a better resolution. The Arnold web is clearly visible, along with the chaotic area at the crossing of the resonances. These pictures show how more and more resonances are highlighted by improving the resolution and increasing the integration time. Notice, in particular, the bottom-right picture which displays a detail *inside* a resonance, showing the auto-similarity of the structure with an emerging Arnold web made up of secondary resonances. Clearly, with an even better resolution and a longer integration time, one would be able to display the Arnold web made up of resonances of the third level, and so forth.

The FMI is therefore a very sensitive indicator of the regularity of the orbit and very efficient in showing the position of the resonances. In order to investigate fine details and to get numerical information, it may be convenient to compute the fundamental frequencies along horizontal or vertical sections. See, for example, Figure 8.12 where ω_L only is reported; the other two frequencies carry basically the same information. The resonances are clearly recognized in the top picture and exactly confirm the position given by the computation of σ_{FMI} in the previous figures. The middle and bottom pictures display further details, and a secondary resonance is clearly enlightened.

Applying the frequency analysis to an orbit starting inside a resonance, one is able to compute numerically the three fundamental frequencies, then the resonance vector satisfying the relation $k \cdot \omega = 0$. For example, $k = (0, 3, 1)$ for the resonance including the point $G = 0.5$, $G_3 = -0.34$, and $k = (1, 2, -4)$ for the resonance including $G = 0.7$, $G_3 = -0.30$, with the argument of pericenter equal to $\pi/2$. In the third window of KEPLER define

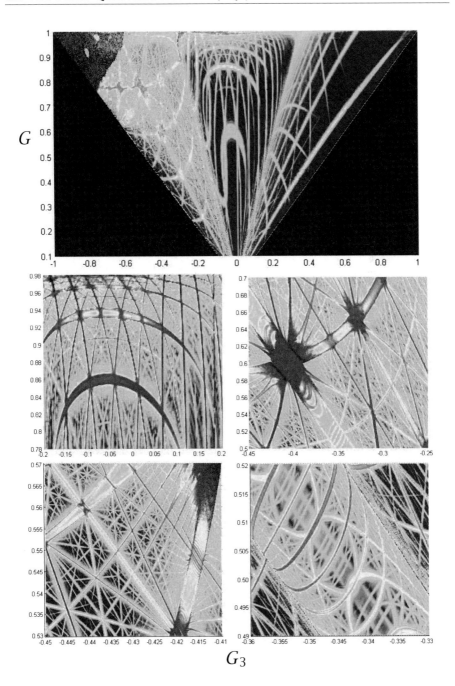

Figure 8.11: FMI for the SQZ crossed problem, with $\mathcal{B} = 0.2$ and $\mathcal{E}_1 = 0.06$. We have chosen $\omega = \pi/2$. Top: the whole phase space. Middle and bottom: details.

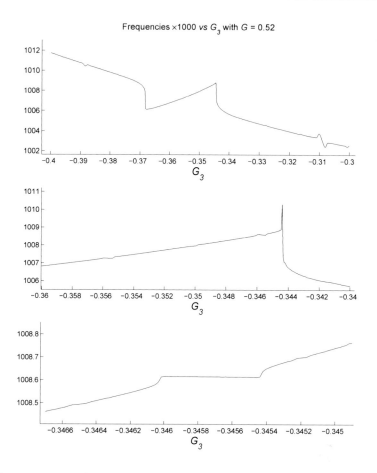

Figure 8.12: Top: frequency ω_L calculated along the section $G = 0.52$ of Figure 8.11; the positions of the resonances agree with those found in Figure 8.11. Middle: a magnified detail, which shows the secondary resonances. Bottom: a further magnified detail, displaying a secondary resonance.

the two unimodular transformations

$$\mathbf{M}_1 = \begin{pmatrix} 1 & 0 & 0 \\ 0 & 3 & 1 \\ 0 & 2 & 1 \end{pmatrix}, \quad \mathbf{M}_2 = \begin{pmatrix} 1 & -1 & 0 \\ 1 & 2 & -4 \\ -1 & -1 & 3 \end{pmatrix}$$

for the two points, respectively; then, by clicking at the end of the computation on "User Functions" and choosing AADelaunayprime_2.m, one can see that the resonance angle librates with a low frequency. Notice that the *second* row of $\mathbf{M}_1$ and $\mathbf{M}_2$ coincides with the resonance vectors k and, consequently, the *second* of the files AADelaunayprime_x.m has been chosen; the first and third rows complete the two unimodular matrices.

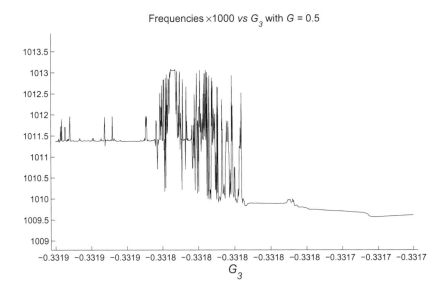

Figure 8.13: Frequency ω_L, calculated along the section $G = 0.50$ of Figure 8.11. Position and frequency range of the stochastic layer surrounding a resonance of Figure 8.11 are shown.

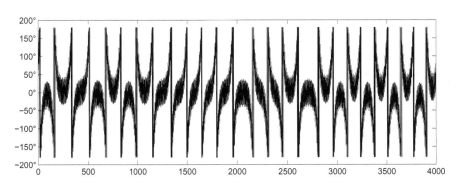

Figure 8.14: Resonant angle vs. revolution number for an orbit starting in the stochastic layer of Figure 8.13. Librations alternate randomly with clockwise and counterclockwise circulations: this is the source of chaos.

Another interesting detail is found in Figure 8.13, which shows a very magnified section of the stochastic layer surrounding the $(0, 3, 1)$ resonance. The dynamical evolution of the resonant angle of an orbit starting here is displayed in Figure 8.14, where circulating and librating motions alternate with each other at random. This is the source of chaos.

In Figure 8.15 we show some typical examples of the different orbits

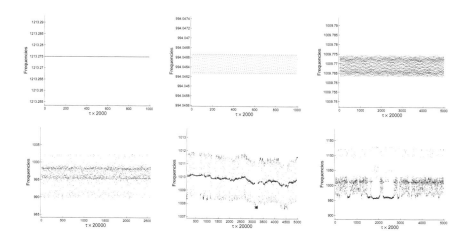

Figure 8.15: Frequency ω_L vs. revolution number for various types of orbits: see the text.

occurring in quasi-integrable Hamiltonian systems. The values of the frequency ω_L are reported for long integration intervals; notice, however, the different scales.

The top-left picture ($2 \cdot 10^6$ revolutions) refers to a KAM orbit, starting in the very neighborhood of the relative equilibrium point, with rotated and normalized values $G = 0.995$ and $G_3 = 0.990$. Notwithstanding the great magnification, the frequency appears constant.

The top-middle picture ($2 \cdot 10^6$ revolutions) refers to a motion starting in a "dark blue zone" of Figure 8.11 (middle-right). The frequency evolution shows a slight oscillation but is very regular, denoting a KAM orbit in the outer neighborhood of a resonance.

The top-right picture (10^8 revolutions) refers to a regular resonant orbit, the rotated and normalized values being $G = 0.5$ and $G_3 = -0.34$. Compare with Figure 8.11 (middle-right).

In the bottom-left picture ($5 \cdot 10^7$ revolutions) the double resonant case is reported. The starting point is $G = 0.6$ and $G_3 = -0.4$, and comparing with the values given by the frequency map (not given here) one sees that the whole chaotic zone at the crossing of the resonance strips is in fact explored.

In the bottom-middle picture (10^8 revolutions) an orbit in the stochastic layer is plotted, starting from $G = 0.5$ and $G_3 = -0.3318$. Looking at figure 8.13, one clearly ascertains that the computed evolution stays permanently within the thin stochastic layer surrounding the resonance, but the most interesting information is the slow and random drift of the whole pattern that only this orbit exhibits, very likely a manifestation of the Arnold diffusion.

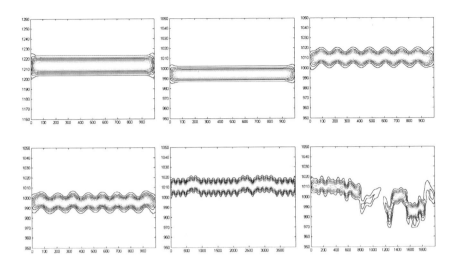

Figure 8.16: Wavelet transform for the first few revolutions of the six orbits of Figure 8.15.

Lastly, in the bottom-right picture (10^7 revolutions) an orbit of Chirikov type is plotted. Notice the remarkable and random variation of the frequency, indicating that the orbit visits a large zone of the phase space.

Figure 8.16 exhibits the wavelet transform of the six orbits of the previous figure, in the same order.

Since Figure 8.11 (top) covers the whole phase space and the equilibrium point we have considered takes up the top-right corner, one may ask: But what about the other three equilibrium points? In the rotated spheres the equilibrium point we have considered lies in the North-North poles, then the remaining three lie in the South-South, North-South, and South-North poles, respectively. For these two last points the signs of the two frequencies are different, thus they may be unstable. In fact, it seems impossible to find a periodic orbit numerically, so their instability is very probable; in Figure 8.11 they are both placed at the vertex $G = G_3 = 0$, which is a singular point in the Delaunay parametrization, encompassing all collision orbits. The point in the South-South poles is instead surely stable and takes up the top-left corner in Figure 8.11 where, however, it appears at first sight immersed in the chaos of the Chirikov zone. But a much more detailed numerical analysis shows in effect a very small KAM zone around this point to which there corresponds a periodic orbit in the physical space; see Figure 8.17. This physical orbit exhibits however two (somewhat surprising) sharp corners, making it very different from an ellipse; the osculating parameters undergo a marked variation, so that the corresponding periodic orbit on $S^2 \times S^2$ will be very large.

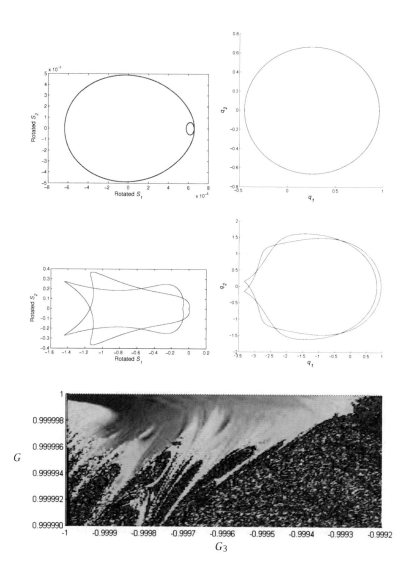

Figure 8.17: Comparing the two stable periodic orbits (100 revolutions are drawn) of the Stark–Quadratic–Zeeman–crossed problem. Top: the case of the equilibrium point in the North-North poles of the two rotated spheres is reported, i.e., the physical orbit (practically an ellipse) on the right and the corresponding very small periodic trajectory of the Pauli vectors on the left. Middle: one sees what happens at the other stable equilibrium point in the South-South poles. The physical trajectory, on the right, displays two surprising corners which make the trajectory, on the left, of the corresponding Pauli vectors very large. Bottom: the FMI at the South-South poles shows a very small ordered zone, with a sharp transition to chaos.

8.1.7 The Generic (SQZ) Problem

Lastly, we consider the full problem: the magnetic and electric fields are both present and neither parallel nor orthogonal. The intersection parabolas with the $\xi_1 \xi_3$ plane are generic: their symmetry axis is always vertical but translated along ξ_1, and the concavity can assume positive or negative values.

In Figures 8.18-8.23 three different cases are considered, which show that the analytical results (bottom-left subpictures) agree with the numerical ones (bottom-right subpictures) very well.

In Figure 8.24 we show how the FMI distribution changes when the angle between the two fields varies from zero (top-left subpicture) to π (bottom-right subpicture), with regular step $\pi/8$.

8.2 The Non-Planar Circular Restricted Three-Body (CR3B) Problem

The CR3B problem is a special case of the general three-body problem: the primary and secondary bodies move in circular orbits about the common center of mass B according to the laws of the two-body dynamics, while a third body of negligible mass moves in their gravitational field. Let 1 be the mass of the primary, $m < 1$ that of the secondary, and d their distance. In the plane of the circular orbits, rotating about B with angular speed $\omega = \sqrt{(1+m)/d^3}$, the two massive bodies are motionless and are supposed to be placed on the q_1 axis at the points $-md/(1+m)$ and $d/(1+m)$, respectively. The Hamiltonian describing the dynamics of the third body is given by

$$H = \frac{1}{2}p^2 + \sqrt{\frac{1+m}{d^3}}(q_1 p_2 - q_2 p_1) - \frac{1}{r_p} - \frac{m}{r_s},$$

where the second term of the right-hand member encompasses centrifugal and Coriolis inertial forces and r_p and r_s are the distances of the small body from primary and secondary mass, respectively.

To reduce the problem to a perturbed Kepler problem, we must suppose: i) the distance d is very large with respect to the distance of the small body from the primary; ii) the mass $m \ll 1$. After shifting the primary in the origin, the Hamiltonian becomes

$$H = \frac{1}{2}p^2 - \frac{1}{q} + \sqrt{\frac{1+m}{d^3}}(q_1 p_2 - q_2 p_1)$$
$$- \frac{m}{\sqrt{(q_1 - d)^2 + q_2^2 + q_3^2}} - \frac{m}{\sqrt{d(1+m)}}p_2,$$

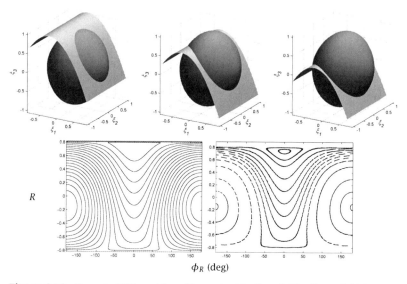

Figure 8.18: Generic SQZ problem (R-case) with $\mathcal{B} = 0.02$, $\mathcal{E}_1 = 0.006$, $\mathcal{E}_3 = 0.0001$, and $G_3 = 0.2$.

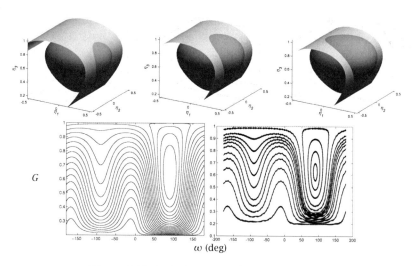

Figure 8.19: As in Figure 8.18 but in the G-case.

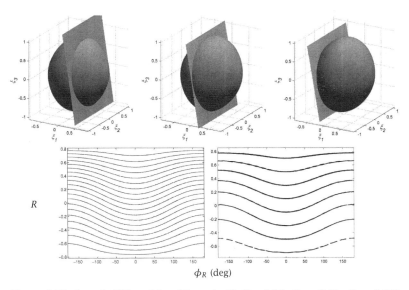

Figure 8.20: Generic SQZ problem (R-case) with $\mathcal{B} = 0.02$, $\mathcal{E}_1 = 0.01$, $\mathcal{E}_3 = 0.001$, and $G_3 = 0.2$.

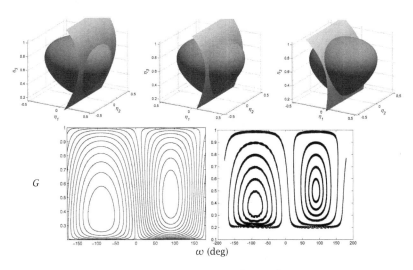

Figure 8.21: As in Figure 8.20 but in the G-case.

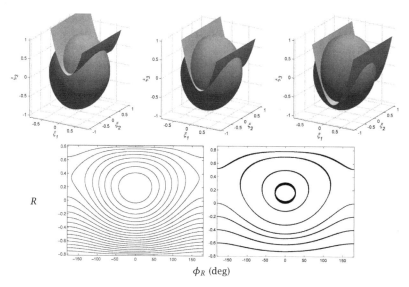

Figure 8.22: Generic SQZ problem (R-case) with $\mathcal{B} = 0.01$, $\mathcal{E}_1 = 0.02$, $\mathcal{E}_3 = 0.0002$, and $G_3 = 0.2$.

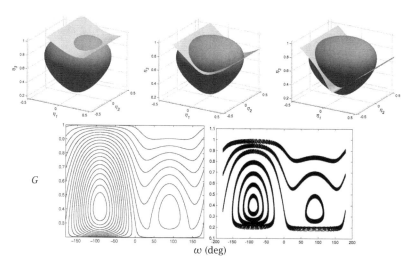

Figure 8.23: As in Figure 8.22 but in the G-case.

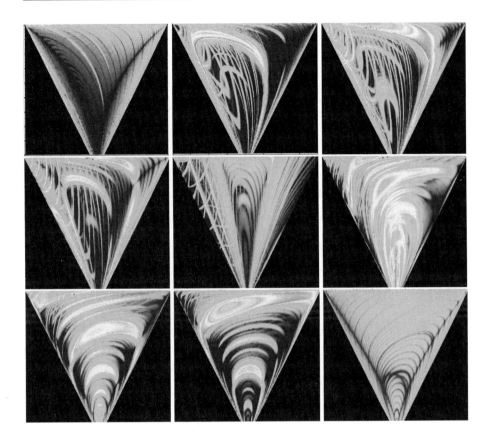

Figure 8.24: FMI for the generic SQZ problem, with $\mathcal{B} = 0.15$ and $\mathcal{E} = 0.045$. The angle between the two fields varies from zero (top-left picture) to π (bottom-right picture), with regular step $\pi/8$. In the first and last pictures the two fields are parallel and the problem is axisymmetric; thus, the vertical component of the angular momentum is a first integral and the Arnold web does not appear.

then, a truncated series development of the potential due to the secondary mass gives

$$K = K_0 + K_1 + K_2,$$

$$K_0 = \frac{1}{2}x(y^2 + 1),$$

$$K_1 = -m\left(\frac{x}{d} + \frac{xx_1}{d^2}\right) + \sqrt{\frac{1+m}{d^3}}x(x_1y_2 - x_2y_1) - \frac{m}{\sqrt{d(1+m)}}xy_2,$$

$$K_2 = -\frac{m}{2d^3}x(3x_1^2 - x^2).$$

Proceeding as in the previous section, we find that the two perturbative terms, averaged with respect to the unperturbed motion, are

$$\frac{\overline{K}_1}{K_0} = \frac{3}{2}\frac{m}{d^2}R_1 + \sqrt{\frac{1+m}{d^3}}\,G_3,$$

$$\frac{\overline{K}_2}{K_0} = \frac{m}{2d^3}\left[\frac{3}{2}(R_2^2 + R_3^2 - G_2^2 - G_3^2) - 6R_1^2\right],$$

up to unessential additive constants of motion. The Poisson bracket $\{\chi, \overline{K}_1 + K_1\}$ may be neglected because the terms surviving to the two subsequent averaging procedures (with respect to the unperturbed motion and to rotation about the vertical axis) are very small. Noticing that the first order term is similar to that of the SQZc problem suggests defining the angle

$$\tan\alpha = \frac{3}{2}\frac{m}{d^2}\bigg/\sqrt{\frac{1+m}{d^3}},$$

from which

$$\overline{K}_1 = K_0\sqrt{\frac{9}{4}\frac{m^2}{d^4} + \frac{1+m}{d^3}}\,(R_1\sin\alpha + G_3\cos\alpha).$$

Rotating the variables and dropping the primes, we get

$$\overline{K}_1 = K_0\sqrt{\frac{9}{4}\frac{m^2}{d^4} + \frac{1+m}{d^3}}\,G_3$$

$$\overline{K}_2 = K_0\frac{3m}{4d^3}[G_1^2\sin^2\alpha - G_2^2 - G_3^2(4\sin^2\alpha + \cos^2\alpha)$$
$$- R_1^2(\sin^2\alpha + 4\cos^2\alpha) + R_2^2 + R_3^2\cos^2\alpha$$
$$- (2G_1R_3 - 6G_3R_1)\sin\alpha\cos\alpha].$$

Because

$$\frac{3m}{4d^3} << \sqrt{\frac{9}{4}\frac{m^2}{d^4} + \frac{1+m}{d^3}},$$

we can take $\overline{K}_1$ as an unperturbed integrable Hamiltonian and $\overline{K}_2$ as a small perturbation. Averaging with respect to the rotations about the vertical axis generated by $\overline{K}_1$ and arranging the terms, we get the final result:

$$\overline{\overline{K}}(\xi) = \frac{1}{2}\cos^2\alpha(\xi_3 + 4\xi_1^2).$$

The Hamiltonian differs from that of the QZ problem only for the concavity value, while the multiplicative factor simply changes the time scale. The critical value is easily calculated, giving $G_3^{\text{crit}} = \sqrt{3/5} = 0.775\ldots$.

The dynamics generated by the normal form reproduces the true one very well, as the reader himself can check with the program KEPLER. To this

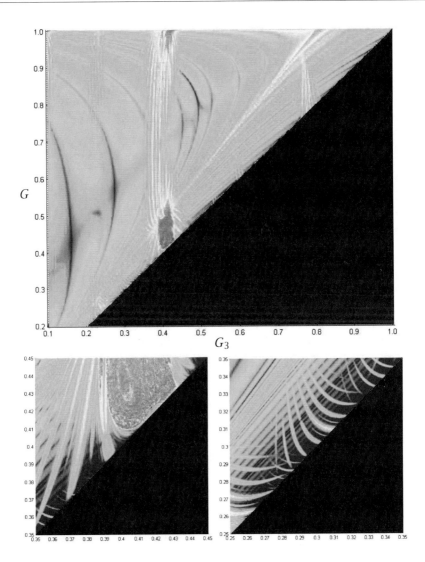

Figure 8.25: FMI for the CR3B problem with $m = 0.05$, $d = 3$ and Total Energy $= -0.5$. The small values of G_3 have not been considered, because in this case the point may sometimes escape to infinity.

end, choose "Circular restricted 3-body" in the menu "Normal form", pick out mass m and distance d, and get the numerical values of the perturbative parameters to insert in the "Perturbation Hamiltonian" panel, along with the approximate values of the two normalized Pauli vectors $\vec{s} = \vec{W}_S/W_S$ and $\vec{d} = \vec{W}_D/W_D$ at the equilibrium point.

Figure 8.25 shows the FMI for the CR3B problem with $m = 0.05$, $d = 3$

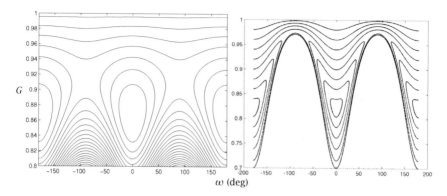

Figure 8.26: Left: action-angle (normal form) for the satellite problem. Right: action-angle (Poincaré section). The oblateness of the primary is 0.05.

and Total Energy $= -0.5$. The line from the origin to the point $G = 1, G_3 = G_3^{\text{crit}}$, which divides the libration from the rotation zone, is clearly visible. Notice that the "ghost undulations" in the top picture do not denote a real phenomenon. They are clearly recognizable because of their "artificial" regularity and because they vary when one increases the integration time, finally disappearing for very long integration time.

8.3   Satellite about an Oblate Primary

For a satellite close to its primary, the principal perturbation arises from the nonsphericity of the planet. We want to study what is called the *main problem*, i.e., we will take only the second zonal harmonic into account; in particular the problem will be axially symmetric around the polar axis.

The Hamiltonian is

$$H = H_0 + \varepsilon H_p, \quad \text{with } H_0 = \frac{1}{2}p^2 - \frac{1}{q} \quad \text{and } H_p = \frac{1}{q^3}\mathcal{P}_2(\cos \vartheta),$$

where $\mathcal{P}_2(x) = \frac{3}{2}x^2 - \frac{1}{2}$ is the second Legendre polynomial and the perturbative parameter ε takes the oblateness of the primary into account. The angle ϑ is the colatitude.

Because of the third power singularity in H_p the problem is not regularizable in the standard manner, and this is basically the source of its difficulties. It was not untill the work of Brouwer (1959) that the second order normal form was calculated. Later Deprit (1981), with the method of the *elimination of the parallax*, radically unraveled the problem. His idea is to factorize the normalizing canonical transformation into the product of two simpler transformations: the first eliminates the anomaly on the plane of the orbit, the latter completes the procedure averaging over the mean anomaly. The normal form of Brouwer is thus recovered, and it turns out that finding the two transformations is strikingly simpler than directly finding the whole one. Moreover, the method can be implemented in a computer program and the average of the perturbative terms can be calculated to higher orders (however, here we will treat only the first two terms); see Deprit (1981) and Coffey *et al.* (1994). We do not enter the somewhat technical details of the procedure but pass directly to the final result. Up to constant multiplicative factors, the second order normal form in the G-case is

$$\overline{H}_p(\eta_1, \eta_3) = \overline{H}_1 + \varepsilon \overline{H}_2,$$

with (G_3 is a first integral and $G = \eta_3$, with G and G_3 normalized to L)

$$\overline{H}_1 = \frac{1}{\eta_3^3} \left(\frac{1}{2} - \frac{3}{2} \frac{G_3^2}{\eta_3^2} \right),$$

$$\overline{H}_2 = -\frac{3}{32} \frac{1}{\eta_3^5} \left(5 - 18 \frac{G_3^2}{\eta_3^2} + 5 \frac{G_3^4}{\eta_3^4} \right) - \frac{3}{8} \frac{1}{\eta_3^6} \left(1 - 6 \frac{G_3^2}{\eta_3^2} + 9 \frac{G_3^4}{\eta_3^4} \right)$$

$$+ \frac{15}{32} \frac{1}{\eta_3^7} \left(1 - 2 \frac{G_3^2}{\eta_3^2} - 7 \frac{G_3^4}{\eta_3^4} \right)$$

$$- \frac{3}{16} \frac{1}{\eta_3^7} \left[\left(1 - 15 \frac{G_3^2}{\eta_3^2} \right) \left(1 - \frac{G_3^2}{\eta_3^2} \right) (1 - \eta_3^2) - 2 \left(1 - 15 \frac{G_3^2}{\eta_3^2} \right) \frac{\eta_1^2}{\eta_3^2} \right].$$

If we consider only the first order term of the averaged perturbation, we find

$$\dot{\omega} = \varepsilon \frac{\partial \overline{H}_1}{\partial \eta_3} = \varepsilon \frac{3}{2\eta_3^4} \left(5 \frac{G_3^2}{\eta_3^2} - 1 \right). \tag{8.3.1}$$

All the points belonging to the intersection between $\overline{\overline{M}}$ and the plane $\eta_3 = \sqrt{5} G_3$ are degenerate critical points. Since the North pole of $\overline{\overline{M}}$ of the G-type has coordinate $\eta_3 = 1$ for every value of G_3, the intersection is nonempty for $G_3 \leq \frac{1}{\sqrt{5}}$. The inclination $i_{\text{crit}} = \arccos \frac{1}{\sqrt{5}}$ is called *critical*: the pericenter of the orbits with $i < i_{\text{crit}}$ circulates in one sense, those with $i > i_{\text{crit}}$ in the other. As G_3 increases toward $\frac{1}{\sqrt{5}}$, the critical circle collapses into the North pole of $\overline{\overline{M}}$. This very crude representation is broken up by also taking the second order term $\overline{H}_2$ into account; the continuous critical circle now disappears, leaving two hyperbolic and two elliptic points only, provided it is not too close to the North pole, i.e., $G_3 << \frac{1}{\sqrt{5}}$. Obviously, to these four critical points one must add the always-present critical points at the poles. For $G_3 >> \frac{1}{\sqrt{5}}$ only these last two critical points are present.

In Figure 8.26 the evolution of G and ω is calculated with the normal form (left) and compared with the true one of the Poincaré section (right), with $G_3 = 0.4$ and $\varepsilon = 0.05$. The two pictures result somewhat different,

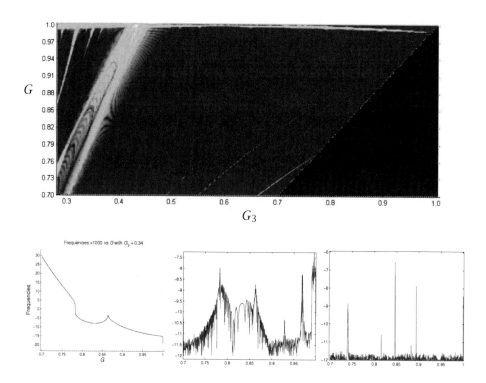

Figure 8.27: Top: FMI for the satellite problem. Bottom-left: frequency map. Bottom-center: FMI for $G_3 = 0.34$. Bottom-right: FMI for $G_3 = 0.7$. The oblateness of the primary is 0.05 and Total Energy $= -0.5$.

though qualitatively the correspondence is passable: in both cases a strip appears, centered about the critical inclination and containing two elliptic and two hyperbolic points. Probably, the discrepancy is due to the neglected higher order perturbative terms, which thus play a significant role.

Figure 8.27 (top) shows the general FMI distribution for $G_3 = 0.4$. Notice in particular the strip of the critical inclination which appears with the expected slope $G/G_3 = \sqrt{5}$, as suggested by (8.3.1). Figure 8.27 (bottom-left) confirms that the frequency of the pericenter motion changes sign when crossing the critical inclination. The other two pictures show the FMI along the vertical sections $G_3 = 0.34$ and $G_3 = 0.7$, respectively, revealing some resonances which surely would escape with a Poincaré section analysis.

CHAPTER 9

The Multi-Body Gravitational Problem

Gravity explains the motions of the planets,
but it cannot explain who set the planets in motion.
God governs all things and knows all that is or can be done.

— I. NEWTON

God? I have no need for that hypothesis.

— P.S. DE LAPLACE

Deducing the motion of bodies interacting gravitationally is probably the most important mechanical problem but also the most difficult. The three-body problem is not integrable, even if the masses are very small but of comparable size, and this fact prevents in general the use of perturbative methods.

In this chapter we will study some important exceptions: the planar three-body problem in its two limit cases, lunar and planetary, which admits a global treatment, and the classical 3-dimensional planetary problem, with one body much more massive with respect to the others.

The reader will find all the software relative to this chapter in the program LAPLACE, described in the third section. The program compares normal form and true motion, performs numerical frequency analysis, and finds the phase space geography of the resonances with the aid of the FMI tool.

9.1   Global Planar Three-Body Problem

The three-body problem admits two well-known limit cases, the *planetary* and the *lunar problems*, which are obtained when a suitable perturbative parameter is very small.

By the *planetary problem* one means the mechanical system consisting of a body of large mass, the "Sun," and other bodies much smaller, the "planets" (in this section we restrict ourselves to two planets) interacting through gravitational forces. The classical approach to the study of its motion is based on series developments, where the quantities playing the role of small parameters in the developments are eccentricities and inclinations of the planets, while the ratio between the masses of the planets and the mass of the Sun is the perturbative parameter. In other words, one studies motions which differ only slightly from the circular coplanar one.

By the *lunar problem* one means the system consisting of a small body, the "Moon," rotating around the "Earth," with a third body, the "Sun," much more distant. Here, the perturbative parameter is the ratio between the distances Moon-Earth and Sun-Earth, but one still studies motions with small eccentricities and inclinations. See, for example, Brouwer & Clemence (1961).

The classical reference on the argument is Poincaré (1892–1893–1899), who was the first to use the action-angle variables of the Kepler problem for reduction of the three-body problem. Jefferys & Moser (1966), Lieberman (1971), and Lidov & Ziglin (1976) studied the problem(s) with finite eccentricities and/or inclinations, but with some limitations. It is only with Féjoz (2001), Féjoz (2002a), and Féjoz (2002b) that a detailed and complete study has been done for the planar case, treating the problem in a global framework. Here we will follow Cordani (2004), which reaches basically the same results though with a different line of reasoning.

While retaining the assumption of smallness of the perturbative parameter, we would like to get rid of the other limitations. Unfortunately, the full 3-dimensional problem appears somewhat difficult to tackle (see the concluding remarks of the section) and we consider here the plane case only, thus with inclinations all vanishing. We will proceed as follows. The system is first reduced to four degrees of freedom thanks to its translational invariance, then, averaging along the unperturbed motion, it is further reduced to two degrees; the averaged Hamiltonian inherits the rotational invariance from the original one and this symmetry results in a further reduction to a system with one degree of freedom, that is hence integrable. As for the perturbed Kepler problem, global conclusions can be drawn by studying the intersections between the reduced manifold and the level surfaces of the perturbative Hamiltonian.

The present section is devoted to this reduction and to the geometric study of the resulting motion, also in the case of large eccentricities. The main technical difficulty is the integration involved in the averaging procedure. Since it cannot be performed exactly, we will proceed with two different approximate methods whose final results agree very well.

(i) Analytical method: since the expression of the averaged Hamiltonian cannot be calculated explicitly, in Féjoz (2002a) the first three terms of its series expansion are given, while further calculations by hand are not practical. We have implemented a procedure on a symbolic manipulator which makes the calculation to any order. The first ten terms are retained and the level curves of the corresponding Hamiltonian are displayed, showing number and location of the critical points. We remark however that the first three terms of the series expansion are already sufficient to give a qualitatively satisfying description of the dynamics, since the other terms do not add further critical points.

(ii) Numerical method: the integrations involved in the averaging process are performed numerically and the level curves of this reduced averaged Hamiltonian are also displayed, showing a very good agreement with the previous case,

We recall some basic expressions on the planar Kepler problem taken directly from Chapter 6 but with various physical constants now made explicit. The 2-dimensional Kepler problem is the Hamiltonian system with phase space $T^*(\mathbb{R}^2 - \{0\})$ and Hamiltonian

$$H_0 = \frac{1}{2m}p^2 - \frac{k}{q}, \quad k = GMm, \quad p = \left\|\vec{p}\right\|, \, q = \left\|\vec{q}\right\|$$

where M is the attractive mass and G the constant of gravitation. The orbit is a conic, whose equation in polar coordinates q, θ is

$$q(\theta) = \frac{G^2/mk}{1 + E\cos(\theta - \varpi)}.$$

G and E are the norm of *angular momentum* and *eccentricity* vector, respectively,

$$\vec{G} = \vec{q} \times \vec{p}, \quad \vec{E} = \frac{1}{mk}\vec{p} \times \vec{G} - \frac{\vec{q}}{q},$$

while ϖ is the *longitude of pericenter* and $f = \theta - \varpi$ is the *true anomaly*. Here and in the sequel we consider the case $E < 1$, which implies that the conic is an ellipse and that $H < 0$. We recall that $\vec{G}$ and $\vec{E}$ are first integrals of the motion.

Another useful parametrization of the orbit is

$$\begin{aligned} X &= a\cos s - aE, \\ Y &= a\sqrt{1 - E^2}\sin s, \end{aligned} \tag{9.1.1}$$

where $a = k/(-2H_0)$ is the *semimajor axis* and s the *eccentric anomaly*, whose relation with time is given by the Kepler equation

$$t = \sqrt{ma^3/k}\,(s - E\sin s).$$

Differentiating this last equation we find

$$\frac{dt}{ds} = q\sqrt{ma/k}. \tag{9.1.2}$$

The action-angle variables are

$$L = \sqrt{mka} \quad \text{and} \quad l = s - E\sin s, \quad G \text{ and } \varpi,$$

so the Hamiltonian

$$H_0 = -\frac{mk^2}{2L^2}$$

turns out to be a function of the action L only; i.e., the problem has only one frequency (total degeneration). The angle l is the *mean anomaly*.

Define the *Runge-Lenz-Laplace* vector $\vec{R} = L\vec{E}$. It satisfies the following properties:

(i) If $\{\cdot,\cdot\}$ is the Poisson bracket, with $\{q_h, q_k\} = \{p_h, p_k\} = 0$ and $\{p_h, q_k\} = \delta_{hk}$, then

$$\{G, R_1\} = -R_2, \quad \{G, R_2\} = R_1, \quad \{R_1, R_2\} = -G,$$

which is the Lie algebra of the group SO(3).

(ii) Once the energy is fixed, the 3-dimensional vectors with components (R_1, R_2, G) span the sphere S^2

$$R_1^2 + R_2^2 + G^2 = L^2,$$

which therefore appears as the space of the orbits.

The property (i) endows $\mathbb{R}^3$ with a Poisson structure, while (ii) states that the space of the orbits S^2 is a leaf of this Poisson structure and is therefore endowed with a symplectic structure.

9.1.1 The 2-Dimensional Secular Problem

Let m_0, m', m'' be the masses of three points undergoing gravitational attraction in the plane. In the standard notation the Hamiltonian is

$$H = \frac{\left\|\vec{p_0}\right\|^2}{2m_0} + \frac{\left\|\vec{p}'\right\|^2}{2m'} + \frac{\left\|\vec{p}''\right\|^2}{2m''}$$

$$- G\left(\frac{m_0 m'}{\left\|\vec{q_0} - \vec{q}'\right\|} + \frac{m_0 m''}{\left\|\vec{q_0} - \vec{q}''\right\|} + \frac{m' m''}{\left\|\vec{q}' - \vec{q}''\right\|}\right).$$

We adopt the Jacobi coordinates

$$
\begin{pmatrix} \vec{Q_0} \\ \vec{Q'} \\ \vec{Q''} \end{pmatrix} = \begin{pmatrix} 1 & 0 & 0 \\ -1 & 1 & 0 \\ -\sigma_0 & -\sigma_1 & 1 \end{pmatrix} \begin{pmatrix} \vec{q_0} \\ \vec{q'} \\ \vec{q''} \end{pmatrix} \quad \text{with inverse}
$$

$$
\begin{pmatrix} \vec{q_0} \\ \vec{q'} \\ \vec{q''} \end{pmatrix} = \begin{pmatrix} 1 & 0 & 0 \\ 1 & 1 & 0 \\ 1 & \sigma_1 & 1 \end{pmatrix} \begin{pmatrix} \vec{Q_0} \\ \vec{Q'} \\ \vec{Q''} \end{pmatrix},
$$

where

$$
\sigma_0 = \frac{m_0}{m_0 + m'}, \quad \sigma_1 = \frac{m'}{m_0 + m'}, \quad \sigma_0 + \sigma_1 = 1.
$$

Clearly, the position vector $\vec{Q'}$ connects m_0 with m', while $\vec{Q''}$ connects the center of mass of m_0 and m' with m'', as required by the Jacobi coordinates. To preserve canonicity, the momenta will change with the inverse transposed matrices

$$
\begin{pmatrix} \vec{P_0} \\ \vec{P'} \\ \vec{P''} \end{pmatrix} = \begin{pmatrix} 1 & 1 & 1 \\ 0 & 1 & \sigma_1 \\ 0 & 0 & 1 \end{pmatrix} \begin{pmatrix} \vec{p_0} \\ \vec{p'} \\ \vec{p''} \end{pmatrix} \quad \text{with inverse}
$$

$$
\begin{pmatrix} \vec{p_0} \\ \vec{p'} \\ \vec{p''} \end{pmatrix} = \begin{pmatrix} 1 & -1 & -\sigma_0 \\ 0 & 1 & -\sigma_1 \\ 0 & 0 & 1 \end{pmatrix} \begin{pmatrix} \vec{P_0} \\ \vec{P'} \\ \vec{P''} \end{pmatrix}.
$$

Choose the reference frame attached with the center of mass of the three points; thus $\vec{P_0} = \vec{p_0} + \vec{p'} + \vec{p''} = 0$. The Hamiltonian becomes

$$
H = \frac{P'^2}{2\mu'} - G\frac{\mu'M'}{Q'} + \frac{P''^2}{2\mu''}
$$

$$
- G\mu'm'' \left(\frac{1}{\sigma_1} \frac{1}{\left\|\vec{Q''} + \sigma_1\vec{Q'}\right\|} + \frac{1}{\sigma_0} \frac{1}{\left\|\vec{Q''} - \sigma_0\vec{Q'}\right\|} \right), \qquad (9.1.3)
$$

where $P' = \left\|\vec{P'}\right\|, \dots,$ and (M'' will be used very soon)

$$
\mu' = \frac{m_0 m'}{m_0 + m'}, \quad \mu'' = \frac{M'm''}{M' + m''}, \quad M' = m_0 + m', \quad M'' = m_0 + m' + m''.
$$

Recall the well-known expansion

$$
\frac{1}{\left\|\vec{y} - \vec{x}\right\|} = \frac{1}{y}\left[1 + \sum_{l=1}^{\infty} \left(\frac{x}{y}\right)^l P_l(\cos\vartheta)\right],
$$

where $\vec{x}$ and $\vec{y}$ are two vectors with $x < y$, ϑ the angle between the two vectors, and $\mathcal{P}_k$ the kth Legendre polynomial. Take $\vec{Q}'' = \vec{y}$ with $-\sigma_1 \vec{Q}' = \vec{x}$ and $\sigma_0 \vec{Q}' = \vec{x}$, respectively, then from (9.1.3) we get

$$H = \frac{P'^2}{2\mu'} - G\frac{\mu'M'}{Q'} + \frac{P''^2}{2\mu''} - G\frac{\mu''M''}{Q''} - G\frac{\mu'm''}{Q''} \sum_{k=2}^{\infty} \sigma_k \left(\frac{Q'}{Q''}\right)^k \mathcal{P}_k(\cos\vartheta),$$

(9.1.4)

$$\sigma_k = \sigma_0^{k-1} + (-1)^k \sigma_1^{k-1},$$

which is the sum of the Hamiltonians of two independent Kepler systems plus a perturbative Hamiltonian

$$H_p \stackrel{\text{def}}{=} -\frac{\mu'm''}{Q''} \sum_{k=2}^{\infty} \sigma_k \left(\frac{Q'}{Q''}\right)^k \mathcal{P}_k(\cos\vartheta).$$

(9.1.5)

Clearly, if $m', m'' \ll m_0$ or $Q'/Q'' \ll 1$ the perturbative Hamiltonian is "small," and the usual techniques of celestial mechanics can be applied.

The secular Hamiltonian is obtained by averaging H_p along the unperturbed motion, i.e., along the two Keplerian ellipses. Once this nontrivial task is done, the averaged perturbative part $\overline{H_p}$ becomes a first integral of the motion, so that it is a function of the first integrals of the two disjoint Kepler problems. In other words, $\overline{H_p}$ becomes a function defined on the space of the orbits, which is the product $S^2 \times S^2$ of the space of the orbits of the two disjoint Kepler problems. More explicitly, $\overline{H_p}$ is expressed as a function of the six variables R_1', R_2', G' and R_1'', R_2'', G'' satisfying the two relations

$$R_1'^2 + R_2'^2 + G'^2 = L'^2, \quad R_1''^2 + R_2''^2 + G''^2 = L''^2. \tag{9.1.6}$$

The two left-hand side members define the *Casimir invariants* of the Poisson structure of $P = \mathbb{R}^3 \times \mathbb{R}^3$, since they have vanishing Poisson bracket with all six variables.

The Hamiltonian now reads as

$$H = -G^2 \frac{\mu'^3 M'^2}{2L'^2} - G^2 \frac{\mu''^3 M''^2}{2L''^2} + G\overline{H_p},$$

and the new phase space is the symplectic manifold $S^2 \times S^2$ on which the constant actions L' and L'' do not generate any dynamics: one can thus take as Hamiltonian only the term $\overline{H_p}$.

The new Hamiltonian system we have defined, with Hamiltonian $\overline{H_p}$ and phase space $S^2 \times S^2$, inherits from the original one the invariance under the action of the group SO(2), which generates simultaneous rotations of the two spheres by the same angle around the vertical axes G' and G'', respectively. $\overline{H_p}$ will therefore be a function of the six invariants:

$$G', \; G'', \; \left\|\vec{R}'\right\|, \; \left\|\vec{R}''\right\|, \quad \vec{R}' \cdot \vec{R}'' = R_1' R_1'' + R_2' R_2''$$

$$\text{and } \; \left. \vec{R}' \times \vec{R}'' \right|_3 = R_1' R_2'' - R_1'' R_2',$$

which are indeed the only quantities having vanishing Poisson bracket with $G_{\text{tot}} = G' + G''$, the generator of the rotations. The averaged problem is therefore a Hamiltonian system with two degrees of freedom and with rotational symmetry, and it is natural to try to reduce out this symmetry, obtaining a one-degree-of-freedom, thus integrable, system.

9.1.2 Reduction of the Symplectic Manifold $S^2 \times S^2$ under the SO(2)-action

We recall that a Hamiltonian system possesses a symmetry if there exists a Lie group $\mathfrak{G}$ which acts on the phase space M preserving its symplectic form Ω and leaving the Hamiltonian H invariant. As a consequence, the components of the moment map $J : M \rightarrow \mathfrak{g}^*$ (where $\mathfrak{g}^*$ is the dual of the Lie algebra of the group $\mathfrak{G}$) are first integrals of the motion. See subsection 2.3.7 on page 58.

In this situation one may try to reduce by this symmetry and obtain a Hamiltonian system with fewer degrees of freedom in the following manner. First, fix a regular value μ (not to be confused with the fictitious masses defined above) of the moment map and consider the submanifold $J^{-1}(\mu) \subset M$, which acquires, by restriction of Ω, a presymplectic structure, i.e., such that the corresponding skew symmetric matrix has a nontrivial kernel. Second, since the isotropy subgroup $\mathfrak{G}_\mu$, i.e., the subgroup of $\mathfrak{G}$ whose coadjoint action does not move μ, acts on $J^{-1}(\mu)$, consider the space of the orbits $M_\mu = J^{-1}(\mu)/\mathfrak{G}_\mu$: so we get rid of the kernel and M_μ acquires a symplectic structure. If, moreover, the $\mathfrak{G}_\mu$-action is free, i.e., without fixed points, M_μ will also be a *differentiable* manifold; otherwise there will exist some singular points. The Hamiltonian H descends naturally to M_μ, defining a reduced Hamiltonian.

In the case we are discussing, $H = \overline{H_p}$, $M = S^2 \times S^2$, and $\mathfrak{G} = \text{SO}(2)$ so that $\mathfrak{G}_\mu = \mathfrak{G}$, since the group is abelian. The difficulty is that the group action has fixed points: the situation is very similar to that already met in Section 6.5, regarding the reduction of the orbit space of the Kepler problem under axial symmetry. As in that case, instead of considering the symplectic manifold $S^2 \times S^2$ we first reduce the Poisson manifold $P = \mathbb{R}^3 \times \mathbb{R}^3$, of which the product of the two spheres is a symplectic leaf, and then restore the two Casimir invariants. Let us view the details.

Consider the SO(2)-action on $S^2 \times S^2$, generated by $J = G_{\text{tot}}$ with $-L' - L'' < G_{\text{tot}} < L' + L''$. The two extreme cases, corresponding to the points

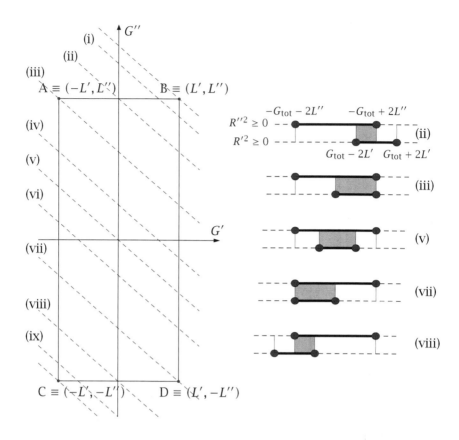

Figure 9.1: Reduction of the orbit space. Left: the admissible values of G' and G'' are those lying on the straight line $G' + G'' = G_{tot}$ and falling on and internally to the rectangle ABCD. When $G_{tot} = \pm(L' - L'')$, i.e., in cases (iii) and (vii) respectively, the SO(2)-action has fixed points: the point D and the point A. Right: the shaded areas show the intervals of the allowed values for ξ_1, where both the inequalities are satisfied at the same time.

B and C in Figure 9.1, are discarded, since the manifolds $J^{-1}(-L' - L'')$ and $J^{-1}(L' + L'')$ degenerate into a point: in both the cases the space of the orbits is made of only two circular orbits with radii a' and a''. In the generic case, having fixed the value μ of G_{tot}, the admissible values of G' and G'' are those lying on the straight line $G' + G'' = \mu$ and falling on and internally to the rectangle ABCD of Figure 9.1. The ordinate of the intersection of the straight line with the vertical axis G'' gives the corresponding value of G_{tot}. When $G_{tot} = \pm(L' - L'')$, i.e., in cases (iii) and (vii), respectively, the SO(2)-action has fixed points: the point D, to which correspond the South pole of the first sphere and the North of the second, and the point A, to which correspond the North pole of the first sphere and the South of the second. It is indeed evident that the poles are not moved by rotations around the vertical axis.

Let us reduce $P = \mathbb{R}^3 \times \mathbb{R}^3$. Define $\psi \in \mathbb{C}^2 - \{0\}$ as

$$\psi = \begin{pmatrix} \psi_1 \\ \psi_2 \end{pmatrix} = \begin{pmatrix} R'_1 + iR'_2 \\ R''_1 + iR''_2 \end{pmatrix}.$$

We have excluded the null element to which correspond circular orbits (points B and C). P is parametrized by ψ, G_{tot}, and $\xi_1 \overset{\text{def}}{=} G'' - G'$. A rotation around the vertical axis leaves invariant G_{tot} and ξ_1, while

$$\text{SO(2)-action}: \psi \mapsto e^{i\varphi}\psi.$$

To reduce this action we must first find the 3-dimensional manifold that parametrizes the space of the orbits of the action. To this end, we apply Proposition 6.3 on page 204; then, putting $G_{\text{tot}} = \mu$, we obtain the reduction

$$P \to \mathbb{R} \times (\mathbb{R}^3 - \{0\}) : (R'_1, R'_2, G', R''_1, R''_2, G'') \mapsto (\xi_1, \xi_2, \xi_3, \xi_4),$$

with

$$\begin{aligned} \xi_1 &= G'' - G', & \xi_2 &= 2\vec{R}' \cdot \vec{R}'', \\ \xi_3 &= 2 \left.\vec{R}' \times \vec{R}''\right|_3, & \xi_4 &= R''^2 - R'^2, \end{aligned} \tag{9.1.7}$$

where $R' = \left\|\vec{R}'\right\|$ and $R'' = \left\|\vec{R}''\right\|$.

By restoring the two Casimir invariants (9.1.6), we pass from the above $\mathbb{R} \times (\mathbb{R}^3 - \{0\})$ to the 2-dimensional reduced manifold M_μ we are seeking. To this end, write the two Casimir invariants as

$$R'^2 = L'^2 - \frac{1}{4}(G_{\text{tot}} - \xi_1)^2, \quad R''^2 = L''^2 - \frac{1}{4}(G_{\text{tot}} + \xi_1)^2, \tag{9.1.8}$$

from which $\xi_4 = L''^2 - L'^2 - G_{\text{tot}}\xi_1$, then substitute (9.1.8) into the obvious relation

$$\xi_2^2 + \xi_3^2 = 4R'^2 R''^2,$$

thus obtaining

$$4\left[L'^2 - \frac{1}{4}(G_{\text{tot}} - \xi_1)^2\right]\left[L''^2 - \frac{1}{4}(G_{\text{tot}} + \xi_1)^2\right] = \xi_2^2 + \xi_3^2. \tag{9.1.9}$$

This is the equation of the reduced manifold M_μ: once L' and L'' are fixed, (9.1.9) describes a family, parametrized by the value μ of G_{tot}, of 2-dimensional manifolds having rotation symmetry about the ξ_1 axis.

In order that the right members of the two relations (9.1.8) be greater than or equal to zero, the inequalities

$$G_{tot} - 2L' \le \xi_1 \le G_{tot} + 2L', \quad -G_{tot} - 2L'' \le \xi_1 \le -G_{tot} + 2L'',$$

must hold simultaneously. In Figure 9.1 the shaded areas show the intervals of the allowed values for ξ_1. Notice that in cases (iii) and (vii) two of the four roots coincide, exactly when the SO(2)-action has fixed points.

Figure 9.2 represents the sections of the reduced manifold M_μ for $L'' = 1$ and $L' = 0.14$. In Figure 9.3 the values $L'' = 1$ and $L' = 0.45$, and in Figure 9.4 the values $L'' = 1$ and $L' = 1/\sqrt{2}$ are considered.

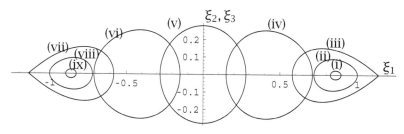

Figure 9.2: The reduced manifold M_μ is a rotation surface around the ξ_1 axis: its sections are represented in the figure, with $L'' = 1$ and $L' = 0.14$. The Roman numerals (i)-(ix) label the values of $G_{tot} = \mu$, as in Figure 9.1.

As expected, M_μ is not smooth at a point in cases (iii) and (vii), due to the fact that the SO(2)-action is not free.

It is interesting to note that, for each of the admissible values of G_{tot}, the reduced manifold appears as a symplectic leaf of a Poisson structure. Indeed, defining

$$C(\xi) \overset{\text{def}}{=} 4 \left[L'^2 - \frac{1}{4}(G_{tot} - \xi_1)^2 \right] \left[L''^2 - \frac{1}{4}(G_{tot} + \xi_1)^2 \right] - \xi_2^2 - \xi_3^2,$$

one checks that the Poisson brackets between the variables ξ are given by

$$\{\xi_i, \xi_h\} = \epsilon_{ihk} \frac{\partial C(\xi)}{\partial \xi_k}, \quad i, h, k = 1, 2, 3. \tag{9.1.10}$$

It follows that $C(\xi)$ is the Casimir invariant, since

$$\{\xi_i, C(\xi)\} = \frac{\partial C(\xi)}{\partial \xi_h} \epsilon_{ihk} \frac{\partial C(\xi)}{\partial \xi_k} = 0 \quad \forall i$$

for the complete skew symmetry of the symbol ϵ_{ihk}. Clearly, the equation of M_μ reads as $C(\xi) = 0$.

We sum up the results of this section in the following theorem.

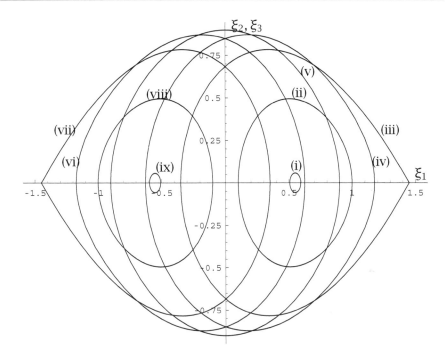

Figure 9.3: As in Figure 9.2 but for $L'' = 1$ and $L' = 0.45$.

THEOREM 9.1 *The symplectic reduction under SO(2)-action of the manifold of the orbits for a fixed energy of the 2-dimensional averaged three-body problem is the symplectic 2-dimensional leaf $C(\xi) = 0$ of a 3-dimensional linear Poisson manifold, with coordinates ξ_1, ξ_2, ξ_3 and Poisson structure (9.1.10). The reduced manifold is homeomorphic to the 2-dimensional sphere and is smooth for all the allowed values of the total angular momentum, i.e., $-L' - L'' < G_{tot} < L' + L''$, except for $G_{tot} = \pm(L' - L'')$ where a pinched point arises.*

9.1.3 The Reduced Motion

Let us now consider the averaged Hamiltonian $\overline{H_p}$: having fixed the three masses, the two semimajor axes, and G_{tot}, it is, at first sight, a function of ξ_1, ξ_2, ξ_3 which we may represent by drawing its level surfaces, satisfying the equation $\overline{H_p}(\xi_1, \xi_2, \xi_3) = $ constant. Every solution point of the Hamilton equations moves on one of these level surfaces, selected by the initial conditions; moreover, it must stay on the reduced manifold M_μ. Hence, the solution point must move on one of the curves determined by the intersection of the manifold M_μ with the family of surfaces given by $\overline{H_p}(\xi_1, \xi_2, \xi_3) = $

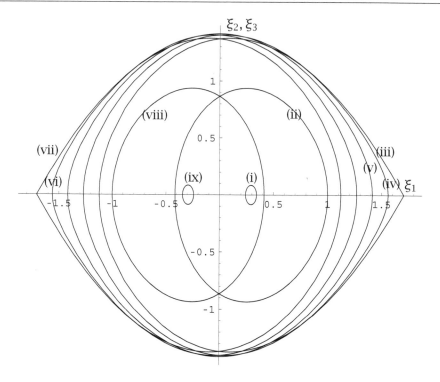

Figure 9.4: As in Figure 9.2 but for $L'' = 1$ and $L' = 1/\sqrt{2}$.

constant.

A closer inspection, however, reveals that $\overline{H_p}$ is a function of ξ_1, ξ_2 only. Indeed, a point on the reduced manifold and relative to some G_{tot}, is characterized by the values of ξ_1 (which fixes the sizes of the ellipses along with G_{tot}) and $\varphi = \arctan(\xi_3/\xi_2)$: as is clear by definitions (9.1.7), this is the angle between $\vec{R}'$ and $\vec{R}''$, and since the interaction energy between the two ellipses is an even function of this angle, it cannot be a function of ξ_3. Notice that $\{\xi_1/2, \varphi\} = 1$. The semidifference of the angular momenta and the angle formed by the two ellipses are therefore a pair of action-angle variables.

The fact that $\overline{H_p}$ is a function of ξ_1, ξ_2 only and that the reduced manifold has rotational symmetry around the ξ_1 axis greatly simplifies the qualitative discussion of the properties of the motion, in particular of the number and position of the critical (or singular) points. We recall that the critical points are the relative equilibrium points on the reduced manifold, which implies that the corresponding physical elliptic trajectories stay unchanged in shape and reciprocal position. The reduced manifold is homeomorphic to a 2-dimensional sphere, so the number of elliptic critical points minus the number of the hyperbolic critical points is always equal to 2 for topological reasons. The simplification arises from the fact that all the qualitative information may be obtained by inspecting the intersection lines of the reduced manifold and of the level surfaces of $\overline{H_p}$ with the plane $\xi_1 \xi_2$.

It is therefore necessary to compute $\overline{H_p}$ explicitly. Unfortunately, the integrations involved in the averaging procedure cannot be performed exactly, so we must proceed with approximate methods. We now describe two different methods, whose final results perfectly agree.

(i) Analytical integration of the approximate Hamiltonian

Our task is to calculate the double integral

$$\overline{H_p} = \frac{1}{(2\pi)^2} \int_0^{2\pi} \int_o^{2\pi} H_p \, dl' \, dl'' \tag{9.1.11}$$

along the unperturbed solution. Here H_p is the perturbative Hamiltonian (9.1.5), with $\vec{Q}'$ and $\vec{Q}''$ evolving along a Keplerian ellipse, while l', l'' are the relative mean anomalies. The basic point consists in reducing the integrand function to a trigonometric polynomial, which makes the integration much easier. To this end, put

$$\left\| \vec{Q}' \right\| = a' \rho' \qquad \qquad \text{where} \quad \rho' = 1 - E' \cos s',$$

$$\left\| \vec{Q}'' \right\| = a'' \left(1 - E''^2\right) \varrho'' \quad \text{where} \quad \varrho'' = \frac{1}{1 + E'' \cos f''}.$$

This suggests taking s' and f'' as integration variables. We must therefore eliminate f' from $\cos \vartheta = \cos(f' - f'' + \varphi)$. Comparing the norm of the radius vector in the Keplerian motion expressed as a function of the eccentric anomaly and of the true anomaly, respectively, one gets

$$\cos f' = \frac{\cos s' - E'}{\rho'}, \quad \sin f' = \frac{\sqrt{1 - E'^2} \sin s'}{\rho'},$$

from which

$$\cos \vartheta = \frac{\cos s' - E'}{\rho'} (\cos \varphi \cos f'' + \sin \varphi \sin f'')$$

$$- \frac{\sqrt{1 - E'^2} \sin s'}{\rho'} (\sin \varphi \cos f'' - \cos \varphi \sin f'').$$

Moreover,

$$dl' = \rho' ds', \quad dl'' = (\varrho'')^2 \left(1 - E''^2\right)^{3/2} df''.$$

The integrand is so expressed as a trigonometric polynomial in s' and f'', and the Hamiltonian (9.1.5) can be calculated in principle to any order; in practice, we stop at $k = 10$. We obtain $\overline{H_p}$ as a function of ξ_1 and

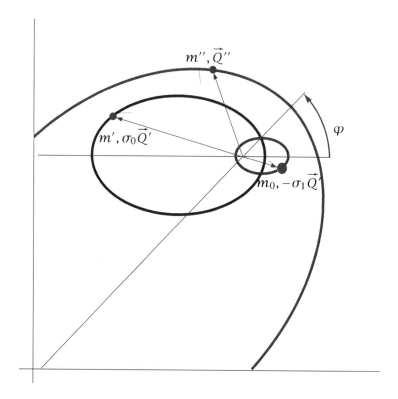

Figure 9.5: The three ellipses described by the averaged motion of m_0, m', m''.

$\cos \varphi$, with the three masses m_0, m', m'', the two semimajor axes a_1, a_2, and the total angular momentum G_{tot} playing the role of parameters. Once parameters are chosen, the program LAPLACE displays the level curves of $\overline{H_p}(\xi_1, \varphi) = $ constant, and from these curves one can infer the motion of the two ellipses described by $\vec{Q}'$ and $\vec{Q}''$: ξ_1, with G_{tot} and a_1, a_2, fixes size and eccentricity of the ellipses, while φ fixes the relative position. Then, the averaged motion of m_0, m', m'' is easily deduced; see Figure 9.5. Take into account that one must add an overall rotational motion to the relative motion, as dictated by the conservation of the total angular momentum G_{tot}.

From a direct inspection of Figures 9.6, 9.7, and 9.8 it is evident that three different cases are possible and the system admits

(i) two elliptic points, or

(ii) one hyperbolic and three elliptic points, or

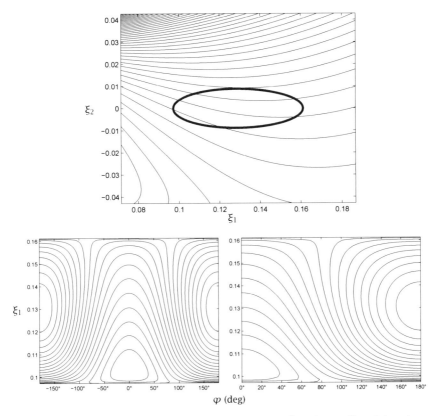

Figure 9.6: The parameter values are: $m_0 = 1$, $m' = 0.1$, $m'' = 0.1$, $a' = 1$, $a'' = 5$, and $G_{tot} = 0.9 \cdot (L' + L'')$. Two elliptic points are present. These and the similar pictures of the subsequent two figures can be obtained with the LAPLACE program. Top: traces of the reduced manifold and of the level surfaces of $\overline{H_p}$. Bottom-left: action-angle variables calculated with the analytical method. Bottom-right: idem but with the numerical method.

(iii) two hyperbolic and four elliptic points.

See the left-bottom picture of Figure 9.6 for the first case, of Figure 9.7 for the second case, and of Figure 9.8 for the third case.

A different visualization, however conveying the same information, is the following. Making the substitution

$$\cos \varphi = -\frac{\xi_2}{2L'E'L''E''},$$

we can express $\overline{H_p}$ as a function of ξ_1, ξ_2. Plotting the level curves of the averaged Hamiltonian $\overline{H_p}(\xi_1, \xi_2) = $ constant (also for $|\xi_2| > 2L'E'L''E''$), one clearly sees that they are topologically equivalent to those of the hyperbolic quadratic form $\xi_1^2 - \xi_2^2 = $ constant for all the admissible values of

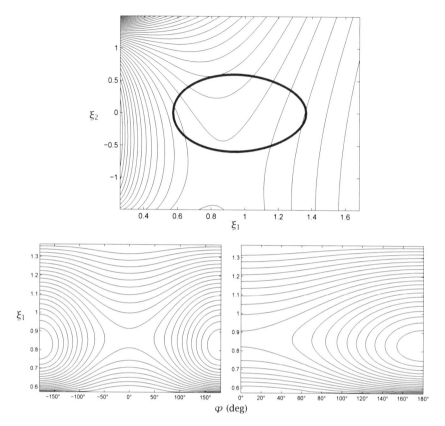

Figure 9.7: As in Figure 9.6, but the parameter values are: $m_0 = 1$, $m' = 0.65$, $m'' = 0.60$, $a' = 1$, $a'' = 5$, and $G_{tot} = 0.8 \cdot (L' + L'')$. Three elliptic points and one hyperbolic point are present.

the parameters. If the saddle point of the quadratic form falls inside the intersection line of M_μ with the plane $\xi_1 \xi_2$, we are in the third case of the above list. If the saddle point is outside, we are in the second case if the curvature radius of the level curves is smaller than that of the intersection of M_μ, and in the first case otherwise. See the top picture of the Figure 9.6 for the first case, of the Figure 9.7 for the second case, and of the Figure 9.8 for the third case.

As one may ascertain with the program LAPLACE, considering also the terms $k \geq 5$ in (9.1.5) does not significantly change the level curves of $\overline{H_p}(\xi_1, \xi_2)$. Let us restrict ourselves to the first three terms and calculate analytically the approximate values of the coordinates ξ_1^{cr}, ξ_2^{cr} of the saddle point. Dropping an inessential multiplicative constant, we get

$$\overline{H_p} = \alpha(\xi_1)\xi_2^2 + \beta(\xi_1)\xi_2 + \gamma(\xi_1),$$

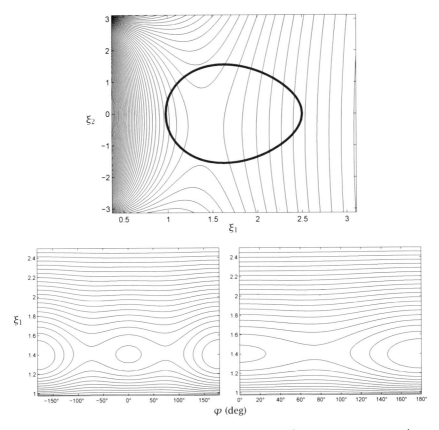

Figure 9.8: As in Figure 9.6, but the parameter values are: $m_0 = 1$, $m' = 0.90$, $m'' = 0.95$, $a' = 1$, $a'' = 5$, and $G_{\text{tot}} = 0.75 \cdot (L' + L'')$. Four elliptic and two hyperbolic points are present.

where we have defined

$$\alpha(\xi_1) = \frac{315}{1024} \frac{\sigma_4}{L'^2 L''^2} \frac{2 + E_1^2}{F_2^7} \left(\frac{a'}{a''}\right)^2,$$

$$\beta(\xi_1) = \frac{15}{64} \frac{\sigma_0 - \sigma_1}{2L'L''} \frac{4 + 3E_1^2}{F_2^5} \left(\frac{a'}{a''}\right),$$

$$\gamma(\xi_1) = \frac{2 + 3E_1^2}{8F_2^3},$$

$$+ \frac{9\sigma_4}{1024} \frac{-25E_1^4 E_2^2 + 30E_1^4 - 20E_1^2 E_2^2 + 80E_1^2 + 24E_2^2 + 16}{F_2^7} \left(\frac{a'}{a''}\right)^2,$$

with

$$E_1 = \frac{R'}{L'}, \quad E_2 = \frac{R''}{L''}, \quad F_2 = \frac{G''}{L''}.$$

One obtains immediately that

$$\xi_2^{cr} = -\frac{\beta(\xi_1)}{2\alpha(\xi_1)},$$

which shows that the sign of ξ_2^{cr} is that of $m_0 - m'$. For the calculation of ξ_1^{cr}, notice that, if the eccentricity of the outer ellipse is small and $a'/a'' << 1$, the dominant term in $\overline{H_p}$ is

$$\overline{H_p} \approx \frac{2 + 3E_1^2}{8F_2^3} = \frac{L''^3}{8L'^2} \frac{5L'^2 - 3G'^2}{G''^3};$$

then, remembering that

$$G' = \frac{1}{2}(G_{tot} - \xi_1), \quad G'' = \frac{1}{2}(G_{tot} + \xi_1) \qquad (9.1.12)$$

and putting

$$\frac{d}{d\xi_1}\overline{H_p} = 0,$$

one finds that the admissible solution is

$$\xi_1^{cr} \approx \frac{4}{3}\left[\frac{9}{4}G_{tot} - \sqrt{\frac{45}{4}L'^2 + \frac{9}{4}G_{tot}^2}\right]$$

and that $\overline{H_p}(\xi_1^{cr})$ is a maximum. Since $\alpha(\xi_1) > 0$, ξ_2^{cr} is a minimum, then the critical point is a saddle point. These approximate values of ξ_1^{cr}, ξ_2^{cr} and the already given description of the reduced manifold M_μ allow one to grasp how the various parameters affect the existence and location of the critical points on the reduced manifold.

The user can check and compare the approximate values with the exact ones given by the LAPLACE program selecting "Normal Form", then "Global planar 3body" and "Vertical section".

(ii) Numerical integration of the exact Hamiltonian

The second method consists in a numerical calculation of the average of the perturbation Hamiltonian. We should thus perform the integrations in (9.1.11), where however the integrand is a series. But, by comparing (9.1.3) with (9.1.4), one immediately checks that

$$\mu'm''\frac{1}{(2\pi)^2}\int_0^{2\pi}\int_0^{2\pi}\left(\frac{1}{\sigma_1}\frac{1}{\|\vec{Q}'' + \sigma_1'\vec{Q}\|} + \frac{1}{\sigma_0}\frac{1}{\|\vec{Q}'' - \sigma_0\vec{Q}'\|}\right)dl'dl''$$
$$+\overline{H_p} = \text{constant}$$

holds, so that we may replace the calculation of $\overline{H_p}$ with that of the double integral in the above expression, at least for what regards the level lines. Obviously, we will consider only those situations for which $\sigma_1 Q'$ and $\sigma_0 Q'$ are smaller than Q'', to ensure the convergence of the numerical integration.

For this calculation, we use the parametrization (9.1.1), with the eccentric anomaly s replacing the mean anomaly l as evolutionary parameter. Due to the rotational invariance of the problem, we may consider the first ellipse in an arbitrary position, the interaction energy depending only on the relative position of the two ellipses. Therefore, for the first ellipse we write

$$\vec{Q'} = \begin{pmatrix} X' \\ Y' \end{pmatrix} = \begin{pmatrix} a'\cos s' - a'\sqrt{1 - \frac{G'^2}{L'^2}} \\ a'\sqrt{\frac{G'^2}{L'^2}}\sin s' \end{pmatrix}$$

and for the second ellipse

$$\vec{Q''} = \begin{pmatrix} X'' \\ Y'' \end{pmatrix} = \begin{pmatrix} \cos\varphi & -\sin\varphi \\ \sin\varphi & \cos\varphi \end{pmatrix} \begin{pmatrix} a''\cos s'' - a''\sqrt{1 - \frac{G''^2}{L''^2}} \\ a''\sqrt{\frac{G''^2}{L''^2}}\sin s'' \end{pmatrix}.$$

By (9.1.2), the integration variables in the averaging procedure may be changed from mean anomaly l to eccentric anomaly s. Modulo an inessential multiplicative and an additive constant, we have

$$\overline{H_p} = \frac{1}{(2\pi)^2} \int_0^{2\pi} \int_0^{2\pi} Q'Q'' \left(\frac{1}{\sigma_1\sqrt{\sigma_1^2 Q'^2 + Q''^2 + 2\sigma_1'\vec{Q'}\cdot\vec{Q''}}} \right.$$
$$\left. + \frac{1}{\sigma_0\sqrt{\sigma_0^2 Q'^2 + Q''^2 - 2\sigma_0\vec{Q'}\cdot\vec{Q''}}} \right) ds'\,ds''.$$

Substituting (9.1.12) and taking G_{tot}, the three masses, and the two semimajor axes as fixed parameters, the integrand becomes a function of s', s'', ξ_1, and φ. It can be integrated numerically, obtaining a grid of values labeled by ξ_1 and φ. Then, we can draw the lines $\overline{H_p}(\varphi, \xi_1) = $ constant, which describe geometrically the relative motion and the eccentricity change of the two ellipses.

These last two tasks (numerical integration and drawing of the family lines) can be achieved with LAPLACE. Some outputs may be viewed in the right-bottom picture of Figures 9.6, 9.7, and 9.8. In all the displayed cases, one notices the very good agreement with the pictures obtained via the analytical, but approximate, method.

To conclude, let us consider the possibility of extending the above work to the 3-dimensional case also. Unfortunately, this does not seem straightforward. In the 3-dimensional case the space of the orbits of the Kepler problem for a fixed energy is $S^2 \times S^2$; hence the manifold to be reduced under the action of SO(3) is $S^2 \times S^2 \times S^2 \times S^2$. Once this reduction is performed, which seems a nontrivial task, we are left with a 4-dimensional symplectic manifold, thus with a two-degrees-of-freedom system. We do not know another first integral, besides the reduced Hamiltonian; thus the system is, very likely, not integrable.

Anyway, the problem surely deserves to be further investigated.

9.2 The 3-Dimensional Planetary Problem

The planetary problem is a particular, but very important, case of the $(N+1)$-body problem: one body, the "Sun," has a mass much larger than the other N "planets" and the forces acting on the system are the gravitational ones.

First, let us consider the case $N = 2$, with m_0 the mass of the Sun and $m', m'' \ll m_0$ those of the two planets; the extension to a generic N is straightforward. We use heliocentric coordinates, thus a non-inertial reference system, with $\vec{q}\,'$ and $\vec{q}\,''$ the radius vectors connecting the Sun with the two planets, respectively. From Newton's law, it is easy to derive the relative equations of motion: see, for example, Brouwer & Clemence (1961, ch. 10, pag. 251, eqs. (4a–4b)). To this end, write the $N + 1$ equations in an inertial frame, then subtract the equation of the Sun from that of every planet, obtaining

$$\frac{d^2\vec{q}\,'}{dt^2} = \nabla_{q'}(U' + F'), \qquad \frac{d^2\vec{q}\,''}{dt^2} = \nabla_{q''}(U'' + F''),$$

$$U' = G\frac{m_0 + m'}{q'}, \qquad U'' = G\frac{m_0 + m''}{q''},$$

$$F' = Gm''\left(\frac{1}{\triangle} - \frac{\vec{q}\,' \cdot \vec{q}\,''}{q''^3}\right), \qquad F'' = Gm'\left(\frac{1}{\triangle} - \frac{\vec{q}\,' \cdot \vec{q}\,''}{q'^3}\right),$$

where $\triangle = \left\| \vec{q}\,' - \vec{q}\,'' \right\|$ is the distance between the planets. The terms F' and F'' are perturbative potentials: if they vanish, we are left with two disjoint Keplerian problems, thus with an integrable system. Moreover, if the two planets do not come too close, the ratio between perturbation and Keplerian forces is of the same order of the ratio between the mass of a planet and the mass of the Sun.

In F' and F'' the *indirect part* appears: $\vec{q}\,' \cdot \vec{q}\,''/q'^3$ and $\vec{q}\,' \cdot \vec{q}\,''/q''^3$, respectively, due to the non-inertiality of the reference system. The indirect part, however, is eliminated in the subsequent averaging (see, for example, Brouwer & Clemence (1961, pag. 508)) so that the equations of motion can be put in Hamiltonian form, with Hamiltonian

$$H = \frac{p'^2}{2m'} - G\frac{m'(m_0 + m')}{q'} + \frac{p''^2}{2m''} - G\frac{m''(m_0 + m'')}{q''} - G\frac{m'm''}{\triangle}. \tag{9.2.1}$$

The secular Hamiltonian is obtained by averaging $\triangle^{-1}$ along the unperturbed motion, i.e., along the Keplerian ellipses. Unfortunately, this is a nontrivial task, which cannot be carried out in a closed form. It requires two preliminary steps. With the first step we will put the expression of the two position vectors in a suitable form, i.e., as a function of an evolutional parameter, closely related to the time, and of five constant parameters characterizing the ellipse. The second step consists in a series expansion of $\triangle^{-1}$ with respect to eccentricity and inclination.

To accomplish the first step, one could use the Keplerian elements of the orbit, but they suffer from the drawback of being singular for orbits which are circular and/or lying on the reference (ecliptic) plane. In contrast, the Poincaré variables are regular for orbits with small eccentricities and inclinations, and are thus well suited for studying the planetary problem of the solar system.

The Keplerian elements of the orbit have a clear geometrical interpretation: semimajor axis and numerical eccentricity fix the size and shape of the ellipse, while inclination, longitude of the ascending node, and argument of pericenter are the three Euler angles fixing the spatial orientation of the ellipse. In contrast, the Poincaré variables are defined on page 190 in a purely algebraic manner and lack a geometrical interpretation. This makes finding the expansion of the two position vectors somewhat involved and awkward, which does not simplify the subsequent series development and averaging process.

We will show that exploiting the geometry of the group SO(3) allows us to write the expression of the Keplerian motion in a very suitable form. Then, we develop $\triangle^{2y}$, $y \in \mathbb{R}$ in such a way that it is immediate, by direct inspection, to detect the terms which vanish under the averaging process. This produces a drastic simplification and allows us to smartly group the surviving terms in a reasonable and very suitable manner, the final result being an even, real-valued polynomial in the Poincaré canonical, heliocentric variables.

9.2.1 SO(3) and Poincaré Variables

SO(3) is a compact simple Lie group: see item (i) on page 46. It is the group of the orthogonal 3×3 matrices with unitary determinant:

$$\mathbf{R} \in SO(3) : \mathbf{R}^t \mathbf{R} = 1, \quad \det \mathbf{R} = 1, \quad \mathbf{R}^t = \text{transposed matrix.}$$

The linear transformation in $\mathbb{R}^3$ induced by $\mathbf{R}$ leaves the Euclidean scalar product invariant, and describes a rotation.

To parametrize SO(3) we use the *canonical exponentiation* of its Lie algebra, which is the algebra of the skew symmetric matrices with product given by the commutator. We utilize this method to find the well-known

Euler rotation formula. Let $\mathbf{R}(\vec{N}, \vartheta)$ represent the rotation of an angle ϑ about a unit vector $\vec{N}$. For example,

$$\mathbf{R}(\vec{e}_1, \vartheta) = \begin{pmatrix} 1 & 0 & 0 \\ 0 & \cos\vartheta & \sin\vartheta \\ 0 & -\sin\vartheta & \cos\vartheta \end{pmatrix},$$

where $\vec{e}_k$, $k = 1, 2, 3$, are the orthonormal vectors of the Cartesian axes. Define

$$\mathbf{G}_k = \left. \frac{d\mathbf{R}(\vec{e}_k, \vartheta)}{d\vartheta} \right|_{\vartheta = 0}.$$

Hence

$$\mathbf{G}_1 = \begin{pmatrix} 0 & 0 & 0 \\ 0 & 0 & 1 \\ 0 & -1 & 0 \end{pmatrix}, \quad \mathbf{G}_2 = \begin{pmatrix} 0 & 0 & -1 \\ 0 & 0 & 0 \\ 1 & 0 & 0 \end{pmatrix}, \quad \mathbf{G}_3 = \begin{pmatrix} 0 & 1 & 0 \\ -1 & 0 & 0 \\ 0 & 0 & 0 \end{pmatrix}$$

satisfy the commutation rules

$$[\mathbf{G}_1, \mathbf{G}_2] = -\mathbf{G}_3, \quad [\mathbf{G}_2, \mathbf{G}_3] = -\mathbf{G}_1, \quad [\mathbf{G}_3, \mathbf{G}_1] = -\mathbf{G}_2,$$

namely the commutation rules of the Lie algebra of SO(3).

We have described how to pass, through differentiation, from the group to the algebra. The inverse procedure is achieved with what is called EXPo-*nentiation*. Indeed, it is immediate to check that

$$\mathbf{R}(\vec{e}_k, \vartheta) = 1 + \vartheta\mathbf{G}_k + \frac{1}{2!}(\vartheta\mathbf{G}_k)^2 + \dots \overset{\text{def}}{=} \text{EXP}(\vartheta\mathbf{G}_k).$$

More generally, define

$$\mathbf{G} = N_1\mathbf{G}_1 + N_2\mathbf{G}_2 + N_3\mathbf{G}_3 = \begin{pmatrix} 0 & N_3 & -N_2 \\ -N_3 & 0 & N_1 \\ N_2 & -N_1 & 0 \end{pmatrix},$$

where N_1, N_2, N_3 are the Cartesian components of a generic unit vector $\vec{N}$, then

$$\mathbf{R}(\vec{N}, \vartheta) = \text{EXP}(\vartheta\mathbf{G}) = 1 + \sum_{j=0}^{\infty} \frac{(\vartheta\mathbf{G})^{2j+1}}{(2j+1)!} + \sum_{j=0}^{\infty} \frac{(\vartheta\mathbf{G})^{2j+2}}{(2j+2)!}$$

represents a rotation of an angle ϑ about $\vec{N}$. To make the above expression explicit, notice that

$$\mathbf{G}^2 = \begin{pmatrix} N_1^2 - 1 & N_1N_2 & N_1N_3 \\ N_1N_2 & N_2^2 - 1 & N_2N_3 \\ N_1N_3 & N_2N_3 & N_3^2 - 1 \end{pmatrix}, \quad \mathbf{G}^3 = -\mathbf{G},$$

from which $\mathbf{G}^4 = -\mathbf{G}^2$, $\mathbf{G}^5 = -\mathbf{G}^3 = \mathbf{G}$, and so forth, so that we obtain the well-known Euler formula:

$$\mathbf{R}(N, \vartheta) = 1 + \sin\vartheta\, \mathbf{G} + (1 - \cos\vartheta)\, \mathbf{G}^2 \tag{9.2.2}$$

$$= \begin{pmatrix} \cos\vartheta & N_3\sin\vartheta & -N_2\sin\vartheta \\ +N_1^2(1-\cos\vartheta) & +N_1N_2(1-\cos\vartheta) & +N_1N_3(1-\cos\vartheta) \\[1em] -N_3\sin\vartheta & \cos\vartheta & N_1\sin\vartheta \\ +N_1N_2(1-\cos\vartheta) & +N_2^2(1-\cos\vartheta) & +N_2N_3(1-\cos\vartheta) \\[1em] N_2\sin\vartheta & -N_1\sin\vartheta & \cos\vartheta \\ +N_1N_3(1-\cos\vartheta) & +N_2N_3(1-\cos\vartheta) & +N_3^2(1-\cos\vartheta) \end{pmatrix}.$$

We recall from (6.1.30) on page 190 the definition of the Poincaré variables for the Kepler motion:

$$\Lambda = \sqrt{mka}, \quad \lambda = l + \varpi,$$

$$\chi_{\mathrm{ecc}} = \sqrt{2(\Lambda - G)}\cos\varpi, \quad \eta_{\mathrm{ecc}} = -\sqrt{2(\Lambda - G)}\sin\varpi,$$

$$\chi_{\mathrm{inc}} = \sqrt{2(G - G_3)}\cos\Omega, \quad \eta_{\mathrm{inc}} = -\sqrt{2(G - G_3)}\sin\Omega.$$

where $\varpi = \omega + \Omega$ is the *longitude of pericenter*. The variables χ and η are *local* canonical coordinates on the symplectic manifold $S^2 \times S^2$, which is the manifold of the orbits with fixed energy. Clearly $\chi_{\mathrm{ecc}} = \eta_{\mathrm{ecc}} = 0$ correspond to circular orbits, while $\chi_{\mathrm{inc}} = \eta_{\mathrm{inc}} = 0$ describe orbits in the ecliptic plane. We immediately obtain

$$G = \Lambda - \frac{1}{2}(\chi_{\mathrm{ecc}}^2 + \eta_{\mathrm{ecc}}^2), \quad G_3 = G - \frac{1}{2}(\chi_{\mathrm{inc}}^2 + \eta_{\mathrm{inc}}^2). \tag{9.2.3}$$

We would express the position vector of a point moving along a Keplerian ellipse as a function of the Poincaré variables, but this is not possible in a closed form: the inversion of the modified Kepler equation is required (see below). Fortunately, it is sufficient for what follows to use σ as parameter, instead of λ. To this end, we proceed with the following geometric construction.

Let us start (see Figure 9.9) from a circular orbit of radius a contained in the ecliptic plane $q_1 q_2$ and let σ be the anomaly of the moving point P. Accordingly,

$$\overrightarrow{FP} = (a\cos\sigma,\, a\sin\sigma,\, 0) \quad \text{and} \quad \left\|\overrightarrow{FP}\right\| = a.$$

Take an angle β, related to the numerical eccentricity E through $\sin\beta = E$, $0 \le \beta < \pi/2$, and rotate about the unit vector $\overrightarrow{u}$ having the same direction as $\overrightarrow{E}$, then project on the ecliptic plane $q_1 q_2$ and finally translate by the vector $-a\overrightarrow{E}$. Consequently, we are interested in $\overrightarrow{E}$ and in $\mathbf{R}(\overrightarrow{u}, \beta)$, the 3×3

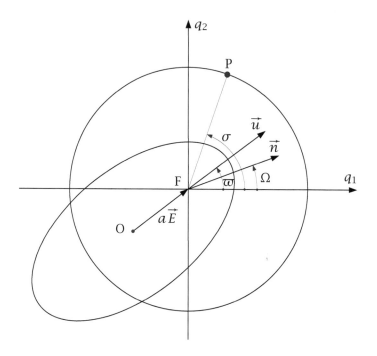

Figure 9.9: The Poincaré variables in the ecliptic plane.

real matrix corresponding to the above rotation. The unit vector $\vec{u}$ has Cartesian components $u_1 = \cos \varpi$, $u_2 = \sin \varpi$, $u_3 = 0$. Since $E^2 = 1 - G^2/\Lambda^2$, we have

$$\cos \beta = \sqrt{1 - E^2} = \frac{G}{\Lambda}.$$

The eccentricity vector $\vec{E}$ has Cartesian components $(E_1, E_2, 0)$, with

$$E_1 = E \cos \varpi = \sqrt{\frac{\Lambda^2 - G^2}{\Lambda^2}} \frac{\chi_{ecc}}{\sqrt{2(\Lambda - G)}} = \sqrt{\frac{\Lambda + G}{2\Lambda^2}} \chi_{ecc},$$

$$E_2 = E \sin \varpi = -\sqrt{\frac{\Lambda^2 - G^2}{\Lambda^2}} \frac{\eta_{ecc}}{\sqrt{2(\Lambda - G)}} = -\sqrt{\frac{\Lambda + G}{2\Lambda^2}} \eta_{ecc}.$$

The *mean longitude* λ is therefore related to the *eccentric longitude* $\sigma = s + \varpi$ by a modified Kepler equation,

$$\lambda = \sigma - \sqrt{\frac{\Lambda + G}{2\Lambda^2}} \left(\chi_{ecc} \sin \sigma + \eta_{ecc} \cos \sigma \right),$$

whose inversion is, in principle, always possible when $E < 1$, but not in a closed form.

Let us now calculate the rotation matrix $\mathbf{R}(\vec{u}, \beta)$. For example, bearing in mind (9.2.3) and the general expression (9.2.2), the entry in the first row and column is

$$\cos \beta + u_1^2(1 - \cos \beta) = \frac{G}{\Lambda} + \left(1 - \frac{G}{\Lambda}\right) \cos^2 \varpi = 1 - \frac{\eta_{ecc}^2}{2\Lambda}.$$

The final expression for the rotation matrix is

$$\mathbf{R}(\vec{u}, \beta) = 1 + \mathbf{A}_1 - \mathbf{S}_1,$$

$$\mathbf{A}_1 = \sqrt{\frac{\Lambda + G}{2\Lambda^2}} \begin{pmatrix} 0 & 0 & \eta_{ecc} \\ 0 & 0 & \chi_{ecc} \\ -\eta_{ecc} & -\chi_{ecc} & 0 \end{pmatrix},$$

$$\mathbf{S}_1 = \frac{1}{2\Lambda} \begin{pmatrix} \eta_{ecc}^2 & \chi_{ecc}\eta_{ecc} & 0 \\ \chi_{ecc}\eta_{ecc} & \chi_{ecc}^2 & 0 \\ 0 & 0 & \chi_{ecc}^2 + \eta_{ecc}^2 \end{pmatrix},$$

where $\mathbf{A}_1$ and $\mathbf{S}_1$ are infinitesimal of first and second order, respectively, for small eccentricities. In order to take into account the projection on the ecliptic plane, we put equal to zero the third row of the rotation matrix. Moreover, the choice of the third column is arbitrary, since the vector on which the matrix acts belongs to the ecliptic plane. The final expression for the position vector is

$$\frac{\vec{q}_{plane}}{a} = (1 - \mathbf{S}) \begin{pmatrix} \cos \sigma \\ \sin \sigma \\ 0 \end{pmatrix} - \vec{E}, \tag{9.2.4}$$

$$\mathbf{S} = \frac{1}{2\Lambda} \begin{pmatrix} \eta_{ecc}^2 & \chi_{ecc}\eta_{ecc} & 0 \\ \chi_{ecc}\eta_{ecc} & \chi_{ecc}^2 & 0 \\ 0 & 0 & 0 \end{pmatrix}, \quad \vec{E} = \sqrt{\frac{\Lambda + G}{2\Lambda^2}} \begin{pmatrix} \chi_{ecc} \\ -\eta_{ecc} \\ 0 \end{pmatrix}. \tag{9.2.5}$$

To also take into account the inclination, let us rotate through an angle i about the unit vector $\vec{n}$ of the node line. By definition,

$$\cos i = \frac{G_3}{G} \quad \text{with } 0 \le i < \pi.$$

To calculate the entries of the rotation matrix $\mathbf{R}(\vec{n}, i)$, we simply notice that it may be obtained from the expression of $\mathbf{R}(\vec{u}, \beta)$ through the substitutions

$$\Lambda \to G, \quad G \to G_3, \quad \chi_{ecc} \to \chi_{inc}, \quad \eta_{ecc} \to \eta_{inc},$$

as is evident from the definitions of the Poincaré variables and the expressions of $\cos\beta$ and $\cos i$. We obtain

$$\mathbf{R}(\vec{n}, i) = 1 + \mathbf{A}_2 - \mathbf{S}_2,$$

$$\mathbf{A}_2 = \sqrt{\frac{G + G_3}{2G^2}} \begin{pmatrix} 0 & 0 & \eta_{\text{inc}} \\ 0 & 0 & \chi_{\text{inc}} \\ -\eta_{\text{inc}} & -\chi_{\text{inc}} & 0 \end{pmatrix},$$

$$\mathbf{S}_2 = \frac{1}{2G} \begin{pmatrix} \eta_{\text{inc}}^2 & \chi_{\text{inc}}\eta_{\text{inc}} & 0 \\ \chi_{\text{inc}}\eta_{\text{inc}} & \chi_{\text{inc}}^2 & 0 \\ 0 & 0 & \chi_{\text{inc}}^2 + \eta_{\text{inc}}^2 \end{pmatrix}.$$

Obviously, in this and the previous formula, one must take into account the expressions (9.2.3).

Finally, we obtain

$$\vec{q} = \mathbf{R}^t(\vec{n}, i)\,\vec{q}_{\text{plane}},$$

where we have taken the transposed matrix, since the rotation matrices act on the basis of the vector space $\mathbb{R}^3$ while we consider the action on the components.

9.2.2 The Secular Planetary Problem

Before calculating $\Delta^{2\gamma}$ explicitly and proceeding to its averaging, we write the position vector $\vec{q}$ in a mixed real-complex form. Let us define the two adimensional complex variables

$$z_{\text{ecc}} = \frac{\chi_{\text{ecc}} + i\eta_{\text{ecc}}}{\sqrt{\Lambda}}, \quad z_{\text{inc}} = \frac{\chi_{\text{inc}} + i\eta_{\text{inc}}}{\sqrt{\Lambda}},$$

and describe the position vector in the ecliptic plane with the complex number q_{plane}. Then, bearing in mind (9.2.4–9.2.5), one can check that

$$\frac{q_{\text{plane}}}{a} = e^{i\sigma} - \frac{1}{4}|z_{\text{ecc}}|^2 e^{i\sigma} + \frac{1}{4}\bar{z}_{\text{ecc}}^2 e^{-i\sigma} - \sqrt{\frac{\Lambda + G}{2\Lambda}}\,z_{\text{ecc}}.$$

In the 3-dimensional case, we determine the position of the moving point with the pair $(q_c = q_1 + iq_2, q_3)$. A direct calculation shows that

$$\frac{q_c}{a} = \frac{q_{\text{plane}}}{a} - \frac{\Lambda}{4G}\left[|z_{\text{inc}}|^2 \frac{q_{\text{plane}}}{a} - \bar{z}_{\text{inc}}^2 \frac{\bar{q}_{\text{plane}}}{a}\right]$$

$$= e^{i\sigma} + X^2 e^{i\sigma} + \bar{Y}^2 e^{-i\sigma} + Z,$$

having defined

$$X^2 = -\frac{1}{4}|z_{ecc}|^2 - \frac{\Lambda}{4G}|z_{inc}|^2 + \frac{\Lambda}{16G}|z_{ecc}|^2|z_{inc}|^2 + \frac{\Lambda}{16G}z_{ecc}^2\overline{z}_{inc}^2,$$

$$Y^2 = \frac{1}{4}z_{ecc}^2 - \frac{\Lambda}{16G}z_{ecc}^2|z_{inc}|^2 + \frac{\Lambda}{4G}z_{inc}^2 - \frac{\Lambda}{16G}|z_{ecc}|^2z_{inc}^2,$$

$$Z = \sqrt{\frac{\Lambda+G}{2\Lambda}}\left[-\overline{z}_{ecc} + \frac{\Lambda}{4G}\overline{z}_{ecc}|z_{inc}|^2 - \frac{\Lambda}{4G}z_{ecc}\overline{z}_{inc}^2\right].$$

The notation X^2, Y^2, Z (with exponent 2 and 1) is to remember the infinitesimal order of the expressions, when, in the case of small eccentricities and inclinations, we must also consider z_{ecc} and z_{inc} small.

For the vertical component we find

$$\frac{q_3}{a} = Ve^{i\sigma} + \overline{V}e^{-i\sigma} + W^2 + \overline{W}^2,$$

$$V = \frac{i\,\Lambda}{2\,G}\sqrt{\frac{G+G_3}{2\Lambda}}\left[-z_{inc} + \frac{1}{4}z_{ecc}^2\overline{z}_{inc} + \frac{1}{4}|z_{ecc}|^2z_{inc}\right],$$

$$W^2 = -\frac{i\,\Lambda}{2\,G}\sqrt{\frac{\Lambda+G}{2\Lambda}}\sqrt{\frac{G+G_3}{2\Lambda}}z_{ecc}\overline{z}_{inc}.$$

Lastly, we shall also need the expression (derived from (6.1.2) on page 177)

$$\frac{q}{a} = 1 - \frac{1}{2}\sqrt{\frac{\Lambda+G}{2\Lambda}}\left[z_{ecc}e^{i\sigma} + \overline{z}_{ecc}e^{-i\sigma}\right].$$

Notice that, in the above formulae, the coefficients of the monomials in z_{ecc} and z_{inc} can be developed in series of even powers of z_{ecc} and z_{inc}, all beginning with the unit term:

$$\frac{\Lambda}{G} = \frac{1}{1-\frac{1}{2}|z_{ecc}|^2} = 1 + \sum_{n=1}^{\infty}\left(\frac{1}{2}|z_{ecc}|^2\right)^n,$$

$$\sqrt{\frac{\Lambda+G}{2\Lambda}} = \sqrt{1-\frac{1}{4}|z_{ecc}|^2} = 1 + \sum_{n=1}^{\infty}\frac{\prod_{k=0}^{n-1}\left(k-\frac{1}{2}\right)}{n!}\left(\frac{1}{4}|z_{ecc}|^2\right)^n,$$

$$\sqrt{\frac{G+G_3}{2\Lambda}} = \sqrt{1-\frac{1}{4}|z_{ecc}|^2 - \frac{1}{2}|z_{inc}|^2}$$

$$= 1 + \sum_{n=1}^{\infty}\frac{\prod_{k=0}^{n-1}\left(k-\frac{1}{2}\right)}{n!}\left(\frac{1}{2}|z_{ecc}|^2 + \frac{1}{4}|z_{inc}|^2\right)^n.$$

(9.2.6)

It is worthwhile to define

$$F(x) \overset{\text{def}}{=} x''e^{i\sigma''} - \alpha x'e^{i\sigma'}, \quad \text{with } \alpha = \frac{a'}{a''} < 1,$$

$$F(1)\overline{F}(1) = 1 + \alpha^2 - 2\alpha\cos(\sigma'' - \sigma') \overset{\text{def}}{=} K(\sigma'' - \sigma').$$

Since

$$\vec{q}'' - \vec{q}' = a'' \left(\frac{\vec{q}''}{a''} - \alpha \frac{\vec{q}'}{a'} \right) \Rightarrow \frac{\triangle}{a''} = \left\| \frac{\vec{q}''}{a''} - \alpha \frac{\vec{q}'}{a'} \right\|,$$

we are led to calculate

$$\left\| \frac{\vec{q}''}{a''} - \alpha \frac{\vec{q}'}{a'} \right\|^2 = \left(\frac{q_c''}{a''} - \alpha \frac{q_c'}{a'} \right) \left(\frac{\overline{q}_c''}{a''} - \alpha \frac{\overline{q}_c'}{a'} \right) + \left(\frac{q_3''}{a''} - \alpha \frac{q_3'}{a'} \right)^2$$

$$= \left[F(1) + F(X^2) + \overline{F}(Y^2) + Z'' - \alpha Z' \right]$$

$$\times \left[\overline{F}(1) + \overline{F}(X^2) + F(Y^2) + \overline{Z}'' - \alpha \overline{Z}' \right]$$

$$+ \left[F(V) + \overline{F}(V) + W''^2 + \overline{W}''^2 - \alpha \left(W'^2 + \overline{W}'^2 \right) \right]^2$$

$$= K + \mathcal{F}_{(2)} + \mathcal{F}_{(1)} + \mathcal{F}_{(0)} + \overline{\mathcal{F}}_{(1)} + \overline{\mathcal{F}}_{(2)}$$

$$= K + \varepsilon,$$

where we have defined

$$\varepsilon = \mathcal{F}_{(2)} + \mathcal{F}_{(1)} + \mathcal{F}_{(0)} + \overline{\mathcal{F}}_{(1)} + \overline{\mathcal{F}}_{(2)}, \text{ with}$$

$$\mathcal{F}_{(2)} = F(1)F(Y^2) + F(X^2)F(Y^2) + F(V)F(V),$$

$$\mathcal{F}_{(1)} = \left(F(1) + F(X^2) \right) \left(\overline{Z}'' - \alpha \overline{Z}' \right) + F(Y^2)(Z'' - \alpha Z')$$

$$+ 2F(V) \left(W''^2 + \overline{W}''^2 - \alpha \left(W'^2 + \overline{W}'^2 \right) \right), \qquad (9.2.7)$$

$$\mathcal{F}_{(0)} = F(1)\overline{F}(X^2) + \overline{F}(1)F(X^2) + F(X^2)\overline{F}(X^2)$$

$$+ F(Y^2)\overline{F}(Y^2) + (Z'' - \alpha Z') \left(\overline{Z}'' - \alpha \overline{Z}' \right)$$

$$+ 2F(V)\overline{F}(V) + \left(W''^2 + \overline{W}''^2 - \alpha \left(W'^2 + \overline{W}'^2 \right) \right)^2.$$

The reason for the particular grouping in these definitions will be clear in a moment.

We now want to calculate the mean value of $\triangle^{2y}$ (in the Keplerian case: $y = -1/2$) when the evolution is the unperturbed one; thus, differentiating the Kepler equation,

$$\frac{d\lambda}{d\sigma} = \frac{dl}{ds} = \frac{q}{a},$$

we obtain (here the overbar denotes "average")

$$\overline{\triangle^{2y}} \stackrel{\text{def}}{=} \frac{1}{(2\pi)^2} \int_0^{2\pi} \int_0^{2\pi} \triangle^{2y}(\sigma'', \sigma') d\lambda'' \wedge d\lambda'$$

$$= \frac{a''^{2y}}{(2\pi)^2} \int_0^{2\pi} \int_0^{2\pi} \left(\frac{\triangle(\sigma'', \sigma')}{a''} \right)^{2y} \frac{q''(\sigma'')}{a''} \frac{q'(\sigma')}{a'} d\sigma'' \wedge d\sigma', \quad (9.2.8)$$

into which we substitute the expressions

$$\left(\frac{\triangle}{a''}\right)^{2y} = (K + \varepsilon)^y = K^y + yK^{y-1}\varepsilon + \frac{1}{2}y(y-1)K^{y-2}\varepsilon^2 + \ldots,$$

$$\frac{q''(\sigma'')}{a''}\frac{q'(\sigma')}{a'} = \left[1 - \frac{1}{2}\sqrt{\frac{\Lambda'' + G''}{2\Lambda''}}\left(z''_{ecc}e^{i\sigma''} + \overline{z}''_{ecc}e^{-i\sigma''}\right)\right] \qquad (9.2.9)$$

$$\times \left[1 - \frac{1}{2}\sqrt{\frac{\Lambda' + G'}{2\Lambda'}}\left(z'_{ecc}e^{i\sigma'} + \overline{z}'_{ecc}e^{-i\sigma'}\right)\right].$$

Again, as for (9.2.7), it is convenient to write this last expression as (here the overbar denotes "complex conjugation")

$$\frac{q''}{a''}\frac{q'}{a'} = 1 + \mathcal{Q}_{(2)} + \mathcal{Q}_{(1)} + \mathcal{Q}_{(0)} + \overline{\mathcal{Q}}_{(1)} + \overline{\mathcal{Q}}_{(2)},$$

$$\mathcal{Q}_{(2)} = \frac{1}{4}\sqrt{\frac{\Lambda'' + G''}{2\Lambda''}}\sqrt{\frac{\Lambda' + G'}{2\Lambda'}}z'_{ecc}z''_{ecc}e^{i(\sigma''+\sigma')},$$

$$\mathcal{Q}_{(1)} = -\frac{1}{2}\sqrt{\frac{\Lambda'' + G''}{2\Lambda''}}z''_{ecc}e^{i\sigma''} - \frac{1}{2}\sqrt{\frac{\Lambda' + G'}{2\Lambda'}}z'_{ecc}e^{i\sigma'},$$

$$\mathcal{Q}_{(0)} = \frac{1}{4}\sqrt{\frac{\Lambda'' + G''}{2\Lambda''}}\sqrt{\frac{\Lambda' + G'}{2\Lambda'}}\left(z''_{ecc}\overline{z}'_{ecc}e^{i(\sigma''-\sigma')} + \overline{z}''_{ecc}z'_{ecc}e^{-i(\sigma''-\sigma')}\right).$$

$$(9.2.10)$$

Our task is therefore to calculate definite integrals of the type

$$\int_0^{2\pi}\int_0^{2\pi} K^{-s}(\sigma'' - \sigma')e^{i(r''\sigma''+r'\sigma')}d\sigma'' \wedge d\sigma', \quad r', r'' \in \mathbb{Z}.$$

It is immediate to prove that these integrals are different from zero if and only if $r'' + r' = 0$: indeed, the coordinate transformation

$$\begin{pmatrix} \sigma'' - \sigma' \\ \sigma' \end{pmatrix} = \begin{pmatrix} 1 & -1 \\ 0 & 1 \end{pmatrix}\begin{pmatrix} \sigma'' \\ \sigma' \end{pmatrix}$$

is unimodular, thus $\sigma'' - \sigma'$ and σ' are still angles on the 2-dimensional torus, so that

$$\int_0^{2\pi}\int_0^{2\pi} K^{-s}(\sigma'' - \sigma')e^{i(r''\sigma''+r'\sigma')}d\sigma'' \wedge d\sigma'$$

$$= \int_0^{2\pi}\int_0^{2\pi} K^{-s}(\sigma'' - \sigma')e^{ir''(\sigma''-\sigma')}e^{i(r''+r')\sigma'}d(\sigma'' - \sigma') \wedge d\sigma',$$

from which the claim follows. A function of the type

$$f(z_{ecc}, z_{inc})e^{i(r''\sigma''+r'\sigma')}$$

is said to have *characteristic* $r'' + r'$.

The following proposition is therefore evident.

PROPOSITION 9.2 *The product of functions of characteristic k and h, respectively, has characteristic $k + h$, and any such function, multiplied by $K^{-s}(\sigma'' - \sigma')$, survives the averaging process if and only if it has vanishing characteristic. The functions $\mathcal{F}_{(k)}, \mathcal{Q}_{(k)}$ and $\overline{\mathcal{F}}_{(k)}, \overline{\mathcal{Q}}_{(k)}$ in (9.2.7) and (9.2.10) clearly have characteristic k and $-k$, respectively.*

It is usual to define the so-called *Laplace's coefficients*

$$\frac{1}{2} b_s^{(r)}(\alpha) \stackrel{\text{def}}{=} \frac{1}{2\pi} \int_0^{2\pi} K^{-s}(\phi) \cos(r\phi) d\phi, \qquad (9.2.11)$$

whose properties are well known when $y = -1/2$: see, for example, Brouwer & Clemence (1961, pp. 471-ff and pp. 495-ff). Laplace's coefficients can be easily computed with a numerical integration; see `LaplaceCoeff.m` in the LAPLACE folder.

At this point, it should be evident what strategy we will follow in calculating the average $\overline{\triangle^{2y}}$.

(i) Substitute (9.2.9) and (9.2.10) into (9.2.8);

(ii) expand the integrand function in (9.2.8), which results in the sum of expressions of the type

$$\frac{1}{h!} y(y-1) \ldots (y-h+1) K^{y-h} (\mathcal{F}_{(2)} + \mathcal{F}_{(1)} + \mathcal{F}_{(0)} + \overline{\mathcal{F}}_{(1)} + \overline{\mathcal{F}}_{(2)})^h$$
$$\times (1 + \mathcal{Q}_{(2)} + \mathcal{Q}_{(1)} + \mathcal{Q}_{(0)} + \overline{\mathcal{Q}}_{(1)} + \overline{\mathcal{Q}}_{(2)}),$$

from which it is immediate to pick out and consequently delete the terms with nonvanishing characteristic, thanks to Proposition 9.2: we obtain a drastic simplification, thus clarifying the reason for the particular grouping in the definitions (9.2.7) and (9.2.10);

(iii) substitute (9.2.6) and expand, retaining the terms up to some fixed infinitesimal order;

(iv) substitute the numerical value of Laplace's coefficients (9.2.11).

At the end of the process we obtain $\overline{\triangle^{2y}}$ as an even, real-valued polynomial in $z'_{\text{ecc}}, z'_{\text{inc}}, z''_{\text{ecc}}$, and z''_{inc}, and thus in the canonical Poincaré variables, whose coefficients are linear combinations of Laplace's coefficients (with, eventually, coefficients that are polynomial functions of the ratio of the semimajor axes).

9.2.3 Linear Approximation: Lagrange–Laplace Theory

The procedure is easily generalized when $N \geq 3$ planets are present. The Hamiltonian (9.2.1) is replaced by

$$H = \frac{1}{2} \sum_{1 \leq j \leq N} \left[\frac{p_j^2}{m_j} - G \frac{m_j(m_0 + m_j)}{q_j} \right] - G \sum_{1 \leq h < k \leq N} \frac{m_h m_k}{\triangle_{hk}}$$

$$\stackrel{\text{def}}{=} H_0 - H_p,$$

$$p_j = \left\| \vec{p_j} \right\|, \quad q_j = \left\| \vec{q_j} \right\|, \quad \triangle_{hk} = \left\| \vec{q_h} - \vec{q_k} \right\|,$$

where the indices $1 \leq j, h, k \leq N$ label the planets. Define the variables

$$z_{\text{ecc}}^h = \frac{\chi_{\text{ecc}}^h + i\eta_{\text{ecc}}^h}{\sqrt{\Lambda^h}}, \quad z_{\text{inc}}^h = \frac{\chi_{\text{inc}}^h + i\eta_{\text{inc}}^h}{\sqrt{\Lambda^h}}, \quad h = 1, \ldots, N,$$

and calculate the averaged $\overline{H_p}$, which results in an even, real-valued polynomial in $\chi_{\text{ecc}}^h, \eta_{\text{ecc}}^h, \chi_{\text{inc}}^h$, and η_{inc}^h.

In order to find the normal form of the nonintegrable Hamiltonian $\overline{H_p}$, one may apply the Birkhoff Theorem 3.10 on page 124, reducing the Hamiltonian to a polynomial in the action variables. However, if we stop at the first terms quadratic in the canonical variables χ and η, the truncated Hamiltonian gives rise to linear, hence integrable equations.

Before we find its explicit expression, we will apply some elementary symmetry considerations, known as "D'Alembert's rules," to prove that the averaged and truncated quadratic Hamiltonian takes the form

$$(\overline{H_p})_{\text{quad}} = \frac{1}{2}\underline{\chi}_{\text{ecc}}^t \mathbf{H}_{\text{ecc}}\underline{\chi}_{\text{ecc}} + \frac{1}{2}\underline{\eta}_{\text{ecc}}^t \mathbf{H}_{\text{ecc}}\underline{\eta}_{\text{ecc}} + \frac{1}{2}\underline{\chi}_{\text{inc}}^t \mathbf{H}_{\text{inc}}\underline{\chi}_{\text{inc}} + \frac{1}{2}\underline{\eta}_{\text{inc}}^t \mathbf{H}_{\text{inc}}\underline{\eta}_{\text{inc}},$$

$$\underline{\chi}_{\text{ecc}} = \begin{pmatrix} \chi_{\text{ecc}}^1 \\ \vdots \\ \chi_{\text{ecc}}^N \end{pmatrix}, \ldots; \quad \mathbf{H}_{\text{ecc}}, \mathbf{H}_{\text{inc}} \text{ quadratic symmetric } N \times N \text{ matrices.}$$

Consequently, the system evolves like two N-dimensional anisotropic harmonic oscillators, whose frequencies are given by the eigenvalues of $\mathbf{H}_{\text{ecc}}$ and $\mathbf{H}_{\text{inc}}$.

Let us assimilate the ecliptic plane with the complex plane of the $2N$ complex variables z. The perturbative Hamiltonian $\overline{H_p}$ is clearly invariant with respect to the following group of transformations.

(i) Rotations about the vertical axis, orthogonal to the ecliptic plane. All the $2N$ vectors z rotate in the plane of the same angle, so that $\overline{H_p}$ must be a function only of their norms and of the scalar and vector products.

(ii) Reflections with respect to the real axis. The angles Ω, ω, and hence ϖ, change sign, the vectors z are reflected with respect to the real axis, and the scalar products do not change sign, contrary to the vector products: therefore, $\overline{H}_p$ does not depend on the vector products.

(iii) Reflections of the orbits of the planets with respect to the ecliptic plane. Ascending and descending nodes swap roles, $\Omega \to \Omega + \pi$, $\omega \to \omega + \pi$, so that ϖ stay unchanged, the vectors z_{inc} change sign, contrary to the vectors z_{ecc}: the perturbative Hamiltonian splits into the sum $H_1(z_{\text{ecc}}) + H_2(z_{\text{inc}})$. Taking into account that in $\overline{H}_p$ only norms $(\chi^k)^2 + (\eta^k)^2$ and scalar products $\chi^h\chi^k + \eta^h\eta^k$ appear, we get the result.

The explicit expression of $\mathbf{H}_{\text{ecc}}$ and $\mathbf{H}_{\text{inc}}$ for $N = 2$ is deduced from the Equations (9.2.9) of the previous subsection. We can write (now the overbar means "complex conjugate")

$$\left(\frac{\triangle}{a''}\right)^{2\gamma} \frac{q'}{a'} \frac{q''}{a''} = K^\gamma \left(1 + Q_{(0)}\right) + \gamma K^{\gamma-1}\left(\mathcal{F}_{(0)} + \mathcal{F}_{(1)}\overline{Q}_{(1)} + \overline{\mathcal{F}}_{(1)}Q_{(1)}\right)$$
$$+ \frac{1}{2}\gamma(\gamma - 1)K^{\gamma-2} \cdot 2\mathcal{F}_{(1)}\overline{\mathcal{F}}_{(1)} + \ldots,$$

where we have omitted all the terms with nonvanishing characteristic or of order greater than 2. Moreover,

$$\mathcal{F}_{(0)} = F(1)\overline{F}(X^2) + \overline{F}(1)F(X^2) + (Z'' - \alpha Z')(\overline{Z}'' - \alpha\overline{Z}') + 2F(V)\overline{F}(V),$$
$$\mathcal{F}_{(1)} = F(1)(\overline{Z}'' - \alpha\overline{Z}'),$$
$$Q_{(1)} = -\frac{1}{2}\left(z''_{\text{ecc}}e^{i\sigma''} + z'_{\text{ecc}}e^{i\sigma'}\right),$$
$$Q_{(0)} = \frac{1}{4}\left(z''_{\text{ecc}}\overline{z}'_{\text{ecc}}e^{i(\sigma''-\sigma')} + \overline{z}''_{\text{ecc}}z'_{\text{ecc}}e^{-i(\sigma''-\sigma')}\right), \text{ with}$$
$$X^2 = -\frac{1}{4}|z_{\text{ecc}}|^2 - \frac{1}{4}|z_{\text{inc}}|^2, \quad Z = -\overline{z}_{\text{ecc}}, \quad V = -\frac{i}{2}z_{\text{inc}}.$$

Take $\gamma = -1/2$ and average over $\sigma'' - \sigma'$. From the elementary identity

$$b_s^{(j)} = (1 + \alpha^2)b_{s+1}^{(j)} - \alpha\left(b_{s+1}^{(j+1)} + b_{s+1}^{|j-1|}\right),$$

we find

$$\mathbf{H}_{\text{ecc}} = \frac{1}{4}G\frac{m'm''}{a''}\frac{a'}{a''}\begin{pmatrix} -b_{3/2}^{(1)}/\Lambda' & b_{3/2}^{(2)}/\sqrt{\Lambda'\Lambda''} \\ b_{3/2}^{(2)}/\sqrt{\Lambda'\Lambda''} & -b_{3/2}^{(1)}/\Lambda'' \end{pmatrix},$$

$$\mathbf{H}_{\text{inc}} = \frac{1}{4}G\frac{m'm''}{a''}\frac{a'}{a''}\begin{pmatrix} b_{3/2}^{(1)}/\Lambda' & -b_{3/2}^{(1)}/\sqrt{\Lambda'\Lambda''} \\ -b_{3/2}^{(1)}/\sqrt{\Lambda'\Lambda''} & b_{3/2}^{(1)}/\Lambda'' \end{pmatrix},$$

whose generalization to a generic N is straightforward: compare with equation (4) of Brouwer & Clemence (1961, page 509) or, for $N = 4$, with the file NormalForm\Jovians.m in the LAPLACE program, variables Hecc and Hinc. The two $N \times N$ matrices are formed by merging all the 2×2 matrices relative to all pairs of planets, getting

$$\alpha_{ik} \overset{\text{def}}{=} \frac{\min(a_i, a_k)}{\max(a_i, a_k)}, \quad a_{ik} \overset{\text{def}}{=} \max(a_i, a_k) \text{ with } k \neq i,$$

$$\mathbf{H}_{\text{ecc}}(i, i) = -\frac{1}{4} G \frac{m_i}{\Lambda_i} \sum_{k=1}^{N} m_k \frac{\alpha_{ik}}{a_{ik}} b_{3/2}^{(1)}(\alpha_{ik}),$$

$$\mathbf{H}_{\text{ecc}}(i, k) = \frac{1}{4} G \frac{m_i m_k}{\sqrt{\Lambda_i \Lambda_k}} \frac{\alpha_{ik}}{a_{ik}} b_{3/2}^{(2)}(\alpha_{ik}),$$

$$\mathbf{H}_{\text{inc}}(i, i) = -\mathbf{H}_{\text{ecc}}(i, i),$$

$$\mathbf{H}_{\text{inc}}(i, k) = -\frac{1}{4} G \frac{m_i m_k}{\sqrt{\Lambda_i \Lambda_k}} \frac{\alpha_{ik}}{a_{ik}} b_{3/2}^{(1)}(\alpha_{ik}).$$

In Figure 9.10 the eccentricity and in Figure 9.11 the inclination of the four Jovian planets in the linear approximation are plotted.

The $N \times N$ matrices $\mathbf{H}_{\text{ecc}}$ and $\mathbf{H}_{\text{inc}}$ satisfy two remarkable properties. The first one is well known and states that $\det \mathbf{H}_{\text{inc}} = 0$, thus one of the eigenfrequencies is null, corresponding to the fact that the planar motion, where all the inclinations are vanishing, is a possible solution. To prove it, define the matrix $\mathbf{D} = \text{diag}(\sqrt{\Lambda_1}, \sqrt{\Lambda_2}, \dots, \sqrt{\Lambda_N})$. Then it is immediate to check that the sum of the rows of $\mathbf{DH}_{\text{inc}}$ (equivalently, of the columns of $\mathbf{H}_{\text{inc}}\mathbf{D}$) is zero, which implies $\det \mathbf{DH}_{\text{inc}} = 0$; since $\det \mathbf{D} \neq 0$, the statement follows.

The latter property is less known and was noticed by M. Herman in some unpublished notes, as reported in Abdullah & Albouy (2001). Consequently, the property is called *Herman's resonance* (perhaps a bit improperly). The diagonal entries of $\mathbf{H}_{\text{ecc}}$ and $\mathbf{H}_{\text{inc}}$ differ only in sign, which entails the *intrinsic* fact that $\text{Tr}\,\mathbf{H}_{\text{ecc}} + \text{Tr}\,\mathbf{H}_{\text{inc}} = 0$ or, equivalently, that the sum of all the $2N$ eigenfrequencies is zero. Moreover, as shown in the same reference, fully expanding $\overline{H_p}$ gives an infinite number of such relations. Unlike the first property, no clear reason is evident for the validity of Herman's resonance, which appears ultimately somewhat mysterious.

9.3 The LAPLACE Program

The LAPLACE program computes numerically the dynamics of a generic solar system, with a central massive star and at most 10 planets. The computing kernel is that of HNBODY: see Rauch & Hamilton (2004). The program is organized in one window with four panels and some menus.

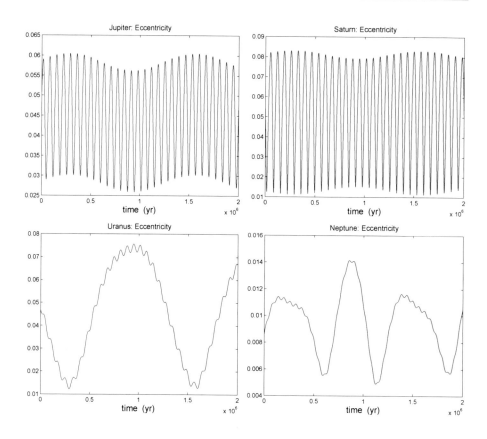

Figure 9.10: Lagrange–Laplace linear theory (blue color): eccentricity of the four Jovian planets. Compare with Figure 9.13.

9.3.1   First Panel: Initial Conditions

By default, mass values and initial conditions are adjusted on our solar system, with 8 planets plus Pluto and one asteroid, but the user can choose other values and even eliminate some planets, leaving blank the relative mass field. The units of measure are: the AU (radius of the Earth orbit), the year = 365.25 days, and the Sun's mass. All angles are expressed in degrees, with 1 deg = $\pi/180$ rad.

9.3.2   Second Panel: Integration

Clicking on "Integrate" starts the numerical integration. The user can choose among three different methods: symplectic (default and strongly recommended), Bulirsch–Stoer, and Runge–Kutta. The first method has a fixed

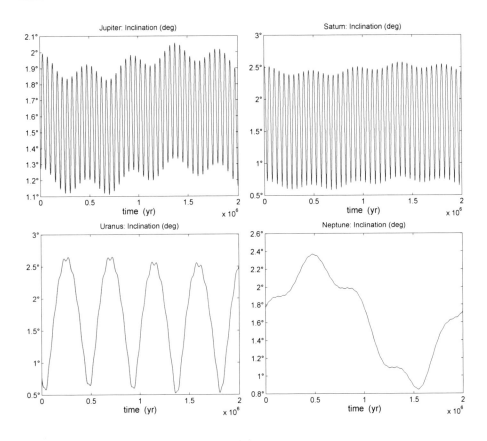

Figure 9.11: Lagrange–Laplace linear theory (blue color): inclination of the four Jovian planets. Compare with Figure 9.14.

integration step, chosen by the user along with the output step. The latter two instead have an adaptive step, which automatically ensures a relative accuracy $= 10^{-12}$, while the output step is still fixed and chosen by the user.

9.3.3   Third Panel: Plot and Frequency Analysis

When the numerical integration process is completed, various fields of the panel are enabled. In the two pop-up menus on the left the user can choose the parameters to plot and the relative planet. If the box "User function" is checked, the user will be prompted to choose a personal script: the file AA_two_planets.m is provided as a template (see also below). Until the button "Clear append" is clicked, the results of subsequent computations are kept in memory and automatically added to the new graphical output.

The right side of the panel is devoted to the frequency analysis of the same planet displayed in the second pop-up menu on the left, and it is very similar to the corresponding part in the first window of the KEPLER program: see items (i) and (ii) at the end of Subsection 7.1.7. The three output frequencies are relative to: semimajor axis-mean longitude, eccentricity-longitude of perihelion and inclination-longitude of ascending node.

9.3.4   Fourth Panel: Frequency Modulation Indicator

When N planets are present, the action space is $3N$-dimensional so that one must pick out a 2-dimensional section plane. The FMI will be computed taking all the actions fixed, with the exception of those spanning the grid on the section plane. The four pop-up menus of the panel allow one to choose the two planets and the corresponding two actions with their range, while all the other actions will keep the constant value displayed in the first panel "Initial conditions". The two planets can be the same; consequently the two actions must be different. The "step number" fields fix the dimension of the rectangular grid on which the FMI is computed. The right side of the panel is equal to the corresponding part in the third/fourth window of the KEPLER program; see Section 7.3.

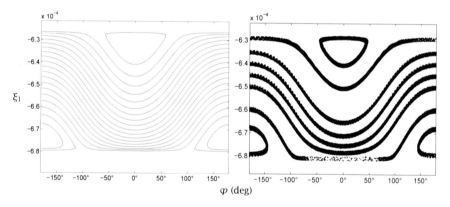

Figure 9.12: Action-angle variables for the planar planetary three-body. Left: normal form output. Right: numerical output. The parameter values are: $m_0 = 1$, $m' = 0.001$, $m'' = 0.0002$, $a' = 1$, $a'' = 3$, and $G_{tot} = 0.98 \cdot (L' + L'')$.

9.3.5   Menu

The menu of LAPLACE is very similar to that of KEPLER except for the "Normal form", which displays two windows. In "Global planar three-body" the user may perform the computations described in Section 9.1: the top, the left-bottom and the right-bottom pictures of Figures 9.6, 9.7, and 9.8 have been plotted with the command "Plot" of the panels "Vertical section",

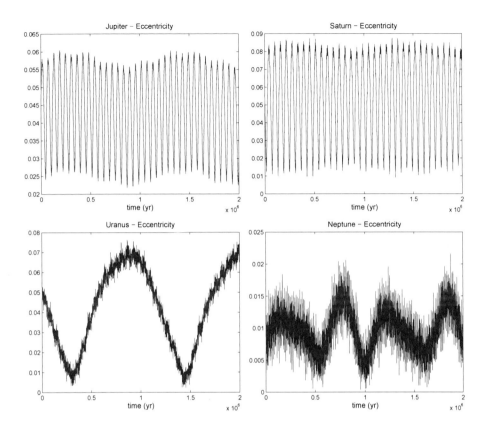

Figure 9.13: "True" eccentricity of the four Jovian planets, computed numerically (red color) with LAPLACE. Compare with Figure 9.10.

"Action-Angle dynamics (analytical)" and "Action-Angle dynamics (numerical)", respectively. "Jovian planets" performs the computations of the Lagrange–Laplace linear theory when Jupiter, Saturn, Uranus, and Neptune are present and displays the result; see Figures 9.10 and 9.11.

9.4 Some Examples

In Figure 9.12 we compare the output given by the menu "Normal Form/Global planar three-body" with that given by the numerical integration of LA-PLACE. In the right picture, we have exploited the ability of the program to plot a user function (here: AA_two_planets.m) keeping in memory the sequential outputs of several computations. Taking into account only two planets (whose names make no difference), first choose the two masses

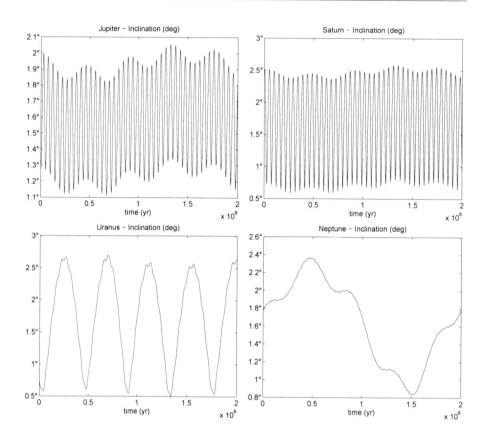

Figure 9.14: "True" inclination of the four Jovian planets, computed numerically (red color) with LAPLACE. Compare with Figure 9.11.

m', m'' and the two semimajor axes a', a''. Then, having fixed the total angular momentum $G_{\text{tot}} = G' + G''$, calculate the list of the two eccentricities

$$E' = \sqrt{1 - \frac{1}{4L'^2} (G_{\text{tot}} - \xi_1)^2}, \quad L' = \sqrt{Gm_0 m'^2 a'},$$

$$E'' = \sqrt{1 - \frac{1}{4L''^2} (G_{\text{tot}} + \xi_1)^2}, \quad L'' = \sqrt{Gm_0 m''^2 a''},$$

with the file GtotXi2ecc.m, letting $\xi_1 = G'' - G'$ vary. Lastly, insert the values of the two eccentricities in the window of LAPLACE and start the corresponding computations.

Figures 9.13 and 9.14 show the dynamical evolution of eccentricity and inclination of the four Jovian planets. The integration has been performed with LAPLACE on 2×10^6 years, in absence of the other terrestrial planets.

The output may be compared with that of the linear and approximate theory of Lagrange–Laplace, showing a good agreement.

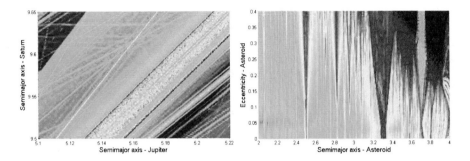

Figure 9.15: Some resonances in solar system; see the text.

Lastly, in Figure 9.15 we show two examples of how resonances are distributed in our solar system. In the computation of both pictures only the four Jovian planets have been taken into account.

In the left picture we let the semimajor axes of Jupiter and Saturn vary in the neighborhood of the true physical values, whereas all the other orbital parameters are kept fixed: the celebrated 2:5 resonance between the revolution frequencies of the two planets is clearly visible; however, many other thin resonances appear whose "deciphering" is not easy.

In the right picture the resonance distribution in the asteroid belt is explored. We fix the inclination of the test body to $2°$ and vary semimajor axis and eccentricity, highlighting all the main, well-known resonances with Jupiter. The most prominent are the 7:2, 3:1, 2:1, 5:3 resonances, located at the 2.3, 2.5, 3.3, 3.7 values of the semimajor axis. Notice that their location is determined essentially by the semimajor axis value, while a high value of the eccentricity cause an enlargement and even an overlapping of the resonance strips: this fact suggests that circular or almost circular orbits are more stable.

The reader may carry out some interesting numerical experiments, comparing the time evolution of the orbit parameters of asteroids starting inside and outside a resonance. For example, with the initial resonant value 2.5 of the semimajor axis, the eccentricity of the asteroid undergoes remarkable oscillations, reaching high values, about $0.25 \div 0.40$ after $5 \times 10^4 \div 10^5$ years and 0.50 after a few million years, while starting with a semimajor axis equal to the nonresonant value 2.45, the eccentricity excursion is bounded by the 0.06 value. This mechanism is very probably responsible for the presence of the so-called *Kirkwood gaps*, since the elongated orbits of the resonant asteroids can lead to a close encounter with Mars. The conjecture is confirmed by considering the resonance 5:2, to which corresponds a semimajor axis equal to 2.82 and a remarkable gap in the asteroidal distributions: the eccentricity reaches very high values, about 0.75. In contrast, consider a semimajor axis equal to 2.685 with its 11 : 4 resonance: the eccentricity is somewhat stable, and in fact it is known that correspondingly there are not any gaps.

CHAPTER 10

Final Remarks and Perspectives

*There is nothing more practical
than a good theory.*
— K. LEWIN

The focus of this book is on theoretical and numerical investigations of quasi-integrable Hamiltonian systems. The goal is to understand the qualitative and quantitative features of the relative dynamics, even for systems with three or more degrees of freedom. By combining analytical, numerical, and geometrical methods, in effect one can also grasp the geography of the resonances, and hence the distribution of order and chaos.

In spite of our efforts to be exhaustive, several topics have surely been omitted, some problems left unresolved, and interesting subjects left undeveloped. We quote a few of them.

Numerical Detection of the Frequency Modulation

As reported in Subsection 5.3.4, the modulation of the fundamental frequencies is an indicator for the presence of a resonance. While this is a well-established fact, its *efficient* numerical detection might be considered as an open problem. The method we have proposed exploits an intrinsic imprecision of the algorithm that computes the frequencies on tori but, as such, produces some spurious phenomena, in particular, the ghost undulations. For that, it would be preferable to develop a program that is able to numerically and directly recognize the typical spectrum of the frequency modulation, as in (5.3.6) on page 162.

Arnold Diffusion

In Arnold (1964) a mechanism is described which underlies an extremely slow diffusion along the resonances, and in Figure 8.15 a numerical example of a such slow motion is exhibited. It seems likely that the mechanism generating this latter is just that proposed by Arnold, but an explicit check or an alternative explanation is lacking. Moreover, it is not clear why there is a preferred direction in the diffusion.

Leaving an Angle in the Normalized Hamiltonian

In constructing the normal form for the perturbed Kepler problem, very satisfying results have been obtained by leaving a permanent angle in the normalized Hamiltonian, thus without completing the normalization procedure which would require the total elimination of the angles. We recall that this allows us to get exhaustive information on the partially normalized dynamics through geometrical tools. It could be that this strategy is also effective in other cases, typically with other totally degenerate Hamiltonians.

Herman's Resonance

While the property $\det \mathbf{H}_{\text{inc}} = 0$, regarding the linearized planetary theory of Lagrange–Laplace, is ultimately due to the conservation of the total angular momentum, and thus to the rotational symmetry of the Hamiltonian, Herman's resonance, stating that the sum of all the eigenfrequencies is zero, appears somewhat mysterious. Notice that this latter property is true *only* in the Keplerian case $y = -1/2$, unlike the first one, which is valid for any y. The situation is strongly reminiscent of the Kepler problem, which admits a further conservation law besides that of the angular momentum. If the Herman's resonance would entail a new first integral in involution with the total angular momentum, the averaged two-planet problem would be integrable and the geometrical method of Section 9.1 extendible to 3-dimensional case.

Transition State Theory

Transition state theory has its origins in early twentieth-century studies of the dynamics of chemical reactions, where it plays a key role. The basic idea is that the phase space can be partitioned into two volumes, the first corresponding to reactants and the second to products. Chemists call the boundary between these two regions the *transition state*. The rate of a chemical reaction is then discussed in terms of the flow across the transition state. These ideas can also be applied to problems in celestial mechanics: the orbits used to design space missions or to study asteroid escape also determine the ionization rates of atoms and chemical-reaction rates of molecules.

The paradigm is an n-degrees-of-freedom Hamiltonian system with an equilibrium point, the linearization about which has eigenvalues $\pm\lambda$, $\pm i\omega_k$, $k = 2,\ldots,n$, where $\lambda, \omega_k \in \mathbb{R}$. Thus we are considering equilibrium points of type: saddle, center,..., center. One can then show that, in the neighborhood of the saddle point, the normal form of the Hamiltonian is

$$H = \lambda q_1 p_1 + \frac{1}{2} \sum_{k=2}^{n} \omega_k (p_k^2 + q_k^2) + F_1(q_1 p_1, q_2, \ldots, q_n, p_2, \ldots, p_n)$$
$$+ F_2(q_2, \ldots, q_n, p_2, \ldots, p_n),$$

where the functions F_1 and F_2 are at least of third order and take into account all the nonlinear terms; moreover, $F_1 = 0$ when $q_1 p_1 = 0$. The first degree of freedom gives the two "reaction coordinates" while the other $2(n-1)$ are the "bath coordinates." The simple expression acquired by the Hamiltonian in the normal form coordinates enables one to construct trajectories showing any possible behavior near the transition state. These trajectories can then be visualized in the original coordinates; see Uzer, Jaffé, Palacián, Yanguas & Wiggins (2002).

Two examples are relevant. The first regards the SQZc system with Hamiltonian (8.1.1) on page 230, where we take $\mathcal{E}_3 = 0$ and $\mathcal{E}_1 > 0$. Indeed, the system admits the equilibrium point

$$q_1 = -\frac{1}{\sqrt{\mathcal{E}_1}}, \quad q_2 = 0, \quad q_3 = 0,$$
$$p_1 = 0, \quad p_2 = \frac{\mathcal{B}}{\sqrt{\mathcal{E}_1}}, \quad p_3 = 0,$$

which is just of the type center, saddle, center; one can so study the hydrogen ionization in crossed electric and magnetic fields.

The latter example concerns the circular restricted three-body problem: see the Hamiltonian (5.4.2) on page 171. The system possesses five equilibrium positions. Two of them, named L_4 and L_5, are linearly stable and can be investigated with the program LAGRANGE; the other three, named L_1, L_2 and L_3, are unstable of type center, saddle, center and are collinear with the two primary bodies. Consider the case Sun-Jupiter-asteroid: L_1 and L_2 are very close to Jupiter and in order to cross Jupiter's orbit, the asteroid can pass only through a narrow bottleneck centered about the planet, thus being forced to "ride" the chaos. We reach the same conclusion for the Earth-Moon-spacecraft system; see Marsden & Shane (2005).

Having very briefly illustrated the key ideas of transition state theory, it is tempting to try to investigate the quoted systems with the aid of the methods described in the course of the book. However, this does not seem straighforward, since the invoked methods and computer programs are not well suited for studying such *unstable* equilibrium points, which, in some sense, lie on the borderline of the applicability field. Therefore, it may be that one must proceed to some adjustments of the code.

* * *

Having arrived at the end of the book, we conclude by expressing the hope to have convinced the reader of the strength and effectiveness of perturbation methods, and of how true is the saying that nothing is more practical than a good theory.

APPENDIX A

What is in the CD?

Let us look through the CD and see how to use the files and programs. We recall that the CD can be downloaded as iso image from the publisher's website by entering the book's ISBN (978-0-8176-8369-6) into http://extras. springer.com/.

1) PhSpGeo.zip

This zipped file is the most important of the CD, since it contains the five MATLAB programs POINCARE, HAMILTON, LAGRANGE, KEPLER and LAPLACE.

Copy the zipped file on your hard disk, unzip, and put the whole folder PhSpGeo where you wish but *preserving the tree structure of the files.* Start MATLAB, navigate to the folder, e.g., KEPLER/Master and type <Kepler> in the MATLAB editor: the relative graphical user interface will open; for the other four programs act similarly. In the following pages the snapshots of the graphical user interface of the five programs are reported.

The code of the programs is written under Windows XP and Windows 7 but has also been tested in some distributions of Linux, where nevertheless the fonts of the graphical user interfaces sometimes appear out of proportion. Try to fix the problem adjusting the numerical value of font_def, in the second row of the ColorFont.m file or by opening, e.g., Kepler.fig in Guide. In Linux be sure that the files AddedODEsolver/bin/main_fmft and Laplace/xxx/hnbody (where xxx is Master or SlaveX) are checked as executable.

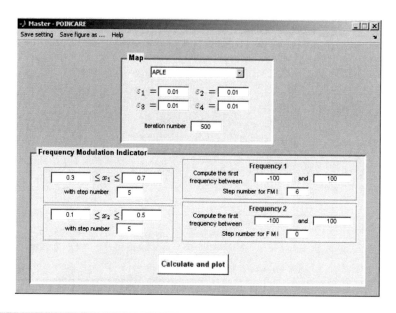

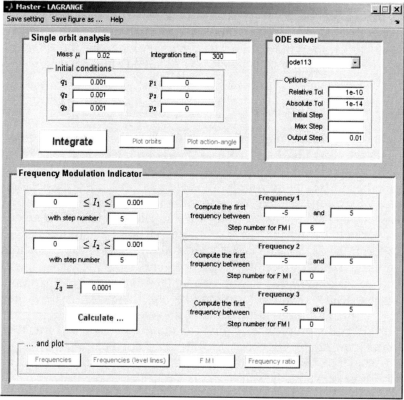

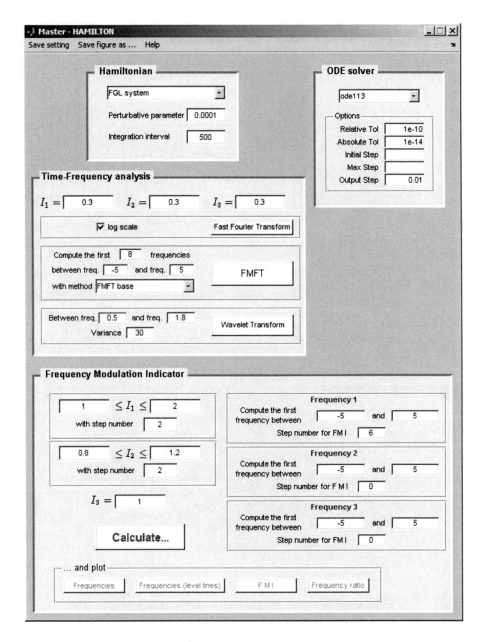

For some esoteric reason, a few labels of the graphical user interface appear incorrectly placed with certain versions of MATLAB. To fix the problem, close and open again the program without closing MATLAB.

The code of the three ODE solvers "IRK_Gauss", "Dop853", and "Odex" is not MATLAB native but requires a compilation. We provide the compiled

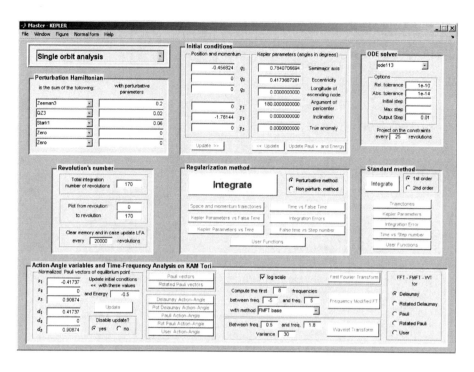

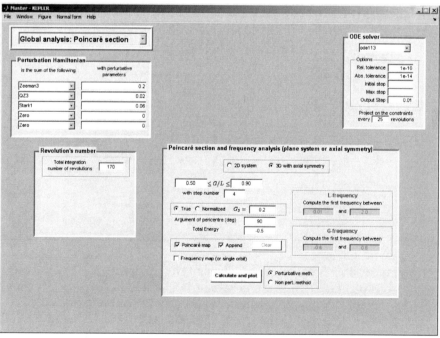

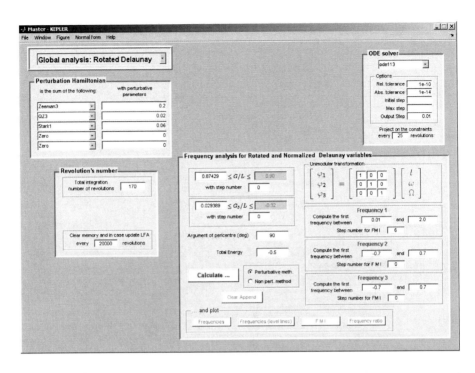

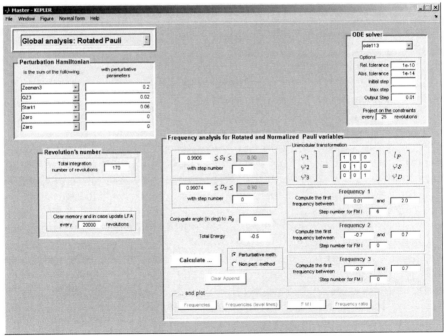

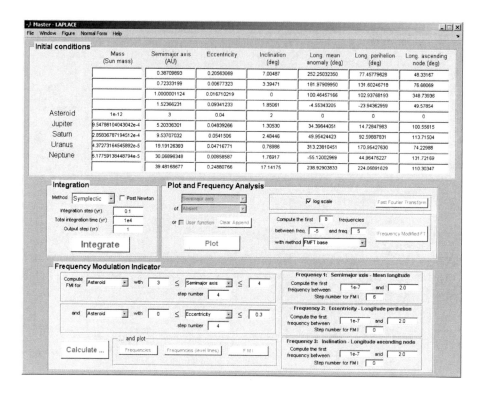

files which should run under Windows32 and some distributions of Linux (both 32 and 64 bit), but the user can also recompile the source code in AddedODEsolver/bin/Source.

Bear in mind that, in the usually very demanding computations of the FMI, the user can exploit a multicore machine. Indeed KEPLER, along with the other four supplied programs, is able to parallelize the computations in the following way. If you possess an n-core machine, in the folder Kepler and beside the subfolder Master create $n - 1$ subfolders Slave1, Slave2, et cetera, then copy the whole program KEPLER identically in every folder: the n folders must differ only in the name. Start MATLAB then KEPLER from the Master folder, set the parameters, and click on "File/Save setting now". *Without closing*, start a new instance of MATLAB, then KEPLER from a folder SlaveX: you will notice that all the buttons of the computations are disabled while the new button "Start Slave" appears. Click on this button and KEPLER will wait for the start of the master. Redo for every slave, and lastly go back to the master and click on "Calculate . . .". The whole work will be automatically shared among the n cores. The final result is displayed by the master. Warning: *do not close any waitbar* during the computation.

Alternatively, in Windows you can create n shortcuts on the desktop, one for the master and the others for the slaves. In the Portraits folder the corresponding icons are available, in color and in a gray scale for Master and Slaves, respectively. In the "Target" field of "Property" you will type something like:

<C:\MATLAB\bin\matlab.exe /r Kepler>

and in the "Start in" field you will type the path of the work folder, for example,

<C:\PhSpGeo\KEPLER\Master> or <C:\PhSpGeo\KEPLER\Slave1>.

2) Visualize_3D.zip

This zipped file contains the MATLAB program sliceomatic by Eric Ludlam which allows you to visualize sets of 3-dimensional data. We provide three examples, regarding the resonance distribution in the action space of the SQZcrossed system of Figure 8.11 on page 249.

Copy the zipped file on your hard disk, unzip, and put the whole folder Visualize_3D where you wish. Then start MATLAB, navigate to this folder and type <Visualize_3D> in the MATLAB editor: you will be prompted to load data for 3D visualization. Enter the SQZcrossed folder and one of the three subfolders LGG3_Total, LGG3_Detail, LSD_Total, then double click on Volume.mat: the graphical user interface of sliceomatic will open (see the figure) and you can begin the exploration by clicking on the slice controllers, then moving the relative arrows.

3) Matlab figures

In the Matlab figures folder the reader may find some pictures in the MATLAB format *.fig. Many of them are the original ones reported in the book, as attested by the name itself; others are unpublished and concern some details. In general they regard FMI computations which require *very* long times of the order of many days or weeks, even with a multicore machine, and may serve as a base to the reader for further explorations. In particular, with the "Mouse track" tool in the "Figure" menu one can record the values of the coordinates of some points in the picture and use them for the computation of the relative orbit, to which a frequency analysis can be applied.

4) Euler.mw

The program EULER, written in the MAPLE language, calculates *analytically*, thus *exactly*, the motion of a point under the gravitational force of two fixed points. For a detailed study of the Euler problem with some examples of the graphic output, see Cordani (2003). The exact integration is useful for a comparison with the numerical integration of KEPLER in order to verify its accuracy.

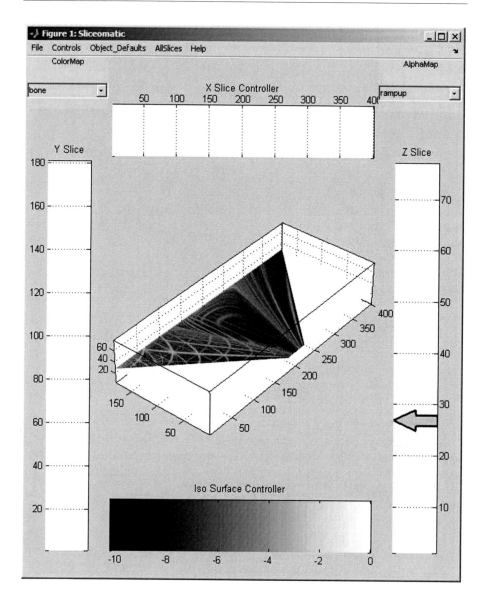

5) Portraits

This folder contains the portraits of the great mathematicians mentioned in the titles of the programs and the respective icons, both in color and in gray scale.

If one has no access to MATLAB

In case the reader does not have access to a MATLAB installation, we provide the compiled version of the five programs POINCARE, HAMILTON, LAGRANGE, KEPLER, and LAPLACE. Four of them are fully working, the exception being KEPLER which, requiring the symbolic toolbox, cannot be integrally compiled. The user is not able to choose the perturbation, as with the interpreted version, but must be content with the two supplied examples, which contain two fixed "hard wired" Hamiltonians. But if one has occasional access to MATLAB, it is possible to recompile KEPLER with a perturbation chosen by the user. Alternatively, one can get the executable from a colleague.

The compiled files are XXX.exe, where XXX is one of the five program names, and run under Windows 7 or Windows XP. In order to run the compiled programs, the user must first install (once and for all) a sort of "scrambled" but free of charge MATLAB, double clicking on the file MCRInstaller.exe, which can be downloaded from the publisher's website by entering the book's ISBN (978-0-8176-8369-6) into http://extras.springer.com/

The five programs can be started with a double click on the file or by typing their name in a DOS window. This latter possibility allows you to get some extra useful information about the progress of the computation and on how it is shared among the various cores.

It is mandatory to also preserve the file structure for the compiled versions; in particular, do *not* move the executables. For the rest, follow the same instructions already given for the interpreted version. We stress that also with the compiled version the user can parallelize the work, sharing the computation among several cores.

Lastly, we call attention to the files XXX.prj, which allow one again to compile the five programs, obtaining in particular the version for Linux.

Bibliography

ABDULLAH, K. & ALBOUY, A. On a strange resonance noticed by M. Herman. *Regular & Chaotic Dynamics*, **6**, pp. 421–432, (2001).

ABRAHAM, R. & MARSDEN, J. E. *Foundations of Mechanics.* Benjamin Cummings, Reading, MA., (1978).

ALFRIEND, K. T. The stability of Lagrangian triangular points for commensurability of order two. *Celestial Mechanics*, **1**, pp. 351–359, (1970).

ALFRIEND, K. T. Stability and motion about L_4 at three-to-one commensurability. *Celestial Mechanics*, **4**, pp. 60–72, (1971).

ARNOLD, V. I. Small denominators and problems of stability of motion in classical and celestial mechanics. *Russ. Math. Surv.*, **18**, pp. 85–191, (1963).

ARNOLD, V. I. Instability of dynamical systems with several degrees of freedom. *Soviet Math. Dokl.*, **5**, pp. 581–585, (1964).

ARNOLD, V. I. *Mathematical Methods of Classical Mechanics.* Springer-Verlag, Berlin, 2nd edn., (1989).

AUSLANDER, L. & MACKENZIE, R. E. *Introduction to Differentiable Manifolds.* Dover Publications, New York, (1977).

BENETTIN, G., FASSÒ, F. & GUZZO, M. Nekhoroshev-stability of L4 and L5 in the spatial restricted three-body problem. *Regular & Chaotic Dynamics*, **3**, pp. 56–70, (1998).

BENETTIN, G., GALGANI, L., GIORGILLI, A. & STRELCYN, J.-M. A proof of Kolmogorov's theorem on invariant tori using canonical transformations defined by Lie method. *Nuovo Cimento*, **79 B**, pp. 201–223, (1984).

BOWMAN, F. *Introduction to Elliptic Functions with Applications.* Dover Publications, New York, (1961).

BROUWER, D. Solution of the problem of artificial satellite theory without drag. *Astron. J.*, **64**, pp. 378-397, (1959).

BROUWER, D. & CLEMENCE, G. M. *Methods of Celestial Mechanics.* Academic Press, New York, London, (1961).

BUTCHER, J. C. *Numerical Methods for Ordinary Differential Equations.* Wiley & Sons, (2003).

CELLETTI, A. *Stability and Chaos in Celestial Mechanics.* Springer-Praxis, New York, (2010).

CHANDRE, C., WIGGINS, S. & UZER, T. Time-frequency analysis of chaotic systems. *Physica D*, **181**, pp. 171-196, (2003).

CHEVALLEY, C. *Theory of Lie Groups I.* Princeton University Press, Princeton, NJ, (1946).

CHIERCHIA, L. & FALCOLINI, C. A direct proof of a theorem by Kolmogorov in Hamiltonian systems. *Ann. Scuola Norm. Sup. Pisa Cl. Sci.*, **21**, pp. 541-593, (1994).

CHIERCHIA, L. & FALCOLINI, C. Compensations in small divisor problems. *Comm. Math. Phys.*, **175**, pp. 135-160, (1996).

CHIERCHIA, L. & GALLAVOTTI, G. Smooth prime integrals for quasi-integrable Hamiltonian systems. *Nuovo Cimento B*, **67**, pp. 277-295, (1982).

CHIRIKOV, B. V. A universal instability of many dimensional oscillator system. *Phys. Rep.*, **52**, pp. 263-379, (1979).

CHOQUET-BRUHAT, Y. *Géométrie Différentielle et Systèmes Extérieurs.* Dunod, Paris, (1968).

CINCOTTA, P. M., GIORDANO, P. M. & SIMÓ, C. Phase space structure of multi-dimensional systems by means of the mean exponential growth factor of nearby orbits. *Physica D*, **182**, pp. 151-178, (2003).

COFFEY, S. L., DEPRIT, A. & DEPRIT, E. Frozen orbits for satellites close to an Earth-like planet. *Celes. Mech. & Dyn. Astron.*, **59**, pp. 37-72, (1994).

CONTOPOULOS, G. & VOGLIS, N. Spectra of stretching numbers and helicity angles in dynamical systems. *Celes. Mech. & Dyn. Astron.*, **64**, pp. 1-20, (1996).

CORDANI, B. *The Kepler Problem. Group Theoretical Aspect, Regularization and Quantization, with an Application to the Study of Perturbations.* Birkhäuser, Basel, (2003).

CORDANI, B. Global study of the 2D secular 3-body problem. *Regular & Chaotic Dynamics*, **9**, pp. 113-128, (2004).

CORDANI, B. Frequency modulation indicator, Arnold's web and diffusion in the Stark-Quadratic-Zeeman problem. *Physica D*, **237**, pp. 2797-2815, (2008).

CORNWELL, J. F. *Group Theory in Physics.* Academic Press, London, (1989).

CRAMPIN, M. & PIRANI, F. A. E. *Applicable Differential Geometry.* London Mathematical Society Lecture Notes Series. Cambridge University Press, Cambridge, UK, (1986).

CROOM, F. H. *Principles of Topology.* Saunders College Publishing, Philadelphia, PA, (1989).

CUSHMAN, R. A survey of normalization techniques applied to perturbed Keplerian systems. In Jones, K., ed., *Dynamics Reported, Volume 1, New Series*, pp. 54-112. Springer-Verlag, New York, (1991).

CUSHMAN, R. H. Reduction, Brower's Hamiltonian, and the critical inclination. *Celestial Mechanics*, **31**, pp. 401-429, (1983).

CUSHMAN, R. H. & BATES, L. M. *Global Aspects of Classical Integrable Systems.* Birkhäuser, Basel, (1997).

CUSHMAN, R. H. & SADOVSKIÍ, D. A. Monodromy in the hydrogen atom in crossed fields. *Physica D*, **142**, pp. 166-196, (2000).

DEPRIT, A. Canonical transformations depending on a small parameter. *Celestial Mechanics*, **1**, pp. 12-30, (1969).

DEPRIT, A. The elimination of the parallax in satellite theory. *Celestial Mechanics*, **24**, pp. 111-153, (1981).

DEPRIT, A., LANCHARES, V., IÑARREA, M., SALAS, J. P. & SIERRA, J. D. Teardrop bifurcation for Rydberg atoms in parallel electric and magnetic fields. *Phys. Rev. A*, **54**, pp. 3885-3893, (1996).

DUBROVIN, B. A., NOVIKOV, S. & FOMENKO, A. *Géométrie Contemporaine.* MIR, Moskow, (1982-1987).

ENGLEFIELD, M. J. *Group Theory and the Coulomb Problem.* Wiley Interscience, New York, (1972).

FASANO, A. & MARMI, S. *Meccanica Analitica.* Bollati Boringhieri, Torino, (1994).

FASSÒ, F. The Euler-Poinsot top: A non-commutatively integrable system without global action-angle coordinates. *J. App. Math. Phys. (ZAMP)*, **47**, pp. 1-24, (1996).

FASSÒ, F., GUZZO, M. & BENETTIN, G. Nekhoroshev-stability of elliptic equilibria of Hamiltonian systems. *Comm. Math. Phys.*, **197**, pp. 347-360, (1998).

FÉJOZ, J. Averaging the planar three-body problem in the neighborhood of double inner collisions. *J. Differential Equations*, **175**, pp. 175-187, (2001).

FÉJOZ, J. Global secular dynamics in the planar three-body problem. *Celes. Mech. & Dyn. Astron.*, **84**, pp. 159-195, (2002a).

FÉJOZ, J. Quasi periodic motion in the planar three-body problem. *J. Differential Equations*, **183**, pp. 303-341, (2002b).

FERRAZ-MELLO, S. *Canonical Perturbation Theories, Degenerate Systems and Resonances.* Springer, New York, (2007).

FOCK, V. A. Zur theory des wasserstoffatoms. *Z. Phys.*, **98**, pp. 145-154, (1935).

FRIEDRICH, H. & WINTGEN, D. The hydrogen atom in a uniform magnetic field–An example of chaos. *Phys. Rep.*, **183**, pp. 37-79, (1989).

FROESCHLÉ, C., GUZZO, M. & LEGA, E. Graphical evolution of the Arnold's web: From order to chaos. *Science*, **289**, pp. 2108-2110, (2000).

GILMORE, R. *Lie Groups, Lie Algebras, and Some of Their Applications.* John Wiley & Sons, New York, (1974).

GIORGILLI, A. Rigorous results on the power expansions for the integrals of a Hamiltonian system near an elliptic equilibrium point. *Ann. Inst. H. Poincaré Phys. Théor.*, **48**, pp. 423-439, (1988).

GIORGILLI, A. New insights on the stability problem from recent results in classical perturbation theory. In Benest, D. & Froeschlé, C., eds., *Modern Methods in Celestial Mechanics*, pp. 249-284. Editions Frontières, Gif-sur-Yvette, France, (1989).

GIORGILLI, A., DELSHAMS, A., FONTICH, E., GALGANI, L. & SIMÓ, C. Effective stability for a Hamiltonian system near an elliptic equilibrium point, with an application to the restricted three-body problem. *J. Differential Equations*, **77**, pp. 167-198, (1989).

GIORGILLI, A. & LOCATELLI, U. Kolmogorov theorem and classical perturbation theory. *Z. Angew. Math. Phys.*, **48**, pp. 220-261, (1997).

GOLDSTEIN, H. *Classical Mechanics.* Addison-Wesley, Reading, MA, (1980).

GUCKENHEIMER, J. & HOLMES, P. *Nonlinear oscillations, dynamical systems and bifurcations of vector fields.* Springer-Verlag, New York, (1983).

GUILLEMIN, V. & STERNBERG, S. *Variations on a Theme by Kepler.* American Mathematical Society, Providence, RI, (1990).

GUZZO, M., LEGA, E. & FROESCHLÉ, C. On the numerical detection of the effective stability of chaotic motions in quasi-integrable systems. *Physica D*, **163**, pp. 1-25, (2002).

HAIRER, E., LUBICH, C. & WANNER, G. *Geometric Numerical Integration–Structure Preserving Algorithms for Ordinary Differential Equations.* Springer-Verlag, New York, (2002).

HAIRER, E., NORSETT, S. P. & WANNER, G. *Solving Ordinary Differential Equations I-Nonstiff Problems (2nd edn.)*. Springer–Verlag, New York, (1993).

HAIRER, E. & WANNER, G. *Solving Ordinary Differential Equations II - Stiff and Differential Algebraic Problems (2nd edn)*. Springer–Verlag, (1996).

HELGASON, S. *Differential Geometry, Lie Groups, and Symmetric Spaces*. Academic Press, New York, (1978).

HONJO, S. & KANEKO, K. Is Arnold diffusion relevant to global diffusion? (2008). URL: *http:// arxiv.org/abs/nlin/0307050*

HORI, G. Theory of general perturbations with unspecified canonical variables. *Publ. Astron. Soc. Japan*, **18**, pp. 287-296, (1966).

ISERLES, A. *A First Course in the Numerical Analysis of Differential Equations*. Cambridge University Press, Cambridge, UK, (1996).

JEFFERYS, W. H. & MOSER, J. Quasi-periodic solutions for the three-body problem. *Astron. J.*, **71**, pp. 568-578, (1966).

KOBAYASHI, S. & NOMIZU, K. *Foundations of Differential Geometry*, vol. I-II. Interscience Publishers, John Wiley & Sons, New York, (1963-1968).

KOLMOGOROV, A. N. Preservation of conditionally periodic movements with small change in the Hamilton function. *Dokl. Akad. Nauk. SSSB*, **98**, pp. 527-530, (1954). Reprinted in *Lecture Notes in Physics* No. 93 (Berlin, 1979) pp. 51-56.

KUMMER, M. On the three-dimensional Lunar problem and other perturbation problems of the Kepler problem. *J. Math. Anal. Appl.*, **93**, pp. 142-194, (1983).

LANCZOS, C. *The Variational Principles of Mechanics*. University of Toronto Press, Toronto, 4th edn., (1970).

LANDAU, L. & LIFCHITZ, E. *Mechanics*. Pergamon Press, Oxford, UK, (1960).

LASKAR, J. The chaotic behaviour of the solar system: A numerical estimate of the size of the choatic zones. *Icarus*, **88**, pp. 266-291, (1990).

LASKAR, J., FROESCHLÉ, C. & CELLETTI, A. The measure of chaos by the numerical analysis of the fundamental frequencies. Application to the standard mapping. *Physica D*, **56**, pp. 253-269, (1992).

LIBERMANN, P. & MARLE, C.-M. *Symplectic Geometry and Analytical Mechanics*. D. Reidel Publishing Company, Dordrecht, (1987).

LIDOV, M. L. & ZIGLIN, S. L. Non-restricted double-averaged three-body problem in Hill's case. *Celestial Mechanics*, **13**, pp. 471-489, (1976).

LIEBERMAN, B. B. Existence of quasi-periodic solutions to the three-body problem. *Celestial Mechanics*, **3**, pp. 408-426, (1971).

LOCHAK, P. Canonical perturbation theory via simultaneous approximation. *Russ. Math. Surv.*, **47**, pp. 57-133, (1992).

LOCHAK, P. Stability of Hamiltonian systems over exponentially long times: the near-linear case. In *Hamiltonian dynamical systems (Cincinnati, OH, 1992)*, vol. 63 of *IMA Vol. Math. Appl.*, pp. 221-229. Springer, New York, (1995).

LUKES-GERAKOPOULOS, G., VOGLIS, N. & EFTHYMIOPOULOS, C. The production of Tsallis entropy in the limit of weak chaos and a new indicator of chaoticity. *Physica A*, **387**, pp. 1907-1925, (2008).

MARSDEN, J. E. & RATIU, T. S. Reduction of Poisson manifolds. *Lett. Math. Phys.*, **11**, pp. 161-169, (1986).

MARSDEN, J. E. & RATIU, T. S. *Introduction to Mechanics and Symmetry.* Springer-Verlag, New York, (1994).

MARSDEN, J. E. & SHANE, D. R. New methods in celestial mechanics and mission design. *Bull. Amer. Math. Soc.*, **43**, pp. 43-73, (2005).

MARSDEN, J. E. & WEINSTEIN, A. Reduction of symplectic manifolds with symmetry. *Rep. Math. Phys.*, **5**, pp. 121-130, (1974).

MATHÚNA, D. O. *Integrable Systems in Celestial Mechanics.* Birkhäuser, (2008).

MAZZIA, F. & PAVANI, R. A class of symmetric methods for hamiltonian systems. *Proceedings of XVIII Congresso AIMETA di Meccanica Teorica e Applicata*, (2007).

MEYER, K. R. & HALL, G. R. *Introduction to Hamiltonian Dynamical Systems and the N-Body Problem.* Springer-Verlag, New York, (1992).

MEYER, K. R. & SCHMIDT, D. S. The stability of the Lagrange triangular points and a theorem of Arnold. *J. Differential Equations*, **62**, pp. 222-236, (1986).

MILNOR, J. *Topology from Differentiable Viewpoint.* The Univ. Press of Virginia, Charlottesville, VA., (1965).

MORBIDELLI, A. *Modern Celestial Mechanics–Aspects of Solar System Dynamics.* Taylor and Francis (free download from http://www.oca.eu/morby/), (2002).

MOSER, J. K. Stabilitätsverhalten kanonisher Differentialgleichungssysteme. *Nachr. Akad. Wiss. Göttingen Math.-Phys. Kl. IIa*, pp. 87-120, (1955).

MOSER, J. K. On invariant curves of area-preserving mappings of an annulus. *Nachr. Akad. Wiss. Göttingen Math.-Phys. Kl. IIa*, **(1)**, pp. 1-20, (1962).

MOSER, J. K. Convergent series expansions for quasi-periodic motions. *Math. Ann.*, **169**, pp. 136-176, (1967).

MOSER, J. K. Regularization of Kepler's problem and the averaging method in a manifold. *Comm. Pure Appl. Math.*, **23**, pp. 609-636, (1970).

NASH, C. & SEN, S. *Topology and Geometry for Physicists.* Academic Press Inc., London, (1983).

NEKHOROSHEV, N. N. Exponential estimates of the stability time for near-integrable Hamiltonian systems. *Russ. Math. Surv.*, **32**, pp. 1-65, (1977).

NEKHOROSHEV, N. N. Exponential estimates of the stability time for near-integrable Hamiltonian systems, II. *Trudy Sem. Petrovs.*, **5**, pp. 5-50, (1979).

OTT, E. *Chaos in Dynamical Systems.* Cambridge University Press, (1993).

PAULI, W. Über das Wasserstoffspektrum vom Standpunkt der neuen quantummechanic. *Z. Phys.*, **36**, pp. 336-363, (1926).

PENROSE, R. Twistor algebra. *J. Math. Phys.*, **8**, pp. 345-366, (1967).

PENROSE, R. Relativistic symmetry groups. In Barut, A. O., ed., *Group Theory in Non Linear Problems.* Reidel Publishing Company, Dordrecht, (1974).

PERKO, L. *Differential Equations and Dynamical Systems.* Springer-Verlag, New York, (1991).

POINCARÉ, H. *Les Méthodes Nouvelles de la Mecanique Céleste. Vol. 1, 2, 3.* Gauthier-Villars, Paris, (1892-1893-1899).

PÖSCHEL, J. Integrability of Hamiltonian systems on Cantor sets. *Comm. Pure Appl. Math.*, **35**, pp. 653-696, (1982).

PRESS, W. H., TEUKOLSKI, S. A., WETTERLING, W. T. & FLANNERY, B. P. *Numerical Recipes in C.* Cambridge University Press, Cambridge, UK, (1992).

QUARTERONI, A., SACCO, R. & SALERI, F. *Numerical Mathematics.* Springer, New York, (2000).

RAUCH, K. P. & HAMILTON, D. P. HNBody symplectic integration package. (2004). **URL:** *http://janus.astro.umd.edu/HNBody/*

RÜSSMANN, H. On optimal estimates for the solutions of linear partial differential equations of first order with constant coefficients on the torus. In *Dynamical systems, theory and applications (Rencontres, Battelle Res. Inst., Seattle, Wash., 1974)*, pp. 598-624. Lecture Notes in Phys., Vol. 38. Springer-Verlag, Berlin, (1975).

SALAS, J., DEPRIT, A., FERRER, S., LANCHARES, V. & PALACIÁN, J. Two pitchfork bifurcations in the polar quadratic Zeeman-Stark effect. *Phys. Lett. A*, **242**, pp. 83-93, (1998).

SALAS, J. P. & LANCHARES, V. Saddle-node bifurcation for Rydberg atoms in parallel electric and magnetic fields. *Physical Review A*, **58**, pp. 434-439, (1998).

ŠIDLICHOVSKÝ, M. & NESVORNÝ, D. Frequency modified Fourier transform and its application to asteroids. *Celes. Mech. & Dyn. Astron.*, **65**, pp. 137-148, (1997).

SKOKOS, C. The smaller (SALI) and the generalized (GALI) alignment index methods of chaos detection: Theory and applications. pp. 1–46, (2007). **URL:** *http://www.imcce.fr/hosted_sites/tempsespace/archives/semin TE-02-04-2007.pdf*

SMALE, S. Differentiable dynamical systems. *Bull. Amer. Math. Soc.*, **73**, pp. 747–817, (1967).

SOMMERFELD, A. *Mechanics.* Academic Press, New York, London, (1964).

SOURIAU, J.-M. Sur la variété de Kepler. *Symp. Math.*, **14**, pp. 343–360, (1974).

SOURIAU, J.-M. Géometrie globale du problème à deux corps. *Proc. IUTAM-ISSIM Symp. on Mod. Devl. Anal. Mech., Atti Acad. Sci. Torino*, **Suppl. 117**, pp. 369–418, (1983).

SOURIAU, J.-M. *Structure of Dynamical Systems.* Birkhäuser, Boston, (1997).

STERNBERG, S. *Lectures on Differential Geometry.* Chelsea Publishing Company, New York, (1983).

TABOR, M. *Chaos and Integrability in Nonlinear Dynamics.* Wiley & Sons, New York, (1989).

UZER, T., JAFFÉ, C., PALACIÁN, J., YANGUAS, P. & WIGGINS, S. The geometry of reaction dynamics. *Nonlinearity*, **15**, pp. 957–992, (2002).

VELA-AREVALO, L. *Time-Frequency Analysis Based on Wavelets for Hamiltonian Systems.* Ph.D. thesis, California Institute of Technology, Pasadena, CA, (2002).

VELA-AREVALO, L. & MARSDEN, J. E. Time-frequency analysis of the restricted three body problem: Transport and resonance transitions. *Classical and Quantum Gravity*, **21**, pp. S351–S375, (2004).

VON MILCZEWSKI, J. & UZER, T. Canonical perturbation treatment of a Rydberg electron in combined electric and magnetic fields. *Phys. Rev. A*, **56**, pp. 220–231, (1997a).

VON MILCZEWSKI, J. & UZER, T. Chaos and order in crossed fields. *Phys. Rev. E*, **55**, pp. 6540–6551, (1997b).

VON WESTENHOLZ, C. *Differential Forms in Mathematical Physics.* North-Holland, Amsterdam, (1978).

WHITTAKER, E. T. *A Treatise on the Analytical Dynamics of Particles and Rigid Bodies.* Cambridge University Press, Cambridge, UK, 2nd edn., (1917).

WHITTAKER, E. T. & WATSON, G. N. *A Course of Modern Analysis.* Cambridge University Press, Cambridge, England, (1952).

WIGGINS, S. *Introduction to Applied Nonlinear Dynamical Systems and Chaos.* Springer-Verlag, New York, (1990).

Index